WITHDRAWN

The Origins of the Turbojet Revolution

Johns Hopkins Studies in the History of Technology

General Editor   Thomas P. Hughes
Advisory Editors   Leslie Hannah and Merritt Roe Smith

The Mechanical Engineer in America, 1830–1910: Professional Cultures in Conflict by Monte Calvert

American Locomotives: An Engineering History, 1830–1880 by John H. White, Jr.

Elmer Sperry: Inventor and Engineer by Thomas Parke Hughes (Dexter Prize, 1972)

Philadelphia's Philosopher Mechanics: A History of the Franklin Institute, 1824–1865 by Bruce Sinclair (Dexter Prize, 1975)

Images and Enterprise: Technology and the American Photographic Industry, 1839–1925 by Reese V. Jenkins (Dexter Prize, 1978)

The Various and Ingenious Machines of Agostino Ramelli edited by Eugene S. Ferguson, translated by Martha Teach Gnudi

The American Railroad Passenger Car, New Series, no. 1, by John H. White, Jr.

Neptune's Gift: A History of Common Salt, New Series, no. 2, by Robert Multhauf

Electricity before Nationalisation: A Study of the Development of the Electricity Supply Industry in Britain to 1948, New Series, no. 3, by Leslie Hannah

Alexander Holley and the Makers of Steel, New Series, no. 4, by Jeanne McHugh

The Origins of the Turbojet Revolution, New Series, no. 5, by Edward W. Constant II

The Origins of the Turbojet Revolution

Edward W. Constant II

The Johns Hopkins University Press
Baltimore and London

This book has been brought to publication with the generous assistance of the Andrew W. Mellon Foundation

Copyright © 1980 by The Johns Hopkins University Press
All rights reserved
Printed in the United States of America

The Johns Hopkins University Press, Baltimore, Maryland 21218
The Johns Hopkins Press Ltd., London

Library of Congress Cataloging in Publication Data

Constant, Edward W
   The origins of the turbojet revolution.
   (Johns Hopkins studies in the history of technology; new ser., no. 5)
   Includes bibliographical references and index.
   1. Airplanes—Turbojet engines—History.
I. Title II. Series.
TL709.3.T83C64   621.43'52   80-11802
ISBN 0-8018-2222-X

For William

"The inertia of society is so stubborn that no one will move against it, if he cannot believe that it can be more easily overcome than is actually the case. And no one will suffer the perils and pains involved in the process of radical social change, if he cannot believe in the possibility of a purer and fairer society than will ever be established. These illusions are dangerous because they justify fanaticism; but their abandonment is perilous because it inclines to inertia."
—Reinhold Niebuhr, *Moral Man and Immoral Society*

Contents

List of Illustrations
ix

Preface
xiii

1
A Model for Technological Change
1

2
Structural Antecedents:
Water Turbines and Turbine Pumps
33

3
Prior Technology:
Steam Turbines, Turbo-Air Compressors, and the First Internal
Combustion Gas Turbines
63

4
Aerodynamics
99

5
Regnant Normal Technology:
Aircraft Piston Engines,
Aircraft Structures, and Alternative Propulsion Systems
117

6
National Patterns in the Pursuit and Utilization of Scientific Knowledge
151

7
The Turbojet Revolution
178

8
The New Normal Technology
208

9
Conclusion
241

A Technical Appendix:
Some Essentials
of the Aeronautic Environment and of Aero-Engines
247

A Note on Bibliography
267

List of Abbreviations
269

Notes
271

Index
303

List of Illustrations

Partial heritage of the turbojet   4
John Smeaton's scale model water wheel test aparatus, 1759   37
The Fourneyron outward-flow water turbine   40
Prony's original sketch of the friction dynamometer, or Prony brake   41
Fourneyron's development of the Prony brake, 1829   42
Hesse's and Pelton's sketches   55
Single wheel of a 25,000 h.p. Pelton double-overhung unit   56
Impeller and diffuser, Reynolds designed Mather-Reynolds centrifugal water pump, 1895   60
Steam turbine types   65
Straight blading of first Parsons steam turbine, 1884   72
Early Parsons portable turbo-generator set, 1886   73
Parsons 67,000 h.p. turbo-generator set, 1923   74
Turbinia at 34½ knots, 1897   75
Ten years' progress in normal technology: *Turbinia*, 2300 h.p., and *Mauretania*, 70,000 h.p.   76
Low-pressure turbine of the *Mauretania*   77
Armengaud-Lemale gas turbine, with 25-stage centrifugal compressor (in three casings) designed by Rateau and built by Brown Boveri, 1907   92
Griffith's RAE turbo-compressor experimental unit, 1928-29   112
Rotor from Griffith-designed nine-stage contra-rotating, contra-flow experimental turbine compressor test unit, built 1939 according to Griffith's 1929 concept   113

Sewing fabric on Boeing MB-3A fighter wing, Seattle, 1921 *130*

Ford Tri-motor, 1926 *130*

Rohrbach Ro II all-metal seaplane with emergency sail kit deployed, 1924 *132*

Wing structure of Rohrbach Ro II all-aluminum seaplane; entire structural load is carried by stressed-skin center box-spar *132*

Boeing Model 200, "Monomail," 1930 *136*

Boeing 247 D, 1934 *136*

Heinkel He 111 C, 1935 *137*

Boeing Model 307 Stratoliner, 1938 *137*

Cylinder components for poppet-valve and sleeve-valve piston aero-engines of comparable power *138*

Langley 30 × 60 foot full-scale tunnel, 1929 *154*

Ackeret's supersonic wind tunnel at Zurich, 1935 *155*

Supermarine S6 B, 1931 *172*

Heinkel He 100, 1939 *173*

Howard Hughes's Lockheed L-14 (1938) *175*

Guillaume patent drawing of a turbojet, 1921 *180*

Frank Whittle at Cranwell in 1925 *181*

Original Whittle turbojet patent drawing, 1930 *185*

Original Whittle W.U. turbojet *189*

Von Ohain's first, privately funded, demonstration turbojet, Bartels and Becker's garage, Goettingen, 1935 *195*

Von Ohain's hydrogen-fueled He S-1 demonstration turbojet for Heinkel, 1936 *199*

The Heinkel He 178, the world's first aircraft to fly purely on turbojet power, 1939 *206*

Gloster E. 28/39, first British jet aircraft to fly *207*

Messerschmitt Me 262 A, 1943 *230*

Junkers Jumo 004 *234*

Gloster F.9/40 "Meteor" *235*

Air flow path in Whittle–Rolls Royce Welland *236*

Wind tunnel model of Lippisch P. 13 supersonic delta wing design, 1944 *237*

Bell P-59 A "Airacomet," 1943 *238*

Lockheed P 80 A *239*

Centrifugal-flow turbojet (with double-sided impeller) *256*

Axial-flow turbojet   *257*
Turbo-prop   *261*
Ram-jet   *263*
Pulse-jet   *264*

Preface

This study began a number of years ago as a naive, if not innocent, inquiry into who "invented" the turbojet. My answer to that question now lies before the reader. The reasons for such a long answer are, I think, well-defined in the first chapter. For those who are not acquainted with the aeronautical environment or with aeroengine types, the technical appendix might be a preferable place to start; alternatively, the subheadings in the appendix should permit it to be used as a glossary.

Since this study began, many people in numerous institutions have shown me extraordinary generosity, tolerance, and encouragement. Among them are: Louis Galambos, Charles E. Neu, and Frank E. Vandiver, all previously of Rice University; W. Burlie Brown, Henry A. Kmen, Herman Freudenberger, and William R. Hogan of Tulane University; Loyd S. Swenson, Jr., of the University of Houston; George H. Daniels, Jr., William Coleman, Carl W. Condit, Donald T. Campbell, Gregory H. Singleton, David Joravsky, George Fredrickson, Michael Radnor, Gerald Zaltman, Robert Church, Salme Steinberg, George T. Romani, and Robert H. Baker, all now or formerly of Northwestern University; and Joel A. Tarr and Ronald Lasser of Carnegie-Mellon University. My personal and intellectual debts to these teachers, colleagues, and friends are of awesome proportion, yet what I have done is my own, and has often defied their best advice and criticism. In addition, several of the heroes of this piece patiently answered letters of inquiry and provided original documentation, among them Sir Frank Whittle, Hans von Ohain, Herbert Wagner, and Clarence L. Johnson.

Support for the research represented here has at various times been provided by a NDEA Fellowship (Tulane University), by the Center for the Interdisciplinary Study of Science and Technology and by the Department of History at Northwestern University, by a United States Air Force Dissertation Year Fellowship (Northwestern), and by the Falk Foundation of Carnegie-Mellon University.

Finally, Thomas Parke Hughes, the editor of this series, and The Johns Hopkins University Press editors, Henry Y.K. Tom, Carolyn Moser, and Mary Lou Kenney, have been especially patient and helpful.

The Origins of the Turbojet Revolution

1

A Model for Technological Change

Time, not reason, separates real from absurd. Nothing is so certain as what is, nothing quite so unsure as what might be. What today is, is yesterday's possibility, a selection out of what-might-have-been's. The schoolchildren of each generation know as commonplace what the finest minds of their fathers' generation only imagined. Ask anyone born before 1950 to make the sound of an airplane, and after he has overcome his incredulity he is likely to emit a noise not unlike that made by a barber's electric shears. Make the same request of someone born after 1950, and the response is likely to be a low, whistling hiss. That younger sound is the final, certain assertion of technological revolution. That subtle change in the way men cognize the sound of an airplane is proof more definite than any statistic, more eloquent than any history, of the reality and the omnipresence of the turbojet revolution.

For forty years, from the first manned, powered flight in 1903 until the middle of the Second World War, the propeller-piston engine combination was the preeminent and, for most of that period, the singular, aircraft propulsion system. But in the generation following the war, the turbojet became the identifying, normative criterion for an airplane. From the start, the significance of the turbojet was clear: its development was "the greatest aeronautical revolution since the Wright Brothers."[1] In succeeding decades, the turbojet attained that ultimate accolade of successful technological revolution—to be acclaimed by pundits an "age," to be broadcast coequally with the "atomic age," the "machine age," the "electronic age." It even lent its name to a mass culture's glittering if tawdry aristocracy, the "jet set." Recognition of the turbojet revolution has become, as common knowledge must, a cliché.

Yet in triumph there is loss. The very completeness of the turbojet revolution has made it so very obvious. That which is, is taken for granted. It has a preordination about it, an implacable logic rarely questioned. It is presumptively true, like the law of gravity or juicy gossip. The tangible revolution in cold steel (or hot, exotic alloys) is apparent and undeniable, and, as such, is simply assumed to be something that "had to be" or whose "time had come." Any conception of the historical process behind the hardware is as rare as the turbojet is pervasive.[2]

This specific lacuna in historical understanding is symptomatic of history's broader failure to confront the process of technological change. Histories of technologies, in the sense of chronologies of devices, abound. Yet plausible reconstructions of how those technologies develop and change are rare. Some general, sociological explications of the nature of technological change, such as those advanced by A. L. Kroeber, William F. Ogburn and Dorothy Thomas, and S. C. Gilfillan, are simply wrong.[3] More contemporary attempts to understand technological change (and its relationship to research and development and to scientific progress) through quantitative techniques, as represented in projects *Hindsight* and *TRACES*, have encountered catastrophic methodological problems.[4] Similarly, the attempt to reduce all technological change to a simple function of economic factors, most notably the work of Jacob Schmookler, demands assumptions (the validity of patent statistics and the equal importance of all patents, for example) that have no empirical foundation.[5] Attempts thus far to present a general interpretation of all technological change have foundered on the great diversity and complexity of that change.

More modest approaches to technological issues have provided more convincing results. John Jewkes, David Sawers, and Richard Stillerman, for example, through a careful examination of numerous individual cases, have demonstrated the tendency for radical innovations to be developed outside the sectors affected.[6] David Hamberg, also through the analysis of case studies, supported Jewkes's, Sawers's, and Stillerman's conclusions.[7] The National Bureau of Economic Research, in its massive study of inventive activity, discovered great diversity in the origins and developmental paths of radical innovations as well as support for the argument that economic growth enhancing improvements in efficiency, productivity, and product performance depend most heavily upon detailed, painstaking, incremental improvements in existing technologies.[8] Meanwhile, confessed historians of technology—notably Edwin T. Layton, Eugene Ferguson, Carl Condit, Thomas P. Hughes, Reese Jenkins, and Derek Price, among others—as well as economic historians such as Hugh G. Aitken and Nathan Rosenberg have made major contributions to understanding aspects of technological change. Certainly these scholars' insights will prove of great value here.

Still, none of these sources provides a comprehensive, middle-level analytical framework readily applicable to the turbojet revolution. That difficulty arises partly from the complex, diffuse, often contradictory character of the turbojet revolution itself:

— "The" turbojet was the multiple, independent invention of only a handful of men—individuals working alone with great insight and even greater determination. Yet the turbojet revolutionized and ultimately redefined a highly sophisticated community of aero-engine manufacturers. Moreover, the turbojet revolution occured within a larger, highly institutionalized aeronautical community.

— The turbojet is heir to two centuries of turbine development, including water turbines, turbine water pumps, steam turbines, internal combustion gas turbines, and piston engine superchargers and turbosuperchargers. (See Figure 1.) Yet the turbojet is a novel, a holistic system with unprecedented performance capabilities.

— Functionally, the turbojet supplanted a spectacularly successful piston engine technology. Yet the turbojet exploited—and was at the same time constrained by—remarkably sophisticated airframe, accessory-system, production, and organizational technologies associated with that conventional propulsion system.

— All of aeronautics stimulated and in turn depended upon aeronautical sciences: aerodynamics and specialized studies of aerostructures, as well as related scientific enquiries into strengths and properties of materials. The turbojet specifically depended upon advances in theoretical aerodynamics. But, like other science-based technological achievements investigated by other historians, the precise nature and extent of that dependence is not obvious.

These paradoxes suggest some of the critical questions involved in understanding the turbojet revolution. Those questions include the nature of widely shared technological traditions, the characteristics of and interrelations among the people who work with those technologies, and the ways in which those technologies change, specifically the roles and relative importance of incremental versus discontinuous or revolutionary changes, and the roles of advances in theoretical science and of testing and experiment in technological change.

To answer these questions with certainty for the turbojet revolution, much less for technology in general, is not now possible. The historical study of technology is, by historical standards, a relatively new undertaking as yet neither blessed nor cursed with orthodox concepts, methodologies, or conclusions. If this circumstance makes conclusions all the more tenuous, it makes the questions asked all the more critical. To phrase the relevant questions precisely, to interrelate those questions

A Model for Technological Change

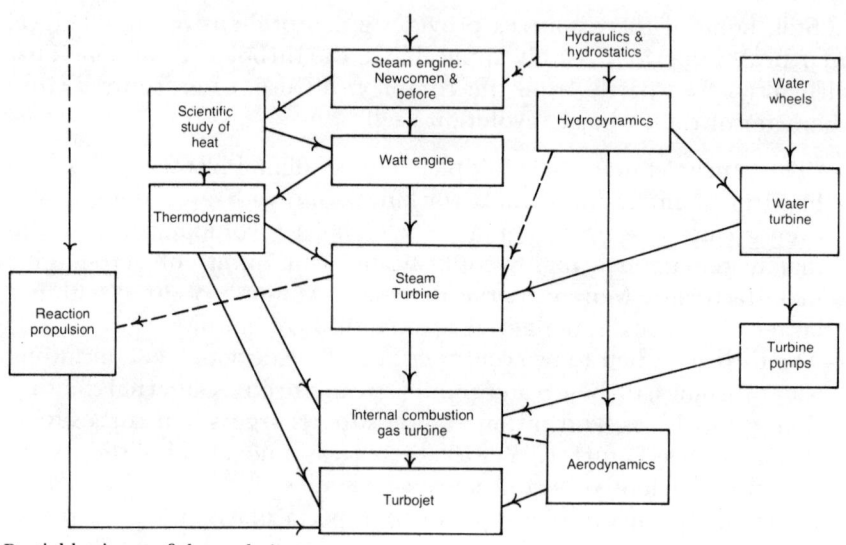
Partial heritage of the turbojet

coherently, this study will utilize an ideal-typical model of the entities and processes involved in the turbojet revolution, and probably in at least some other episodes of major modern technological change. No claim is made for the model beyond heuristic value. The purpose here is not ultimate truth but the beginning of historical understanding.

The use of models in history is controversial.[9] Yet all historians, all people, employ "models" of reality.[10] As E. H. Carr has so persuasively argued, history contains no "facts," only "historical facts"[11]—a quantitatively modest subset from among known and inferred events and processes which historians themselves have selected and ordered into supposedly comprehensible representations of a specified universe. The biographer of a great man presumes that his subject was not altogether unlike his portrayal of him and that the great man's acts somehow made a difference to his fellows and heirs. The historian of ideas supposes that the entities he imagines exist and matter; political or economic historians assume the theoretical constructs they employ have an independent validity that serves to enhance the analysis of specific historical episodes.

The model proposed here is different from those of more conventional history in two respects: it is explicit rather than implicit, and it is tentative. It is explicit for clarity and for candor. It is tentative because it is offered in a relatively new discipline not yet possessed of well-winnowed concepts. Use of the model is founded upon the premise that if there exist historical processes such that only the consequences of those processes appear directly in the historical record, mere positivist recitation of a sequence of such consequences is not adequate. Rather, the model is intended also to portray underlying social processes.

The model is conceived to be ideal-typical in the Weberian sense: a qualitative, artificial, internally consistent construct the elements and interrelations of which may be used to describe and analyze but not to replicate historical reality.[12] These artificial elements and interrelations are not deemed to be real, any more than "class," "personality," "value," or "paradigm" are real, objective entities in the empirical world. Rather, one seeks real-world analogues to the analytical construct. A person is not a personality, yet the behavior of persons may be fruitfully described as if they were personalities. Similarly, a group of historical persons may be analyzed as if it were a class or the historical development of a scientific idea may be traced as if it were a paradigm.

To say that a:b ~ c:d is not say that a = c. To say that piston aero-engines represented a tradition of technological practice is not to assert the existential identity of the piston engine and the tradition of practice. Neither does the piston engine necessarily possess all of the properties ascribed to a tradition in the model and none other, nor does a tradition necessarily exhibit exhaustively all the characteristics of the piston engine and none other. Rather, one can assert that the relationship, say, between an ideal-typical technological tradition of practice and an ideal-typical community of technological practitioners is like that between the piston aero-engine and the community of aero-engine manufacturers. The model is an heuristic analogue, not a carbon copy.

A model so used cannot deductively prove a history so generated; application to one or a few (or even very many) historical instances cannot inductively prove a model. Any model or any theory always poses a danger of circular selection-corroboration. Yet the model proposed here can be judged intuitively plausible or not; it can be examined for internal consistency; it can be evaluated in terms of the ideas and prescriptions of other students of technology; it can be carefully compared to analagous models and concepts from other areas. Most importantly, the model's efficacy in making the historical complexities of the turbojet revolution clearer should become evident. As Max Weber himself emphasized, it is only expedience, efficacy in the explication of "cultural reality," that justifies the utilization of ideal-typical models in the first place.[13]

The central thesis here is that major advances in theoretical aerodynamics led a handful of men to radically new assumptions about possible aircraft speed and about possible gas turbine component efficiencies, which for them implied the turbojet. Making that case, however, will require analysis of the many traditions and circumstances that impinged upon the creation of the turbojet. It is to keep that analysis sorted out that the model has been constructed. It is to be hoped that explicit definition and application of the model will permit investigation and comprehension of the diffuse and multifaceted sequences of events that were the turbojet revolution at a level and with a precision not available through more conventional historical techniques. The need for more refined analytical methods in dealing with the turbojet revolution is obvious. For example, as devices, the turbosupercharger and the turbojet are quite similar. Yet the turbosupercharger represented perhaps the highest development of the piston aero-engine tradition, while the turbojet was created to meet radically different performance assumptions. Utilization of the proposed model will help prevent historical misconstruction of the relationship between these two mechanically similar devices and should help resolve an array of other equally problematic questions of historical interpretation.

The debt of the model put forward here to Thomas S. Kuhn, Donald T. Campbell, Karl Popper, and other scholars of science and technology is obvious. Many of the basic concepts will be recognized as theirs. Yet this model is not rigorously Kuhnian or Popperian. It concerns technology, not science. It is, most of all, tentative, expedient, and corrigible.

Technology is knowledge, and as such is qualitatively no different from any other knowledge that is the property of any other biological entity. Donald T. Campbell has argued persuasively that all knowledge processes, from those of the simplest one-celled organism to those of a Nobel laureate, rest equally upon mechanisms of random variation and selective retention.[14] For the simple organism, behavior is random within the limits prescribed by its genetic coding (which itself represents considerable acquired knowledge about past environments). Selection of behaviors is direct and abrupt: dumb "ideas" die. Successful ideas (variants of behavior), if they are to persist, must be retained genetically—that is, must have genetic origin. The environment directly selects successful mutants.

More complex organisms, those with developed senses, can explore the environment vicariously rather than directly. The sensory organs themselves, the perceptional apparatus, represent considerable knowledge about the environment, being also well-winnowed products of blind variation and selective retention. Human sensory systems and mental faculties likewise have resulted from an evolutionary process. Although human perception is demonstrably fallible, it does constitute a well-winnowed, highly effective means of apprehending those facets of the environment that have had survival value. Similarly, human science and technology represent further vicarious exploration of the environment and are not epistemologically different from other knowledge. Science and technology are also well-winnowed, fallible, and corrigible. Although no specific degree of approach to ultimate truth may be asserted for science, it can be asserted confidently that science does contain some truth about the environment. Similarly, although no approach to perfection may be shown for technology, technologies do work.

According to Campbell's paradigm, selective retention processes comprise a "nested hierarchy" of mechanisms, beginning, at the simplest level, with direct environmental elimination of unsuccessful variants, and progressing up through vicarious selection by internalized criteria (the operation of eyesight, taste, or instinct) to conscious, hopefully nonlethal, vicarious exploration (purposeful trial and error). Each level in the hierarchy of processes itself contains in its "own operation a blind variation and selective retention process at some level."[15] This epistemological paradigm is compatible both with the "introspectionist" models of scientific creativity, which usually postulate some "unconscious," purely random association of ideas from which the mind's aesthetic sensibility selects some subset for conscious appraisal, and with the more structured (but still random at some hierarchical level) search processes posited by advocates of artificial intelligence.[16]

The conjecture and testing of scientific theories represents perhaps the highest, most abstruse development of the fundamental blind-variation–selective retention model. Significantly, both the standards for legitimate scientific conjecture (variation) and the standards for legitimate testing (selective retention) are normally defined by a well-winnowed, socioculturally selected scientific community. To whatever degree a scientific conjecture actually confronts the environment (nature), that confrontation is mediated by community behavioral prescriptions.[17] Technological variation-retention is similar, but does differ at important points. Normal technological activity, like normal scientific activity, is dominated by community practice. Radical technological conjecture, like radical scientific conjecture, frequently violates the accepted rules. But technological selection differs from scientific selection. Technological systems directly, not vicariously, explore the environment: planes crash, engines explode. Technological systems thus face direct elimination by an environment unmediated by background scientific theory in a sense that scientific conjectures ordinarily do not. Furthermore, as noted in detail below the selective processes for technology, because of interaction among technologies, may be considerably more complex than those for science. Finally, if the portrayal of science-technology interrelations advanced below holds, technology may constitute, in effect, a less vicarious exploration of the empirical universe described by (well-winnowed) scientific theory. This selective retention function of technology for science may be valid both in a physical sense (which technological systems based on which theories work) and in a sociocultural sense (which scientific theories or activities or communities are perceived by the society at large to be valid and legitimate).[18]

That both science and technology are epistemologically similar and not fundamentally different from other knowledge, that both are products of evolutionary variation-retention processes, bounds inquiry into the nature of changing technological knowledge and suggests avenues of investigation likely to be fruitful. These considerations, however, do not provide a comprehensive analytical framework.[19] For that purpose, a more highly articulated, middle-level model for technological change, a model less general than evolutionary epistemology but in no way contradictory to it, is necessary. The model here presented attempts to meet that need.

If technology is epistemologically similar to other knowledge, its goals and values frequently are not. Defined in this dimension, technology may be considered purposive knowledge about things, products, or processes—in distinction to science, which ostensibly seeks phenomenological truths about the universe, or art, which seeks only subjective aesthetic satisfaction. A broad spectrum of other definitions of technology are equally agreeable: Edwin Layton's distinction between science as "knowing" and technology as "doing,"[20] or even the anthropologists' general conception of culture as "man's extrasomatic means of adaptation." In the model presented here, the epistemological and especially the sociological dimensions of technology, and the interactions between them, are more important than definitions.

The Community Structure of Technological Practice

Technological knowledge comprises traditions of practice which are properties of communities of technological practitioners. While extensive research has been done on "invisible colleges," research fronts, and the community structure of science, there has been little analogues sociological or historical investigation of technological practice.[21] For technology, basic work on communication patterns and gradients, career paths, and professional mobility, age, and educational levels has yet to be undertaken. The existence of specialized, differentiated, well-defined communities of technological practitioners, however, may be inferred from the evidence of identifiable industry, sector, and product groupings. Standard Industrial Classification codes, for example, are a crude indicator for such communities. For well-institutionalized technologies, other indicators are engineering societies and special interest groupings within them. Specific educational programs (aeronautical engineering, petroleum geology) also demark esoteric communities of practitioners with stringent entry requirements. Nevertheless, just as membership in a scientific invisible college is attained by education combined with actual practice, often initially under the tutelage of a senior researcher, so membership in technological communities seems normally to depend upon both education and direct, practical experience in the field in question.

There are two sources of ambiguity in defining these communities of technological practitioners. First, to a degree more pronounced than in science, technologies and their relevant communities are hierarchical. For example, the aeronautical community writ large is composed, at the least, of manufacturers, of civil and military users, of governmental and community agencies (airport authorities, for instance), and of industry, government, private nonprofit, and university-related aeronautical sciences organizations. Manufacturers, in turn, comprise communities of practitioners specializing, usually, in airframes, power plants, or accessory systems (fuel systems, landing gear, electronic control and NAV/COM systems). Other highly specialized suppliers and subcontractors provide esoteric components and materials to the manufacturing organizations. Thus the aeronautical community is composed of a multilevel hierarchy of subcommunities.

Second, community membership at any level within the hierarchy may be nonexclusive or may overlap other technological communities. For example, among the piston aero-engine manufacturers who will be prominent here, the Allison Division of General Motors and the aero-engine divisions of Rolls-Royce, Diamler-Benz, and B.M.W. all belonged to corporations with major commitments in the automotive industry. Other firms—Junkers, for instance—were also airframe manufacturers, while still others, such as the Pratt and Whitney division of United Aircraft, were almost exclusively aero-engine manufacturing firms. For these reasons, definition of a community of practitioners, say the community of piston aero-engine manufacturers, involves a certain arbitrariness.

Utilization of a community of practitioners as a primary unit of historical analysis nevertheless does promise to generate basic insights for the history of technology. A community of technological practitioners, moreover, may be analyzed in turn as an aggregation of individuals or of firms, just as a scientific specialty may be analyzed as an aggregation of individuals or of labs and departments. While much of the data presented in this historical reconstruction of the turbojet revolution will concern the activities of specific individuals, their work must be considered in context, must be compared or juxtaposed to that of the relevant communities to which those individuals did or did not belong. Certainly the middle-level resolution offered by such a community focus seems notably superior under current circumstances to further investigation of isolated individual inventors or firms or to continued attempts to draw conclusions from masses of homogenized data. Major discontinuities in technological practice occur within communities of practitioners, and the dynamics of that process of change can be better studied at the community level than at individual, firm, national, or industrial aggregate levels.[22]

Traditions of Practice

Technological traditions of practice comprise complex information physically embodied in a community of practitioners and in the hardware and software of which they are masters. Such traditions define an accepted mode of technical operation, the conventional system for accomplishing a specified technical task. Such traditions encompass aspects of relevant scientific theory, engineering design formulae, accepted procedures and methods, specialized instrumentation, and, often, elements of ideological rationale.[23] A tradition of technological practice is proximately tautological with the community which embodies it; each serves to define the other. Traditions of practice are passed on in the preparation of aspirants to community membership. A technological tradition of practice has, at minimum, a knowledge dimension, including both software and hardware, and a sociological dimension, including both social structure and behavioral norms. These aspects together define the tradition itself.

Normal Technology

Normal technology—what technological communities usually do—comprises the improvement of the accepted tradition or its application under "new or more stringent conditions." It is technological development. It is, like Thomas Kuhn's normal science, "puzzle solving." That is not to say that the problems of normal technology are easy or obvious of solution, just that their solution is sought within the limits of the received tradition and that their solution is presumed to exist within those limits. Normal technology generally is requisite to technological progress. Only a concentration on a functional normal technology can avoid a hopeless dispersion of scarce resources. Normal technology therefore constitutes the bulk of all technological activity, whether measured in terms of funds or in terms of engineering time and talent.

To describe normal technology as narrow is not to say, however, that it is unchanging, that it does not progress. Quite the contrary: the many subsystems that make up a total system undergo relatively frequent change, and indeed, change may be revolutionary at that lower level: a new bearing, a new valve-hardening or valve-cooling technique, a different radiator, a higher-octane fuel, a new fuel-injection system. There is always some way in which an existent technology can be improved, and it usually is. The essence of normal technology at a given level is articulation. The evolution of ever more precise design techniques, the development of new configurations, the establishment of whole lines of development (like the Rolls-Royce Eagle-Kestrel-Merlin-Griffon line, which spanned a good thirty years of aircraft piston engine history)—all constitute normal technology. The rewards of such enterprise are great. For example, between 1925 and 1945, service aircraft piston engine power increased about tenfold, from under 350 h.p. to over 3,500 h.p. Similarly, piston-engined aircraft speed doubled, from about 200 mph to well over 400 mph. Such development is achieved at great cost and by great effort. It is what is most often thought of, with some justice, as technological progress. It is normal technology.

Specific community dominance of this normal technology is obscured somewhat by interface requirements among the practices of various communities and subcommunities. Each community or subcommunity of practitioners is pretty much free to pursue its own ends, provided essential interface requirements are met: provided the cost and performance boundary conditions set by the expectations of other communities concerned with other aspects of the same overall system are not violated. For example, the practice of aero-engine makers was bounded by the power, weight, fuel consumption, and reliability expectations of the airframe community, which in turn were shaped by the values and expectations of military and civilian users, whose own desiderata were partially determined by standards maintained by still other communities —the availability of certain fuels and long runways, for instance.

Edwin Layton has noted the importance of acts of creative transformation in transferring knowledge from scientific to technological social systems,[24] while Hugh G. Aitken has begun formulation of a more general information-transformation model for the interaction of science, technology, and economics.[25] Similar but continuous translation or transformation processes also seem necessary among levels within the same broad community of technological practitioners. As each community pursues development of its own tradition, boundary conditions must be continually restated, redefined, and renegotiated.

In instances of major change, these considerations become critical. Radical changes in one subsystem almost always require commensurate

A Model for Technological Change

changes in the rest of the system. For example, early jet aircraft, while offering much superior speed and altitude performance, imposed severe penalties in fuel consumption, required very long runways, and demanded redesign of numerous aircraft subsystems. To an extent, subcommunities seem to hold a veto over significant deviations from standard practice on the part of any one of their number. Thus the process of radical change in any one system (engines) requires translation of its consequences into the interest frame of each of many relevant communities and the persuasion of each of them that the overall gains to be had from the new system outweigh its costs to them. These conditions would go far to explain technological persistence—the detailed development of community practice until some new system offers a very steep performance gradient compared to older practice. Not only does the inertia or vested interest of any one community create internal resistance to change, but interface with all other relevant communities or subcommunities may generate powerful external opposition to such change.

Yet such hierarchical structure both in technological practice itself and in the technological communities which are approximately isomorphic with that practice does give technology immense advantages for overall rapid evolution. Herbert Simon has argued persuasively that all complex hierarchical systems which are "nearly decomposable" into subsystems have many stable states and thus are much more efficient in evolving toward greater internal complexity and better adaptation to external environments.[26] The isomorphism between technological practice, communities of practitioners, and technological design processes reflects this fundamental evolutionary and epistemological principle.

Technological Change: Functional Failure

No technology, however sophisticated or highly developed, is ever perfect. In Henry Ford's words, "If we have a tradition, it is this: Everything can always be done faster and better."[27] Every technical process has real costs which are hypothetically subject to reduction. The need for technological change is inherent in the scarcity of all economic goods and implicit in the human condition.

For these general circumstances to be perceived as a problem within a specific tradition of practice, however, commonly requires one of two situations. First, articulation or application of a conventional system under "new or more stringent conditions"[28] may result in system failure. For example, as aircraft flew to higher altitudes during the First World War, the normally aspirated (unsupercharged) piston engine quickly encountered severe performance limitations: it failed under more demanding conditions. Second, technological disequilibrium, of either the intersystem variety (improved looms demanded improved spinning) or intrasystem variety (improvements in one machine component demand improvements in other components), can result in a pressing need for change.[29] In either case, a conventional system fails to adequately perform its traditional function.

The solution to these problems is usually sought through intensive development of the conventional system. If, however, problems persist, a search for other, more radical solutions may begin. Such a search is, at that level, *ad hoc* and random. Although it may have all the appearances (equipment, laboratories, personnel) of what commonly is called inductive science, such a search is blind, without success postulated, without a solution presumed to exist on any evidence save, possibly, an inquirer's hunch.[30]

But for the relevant community to recognize a given instance of functional failure as a potentially soluble problem rather than as just a limitation or pervasive nuisance, that community must associate the problem with some candidate solution, conventional or radical. Thus, recognition of the deficient performance of the piston engine at high altitudes as a soluble problem rather than as a challenge in the same category as levitation came with the first proposals for supercharging. The community's selection of a solution to functional failure, moreover, depends upon *ex post* competition among proposed alternatives.

While functional failure affects the practice of a specific community, neither initial recognition of the problem nor creation of alternatives necessarily comes from within that community. Usually, proponents of new systems are well acquainted with the technological tradition in question, although they often are not practitioners. For example, the men who built the first steam turbines were not employed in reciprocating engine manufacture; nevertheless, several had experimented with reciprocating steam engines. Whether initial recognition of a problem is indigenous or exogenous to a given community depends on the hierarchical level considered and, to a degree, on where the community boundaries are arbitrarily drawn.

Technological Co-evolution

Although perhaps strictly speaking a species of change induced by functional failure, one especially potent process of recognizing technological problems and selecting solutions takes on a more patterned, organized form and therefore warrants special attention: technological co-evolution.

A Model for Technological Change

Concepts of co-evolution are just now finding theoretical expression in evolutionary biology, but the idea is clear: biological co-evolution "refers to the joint evolution of two (or more) taxa that have close ecological relationships but do not exchange genes, and in which reciprocal selective pressures operate to make the evolution of either taxon partially dependent upon the evolution of the other."[31] Transposed to technology, the concept of co-evolution implies that the development of one set of devices may be linked intimately to the development of other devices within the same macrosystem, and that the two sets of devices may exert powerful, mutually selective pressure on each other. For example, as noted below, the direction of both water turbine and steam turbine development was highly responsive to the demands of electrical power generation; the development of dynamos, in turn, was dependent upon the characteristics of water and steam turbines. Similarly, new airframes partially defined the evolutionary environment of aircraft piston engines, while creation of the turbojet ultimately would compel a revolution in airframe design and construction.

Because inventive activity is self-consciously purposive—that is, normally has a well-specified goal or function—the distinction between mechanisms of variation and of retention is not as clear in technology as in nonhuman biological systems. Nevertheless, to the extent that learned processes of *ex ante* evaluation (those directing inventive activity) represent the results of vicarious exploration of the environment in Campbell's sense, they remain formally mechanisms of selection, not variation. For instance, designing a steam turbine to meet desired dynamo characteristics represents vicarious selection from among possible turbine variants.

Technological co-evolution in this sense implies more than either technological disequilibrium or "reverse salients in an advancing technological front," the image Thomas P. Hughes has used to portray severe problem areas that hold up the rapid advance of an entire technology and serve to attract the attention of concerned technologists.[32] Technological co-evolution implies, first, specificity: the direction of development of a given technology (steam turbines) is linked to some other specific co-evolving technology (power generation and transmission). The evolution of one is partially dependent on the evolution of the other. Second, technological co-evolution implies a hierarchy of retentive or selective processes. The fate of a given invention—its developmental direction—not only depends on its competition with alternative devices performing the same or similar functions and on its co-evolution with a specific other technology, but also depends on the evolutionary success or failure of the higher-level macrosystems of which it is a part. If electrical power had lost out to gas lighting and pneumatic power distribution, the history of steam turbines and of Pelton water wheels would be different. Even at a purely technological level, selective retention is as critical as sources of variation. Indeed, in complex, progressing technologies, the two routines feed back into one another circularly and cumulatively, which perhaps explains why the distinction between the two is not easily made. Co-evolution nonetheless does characterize a fundamental process of technological change.

Science and Technology: Presumptive Anomaly

Presumptive anomaly occurs in technology, not when the conventional system fails in any absolute or objective sense, but when assumptions derived from science indicate either that under some future conditions the conventional system will fail (or function badly) or that a radically different system will do a much better job. No functional failure exists; an anomaly is presumed to exist; hence presumptive anomaly.

Formulation of an alternative technological system is based upon such scientific inference; and a specific, remedial technological problem may exist only in light of such insight. The search for technological solutions and technological experimentation is directed by this scientific perception (although perhaps empirical or blind within the bounds set by the perception). For technological creation based on presumptive anomaly, a proposed new system draws its validity only from conditions visualized in the presumptive anomaly, and vis-à-vis any other set of assumptions from any other source, may have no particular validity.[33]

Presumptive anomaly would seem to represent one direct causal link between theoretical science and technological practice.[34] This concept of presumptive anomaly has greater, more explicit explanatory power than do other concepts with which other investigators have tried to define science-technology interaction (the use of packed-down, generation-old theory in engineering design, some nebulous benefit of scientific education, or the awkwardly simple assertion that scientific and technological accomplishment just go together).

The turbojet revolution is thought to represent a preeminent example of presumptive-anomaly-induced radical technological change. In bare outline, insights emerged from aerodynamics during the last half of the 1920s that produced three scientifically valid assumptions that together created presumptive anomaly for conventional aero-engine practice. First, aerodynamics suggested that by proper streamlining, aircraft speeds could be increased at least twofold from then-current levels, about 200 mph. Such an increase would bring aircraft speed into a regime in which some form of jet or reaction propulsion might become feasible. Second, the application of aerodynamic theory implied new modes of design for axial compressors and turbines and induced some men—a very few—to believe that vastly more efficient gas turbines could be built. Third, by the middle 1920s, aerodynamics indicated that its own laws underwent violent change as the velocity of objects through the air approached the speed of sound.[35] The incapacity of conventional aerodynamic theory to describe such conditions clearly implied that the propeller, the other half of the piston engine–propeller system, would not function at near-sonic speeds. The conventional technology had not faltered or failed by any means or measure: it still held out a great deal of development yet to be done; it still promised and in the event delivered greatly increased performance. But the insights of aerodynamics indicated that in the future the conventional system would, under the last assumption, fail absolutely, and under the first two, probably become inefficient relative to alternative competing systems. Presumptive anomaly had emerged for conventional practice.

It should perhaps be noted that the existence of presumptive anomaly does not depend upon community boundary definition.[36] Presumptive anomaly describes a relationship only between scientific theory and certain technological practice. For example, if the aero-engine community is taken as the unit of analysis for the turbojet revolution, then presumptive anomaly originated outside that community. On the other hand, if the entire aeronautical community, including aero-engine and airframe manufacturers as well as aerodynamicists, is taken as the appropriate unit of analysis, presumptive anomaly for aero-engine practice arose within the relevant community. The concept of presumptive anomaly is relatively insensitive to specific community definitions and will retain its analytic power despite difficulties in precisely drawing community boundaries.

Technological Revolution: Competition and the Community

Once an alternative technological system is formulated, the afflicted community of practitioners must decide whether to adopt the new system or to continue developing the conventional system. As noted, normal technology, even if plagued with problems, is not easily abandoned. Decisions to adopt a new technology depend upon such obvious factors as perceived costs, efficiency, and risks, as well as upon such other characteristics as the ease with which the new system can be explained and understood, its compatibility with its intended environment, technical and nontechnical, and the extent to which it can be easily tried.[37]

The process of technological change founded upon presumptive anomaly, however, is more abstruse. Because presumptive anomaly is science-based, it is most likely to be recognized initially, and in some cases, singularly, by those very close to the intellectual foundations of their respective fields. This circumstance explains why young outsiders, those who have been recently exposed to the leading edge of relevant science and who are not yet fully committed to the conventional technology, tend to be the instigators of presumptive-anomaly-based technological revolution. In addition, because of the esoteric source of presumptive anomaly and its essentially hypothetical and therefore disputable nature, its recognition tends to be even more isolated and, relative to the community, unimpressive than alleged functional failure.

Furthermore, to be at all credible, to imply a testable new system and therefore a likely candidate for adoption, presumptive anomaly must be expressible quantitatively either as it comes from science or as a part of the proposed technology. The presumptive anomaly must be such that it is possible to tell whether it is or is not met by a proposed system. Unfortunately, the scientific insight from which a presumptive anomaly is derived may and usually does contain simplifications or idealizations which render it not directly applicable to technological reality. For example, classical hydrodynamics postulated an ideal, inviscid, irrotational fluid; classical subsonic aerodynamics dealt only with an incompressible gas and did not consider interference effects. Furthermore, rarely if ever does presumptive anomaly cover a complete system. Usually it applies only to one or a few critical variables or elements within the system. In the case of the turbojet, while the requisite aircraft speeds and turbine component efficiencies could be validly deduced from aerodynamic theory, neither the combustion intensities necessary nor the high-temperature materials demanded existed in practice or were clearly implied in any known theory. For those who percieve presumptive anomaly, these lacunae do not make it any less compelling or, in the end, any less valid. But proponents of new systems are put in the unenviable position not only of having to transform theory so that it applies to technology but also of often having to argue *de novo* for certain features of their proposals without backing from either existent scientific theory or technological practice.

Competition between a conventional system and new systems, whatever their derivation, is further muddled by a variety of factors. First, neither the cost nor the performance of any new system can be accurately evaluted *ex ante*. No purely rational choice between conventional and alternative systems is possible until exhaustive attempts have been made to refine the conventional system and considerable effort has been devoted to developing the new system. Given limitations of time and resources, concurrent pursuit of both those programs is usually impossible. Thus, choice often must be made *ex ante*, before the old system is further refined or the new system well developed, only on the basis of the perceived subjective promise or the relative aesthetic appeal of the two systems. Such choices are exacerbated by the tendency of old and new systems to exhibit their greatest virtues along different dimensions —efficiency versus speed, for example. Thus, adoption of a new system often implies not only a new device but also a new value system, a new set of criteria by which technology is judged. This is not to suggest that new and old systems are in any logical sense incommensurable. Nevertheless, what the community regards as essential, invaluable, and normal in the old system may be regarded by proponents of a new system as trivial and expendable. Even if they speak the same language, protagonists of revolution and defenders of the faith rarely hold the same things dear.

Second, although an anomaly and a given new system may be intimately associated in a proponent's mind, it does not follow that the association is exclusive. Often, the same or very similar anomalous perceptions separately realized by several individuals or groups call forth different alternatives. For each individual or group, perception of the anomaly and formulation of a specific alternative are intertwined. But the community may face a variously asserted anomaly together with a number of competing alternative systems. Furthermore, although a specific new device may be recognized as implying a new technological tradition, as having a new set of basic principles compared to the conventional system, the new system usually emerges not as one consensual design but as a group of mutually competing designs not all of which may share all of the characteristics later to be seen as central to the new tradition. The derivation of a new tradition of practice implies competition between specific designs to, in effect, define that tradition.

These debates over the definition and adoption of a new system may be clouded by its relationship to prior technology. The turbojet, for example, was clearly founded upon a vast store of experience with steam turbines plus less fortuitous earlier experiences with internal combustion gas turbines. Moreover, in addition to technological heritage, there was a remarkable persistence of the same organizations and, in some cases, of the same men in the creation of the steam turbine, the first gas turbines, and finally the turbojet engine. Those commonalities do not imply, however, that there was a simple one-to-one transfer from prior technology to the turbojet. Instead, there was a complex transformation of prior relevant technology, a transformation shaped by the insights of aerodynamics and distorted by the special requirements of any airborne mechanical system. That transformation not only produced a completely new aeronautical tradition but also in the end essentially altered the prior technology itself. Clearly, exploitation of prior technology cuts two ways. Use of prior technology, either directly or as a model or guide, does immensely simplify the search for problem or subproblem solution: it is highly efficient for design.[38] Yet to the extent that a prior technology shapes the formulation of systems in a completely alien field, its use can only exacerbate an already abstruse decision process.[39]

Ideally, competition between a conventional system and an alternative can have one of two outcomes: nonadoption (either the alternative fails, or else attempts to refine the older system are perceived by the relevant community to have been successful, thus negating arguments for radical change) or adoption (the community adopts the new system and begins a new normal technology), which is the equivalent of technological revolution. This ideal dichotomy may not be fully representative of reality. For some time, old and new systems may be pursued in parallel: indeed, for some uses, piston aircraft engines are still preferable to turbines. More importantly, the relevant community itself may be redefined radically by the adoption or triumph of a new tradition. Technological revolution in fact rarely seems to result in the one-to-one conversion of an existing community. More often, community boundaries are changed: some members drop out; others are added.[40] Certainly the turbojet revolution will be seen to have entailed the extinction of some aero-engine community members (Curtiss-Wright) and the addition of some new members who previously had not manufactured any form of piston engine (General Electric, Westinghouse, Heinkel). Whatever its outcome, technological revolution has a significant time dimension. It is not a discrete critical event, but a dynamic process sometimes requiring decades to unfold. Development is not done overnight nor is competition resolved instantaneously.

For the instances of successful change of primary interest here, technological revolution can now be more precisely defined. Technological revolution is the professional commitment of either a newly emerging or redefined community to a new technological tradition. Community commitment need be neither total nor singular. The criterion for technological revolution is that the leading edge of technological progress be based upon development of a new system. Technological revolution has occurred when a new tradition of practice comprising a new normal technology is initiated. This concept of technological revolution conflicts with more conventional conceptions. Here, technological revolution is placed much earlier. The revolution occurs not when the new system is operational, not when it is universally accepted, not even when it first works, but when it is accepted by even a significant minority of the relevant community as the foundation for new normal practice. Technological revolution, the essential radical change, comes quite before and quite differently from later development. Technological revolution is here defined only in terms of a community of technological practitioners: it has nothing to do with social significance, economic impact, or tawdry advertising. A technological revolution occurs when a community of practitioners embraces a new tradition, whether that community consists of three little old wine makers or of every major aero-engine manufacturer in the world.

Those studies which have tried to identify technological revolution with patents, invention, or investment are easily reconciled. Clearly, an increasing number of inventions (especially of the highly specific sort likely to be patentable), massive doses of investment, proliferation of new types and models, are all characteristic of a recently initiated new normal technology. Even after the new system itself is working quite well, it continues to engender a mounting swell of second-generation and interface problems that provide the hugely expensive but still rewarding challenges of normal technology. Those who look at patent statistics or investment see the *consequences* of technological revolution, not the revolution itself.

Science and Technology: the Method

The work of Karl Popper, Thomas Kuhn, Imre Lakatos, Donald Campbell, and others suggests that the success of the physical and biological sciences since Galileo has in part derived from a unique and powerful methodology. That methodology comprises the bold conjecture of theoretical systems—their basic entities and the relationships among them—followed by the rigorous testing and refinement of those conjectured systems. In a sense that cuts right across the relationship between advances in scientific theory and technological presumptive anomaly described above, indeed in a sense that cuts orthogonally across the entire community model presented above, the application of this scientific method to technology would seem to have become increasingly pervasive and effective since, at the latest, the beginning of the nineteenth century.

The use of scientific method in a technological context has had at least three major dimensions. First, scientific method accurately typifies the conjecture, testing, and progressive development of large-scale, complex technological systems.[41] Second, application of this scientific method by technological practitioners led to the creation, utilization, and intensive development of ultimately independent, hierarchically more general traditions of technological testability. These traditions in turn are the central mechanism both for the incremental improvement of conventional systems and for the comparative evaluation of alternative systems. Third, associated both with the conjecture and testing of complex systems and with the creation of quasi-independent traditions of testability are fundamental social norms governing the behavior of technological practitioners which are very close in structure, spirit, and effect to the norms governing the behavior of scientists.

Numerous scholars have commented on the use of scientific method, in distinction to scientific knowledge, in technology,[42] but the method supposedly borrowed has commonly been misconceived as that of mere trial-and-error experimentation. While the approaches of any number of technological innovators have been overtly empirical and have been described by themselves and by their biographers in fashionably inductive terms,[43] in fact those trial-and-error techniques frequently were applied within the context of previously conjectured holistic systems. Even so practical a Baconian as Thomas Edison conducted his search for a practicable incandescent light filament within the context of a well-conceived, total electrical generation-distribution-utilization system and with an extraordinarily precise *a priori* idea of just what properties such a filament had to have.[44] Clearly, Edison's search process embodies Campbell's conception of a "nested hierarchy" of variation-retention processes: Edison's search was random only at the level of filament materials, but highly structured at the hierarchically higher level of systems design.

It may well be that the application of this method of bold total-systems conjecture and rigorous testing (or of variation and selective retention) to large-scale, complex, multilevel systems beginning in the nineteenth century created a fundamentally novel category of knowledge and knowledge processes distinct both from science proper and from craft technology. What men of the nineteenth and twentieth centuries have persisted in terming "science," "scientific," "empirical," "practical," "applied science," or "inductive method" really involves this conjecture, design, development, testing, and evolution of complex technological systems.

The key element in this application of scientific method to technology is testing. At its simplest, testing consists of running a complete system—an engine for example—until something breaks, redesigning or strengthening that part, and continuing until something else breaks. At a more sophisticated level, testing involves construction of complex test rigs that are themselves major technological achievements. With such specialized apparatus, data can be collected on the behavior of individual components of systems at various systems performance levels and conditions. Although historians of technology have largely neglected the creation, evaluation, and application of these instruments of technological testing, reports of their construction and usage dominate much of the technological literature.[45] Certainly the ability to test systems that are holistic (have novel properties or capabilities not derivable from their components separately) and multivariate (have many components or subsystems each of which is complex, yielding an overall system with a vast array of interrelated performance parameters) has been critical to progressive, cumulative technology at all levels and is essential to the choice between conventional and radical systems.

The efficacy of such "scientific" technological testing still remains somewhat compromised by the differentiated community structure of technological practice. Particular techniques, technologies, and practices relating to testability ordinarily will be commensurate with, and to some degree will define, a specific community's normal technology. The dominant system and conventional modes of testability usually are mutually reinforcing. To overthrow a conventional system frequently requires, in addition to the formulation of an alternative system and the definition of new performance parameters, the creation of new or much refined testing techniques. Testability, then, can cut several ways. It can buttress conventional practice. Or it can overthrow conventional practice: in technology as in science, one of the quickest ways to impugn a normal technology is to subject it to a novel experimental test which reveals theretofore unseen deficiencies. But such a strategy—used, for example, by Auguste Rateau in developing and arguing for his centrifugal air compressor—can also lead to further development of the conventional system. Traditions of testability, and the techniques that express them, thus frequently become enmeshed in, rather than resolve, competition among systems.

Ultimately, however, some techniques of testing, together with associated norms, do become independent traditions of technological testability more general than any particular technological system. For example, the initial development of the Prony brake, or dynamometer, was integral to the development of the first successful water turbines. Yet utilization of dynamometers soon became *de rigeur* across the whole spectrum of prime mover development. Construction and utilization of dynamometers became an independent, dynamic tradition in itself, and ultimately produced its own internal varieties of technological revolution, as embodied in hydraulic and electrical dynamometers.[46] Moreover, the behavioral norms governing the proper design and use of these devices, and data collection and reporting from them, became generalized and binding upon all legitimate practitioners of "scientific" technology regardless of their specialized community affiliations. These traditions of technological testability, then, are the critical link between technological systems traditions as properties of specialized communities of practitioners and those practitioners' application of scientific method to technology. Indeed, such traditions of testability are cemented into the practice of individual technological communities as norms or values and are perhaps one of the key elements common to all such specialized communities.

The distinction between this modern, progressive, "scientific" technology and older craft traditions, however, requires careful definition. Craft technologies do progress, but slowly. The winnowing process for variants in a craft technology is imprecise if not haphazard. Except in extraordinary circumstances, when a new system is overwhelmingly superior, the isolated construction of even advanced designs in a craft context may not lead to cumulative progress. A tradition of technological testability, in contrast, provides both a means and a social imperative to rapidly, accurately, and publicly compare alternatives.

Evolution of this critical capability to test complex systems involves not only development of testing techniques or hardware, but also implies evolution of behavioral norms governing testing procedure. The development of the technologies of primary interest here, from water turbines to turbojets, happens to encompass also the evolution both of techniques of testing and of rigorous community norms governing development of and data reporting from those techniques.

The critical role of norms in technological testability highlights a fundamental bifurcation in the behavior of technological practitioners which mirrors the behavioral bifurcation Ian Mitroff found among project Apollo moon scientists.[47] Mitroff set his own analysis against the backdrop of the normative structure of science as defined by Robert Merton, Elinor Barber, Norman Storer, and others. Those investigators argue that the source, at least in part, of the awesome epistemological power of science lies in the behavioral norms the social system of science imposes upon its practitioners: faith in rationality, emotional neutrality, communism, disinterestedness, and skepticism.[48]

Among the scientists he investigated, however, Mitroff also discovered a set of counternorms. He found the scientists judged by their peers to be the most brilliant and creative frequently to be those also judged most opinionated, egotistical, and resistant to counterevidence or counterargument. Such scientists orient their behavior toward an alternative set of counternorms: emotional commitment, particularism, interestedness, and dogmatism. Mitroff, drawing upon another of Merton's concepts, suggests that the powerful creative force found in modern science derives from the dynamic alternation of norms and counternorms in the behavioral orientation of individual scientists. Although such dynamic alternation in individual behavior may indeed be critical to the functioning of science, it does not vitiate the objectivity of the whole scientific

A Model for Technological Change

enterprise, which derives from specific holistic properties of the social system of science, especially its insistence (at least ostensibly) upon replicable experimental corroboration.[49]

A similar structure of norms and counternorms for the behavior of individual practitioners, plus a social systemic demand for replicable results, would seem to appertain to technology. Technological practitioners are required to be objective, emotionally neutral, rational, and honest. Yet technological practitioners often are—and protagonists of technological revolution usually are—passionate, determined, and irrationally recalcitrant in the face of unpleasant counterevidence bearing on their pet ideas. Indeed, without such behavior, radical changes such as the turbojet revolution would be unlikely to occur. Furthermore, the social system of technology, like that of science, reserves its greatest reputational (but not necessarily financial) rewards for successful rebels, for those truculent individuals who actually bring off a technological revolution.

Thus, the scientific methodology underlying modern technological practice, the dominance of specific technological traditions within specific communities of practitioners, separate traditions of technological testability, and the norms and counternorms governing practitioners' behavior form a single, internally interacting "sociotechnical system."[50] The social system of technology, like that of science, can thereby functionally harness those highest and yet, paradoxically, potentially most destructive of human traits, ego, emotion, and commitment.

The basic model for technological change presented thus far refers specifically to middle-level community practice—to the development and transmutation of technological knowledge at that level. In this sense, this ideal-typical model describes a horizontal slice through all technological practice, together with orthogonal traditions of technological testability which permeate all levels. One of the costs of utilizing such a model is that it partially obscures other critical factors which influence technological change but which lie outside the model's principal focal plane. Three sets of such factors require consideration here: the process of engineering design, economics, and patents, and cultural influences.

Engineering Design

Engineering design is the process by which technology as social knowledge is transformed into technology as artifact. That process is always multilevel and hierarchical, and this complexity in part may be responsible for obscuring the community nature of technological practice.

Engineering design is the overt, manifest behavior actually observed in communities of technological practitioners. Through engineering design, systems concepts, whether normal or revolutionary, are translated into hardware. The core of such design is hierarchical decomposition of the overall design problem into more manageable subproblems. This partitioning, as suggested above, usually is isomorphic both with subsidiary, more specialized technological traditions of practice and with related subcommunities of practitioners, although often there may be multiple ways of partitioning any given problem. Herbert Simon, in general terms, emphasizes the nested-hierarchical nature of problem decomposition, as well as the interactions among different levels, and the resulting indeterminance of whether ultimate design is done from "the top down" or vice versa:

> One way of considering the decomposition, but acknowledging that the interrelations among the components cannot be ignored completely, is to think of the design process as involving first the generation of alternatives and then the testing of these alternatives against a whole array of requirements and constraints. There need not be merely a single generate-test cycle, but there can be a whole nested series of such cycles. The generators implicitly define the decomposition of the design problem, and the tests guarantee that important indirect consequences will be noticed and weighed. Alternative decompositions correspond to different ways of dividing the responsibilities for the final design between generators and tests.[51]

These hierarchical design processes represent a relatively efficient, if essentially satisficing, strategy for searching the very large spaces implied by large, complex systems.

More specific investigations by David L. Marples and P. J. Booker also emphasize the hierarchical nature of the design process as well as the mutual interactions among different levels. Marples and Booker, however, also note the importance of explicitly technological variables, such as the very close relationship of even novel design solutions to previous practice and the major roles played by time, relative certainty, costs, and marketing considerations in design of components at all systems levels.

Marples advances a decision-tree model of the design process and argues that existent technical capabilities and higher-level design are reciprocally interacting:

> In fact, the result of all engineering design activities is "hardware"; manufacturing techniques, standard parts and assemblies constitute a set of means which have been previously devised for carrying out a variety of design decisions made at a higher level of abstraction. All new high-level decisions are

A Model for Technological Change

made in the knowledge of the existence of these means. All new detail designs and methods of manufacture increase the means available for carrying out more high-level decision.[52]

Marples visualizes two functions for design: the search for possible subproblem solutions and their rapid and accurate evaluation (including the possible generation of lower-level subproblems and the estimation of likely effects on higher-level decisions already taken). Marples further suggests that a hierarchy of decision criteria are called upon at each level of the design process.

Booker, somewhat more elaborately, visualizes design as a dynamic learning process comprising initial problem specification, search of direct and remote precedents for possible solutions, evaluation and breakdown into subproblems for each candidate solution, choice, design itself, testing, and repetition of the cycle.[53] Booker also notes a preference for familiar designs over novel solutions and the importance of commercial considerations, especially perceived development potential, in design selection. Booker stresses the holistic nature of engineering design: "No design exists in complete isolation from other designs, neither, except in unusual circumstances, are designs built upwards by adding together separate elements. Every design must start with a complete idea, however indefinitely conceived, and this is successively broken down into smaller parts which receive greater attention. The definition of the whole is intimately bound up with all of its constituent parts."[54]

Within these hierarchical design processes, both Marples and Booker discover a preeminent role for what Marples terms "classic research," studying technological phenomena "under more controlled conditions" in search for solutions to intractable design problems. Booker finds that "forced invention" arising from the nature of complex systems is one of the primary sources of design progress:

> During the progress of the design it may often turn out upon closer examination that something is not attainable at some level or another. This will force a change either at that level or at some level above it. Since it becomes increasingly difficult and costly to go back and change a decision at a higher level, great creative effort will be applied in solving the problem at the lowest level—resulting in forced invention.[55]

These two types of technological investigation—basic information-gathering about a system, process, or material as it behaves in the artificial environment of a man-made design, and forced invention to meet an urgent, unforeseen need within a complex system—may account for the preeminence of "mission-oriented research" in technological practice. Clearly, such mission-oriented research has no *prima facie* relationship to normal or revolutionary technology as defined in our model. Mission-oriented research is equally applicable to either system. Similarly, presumptive anomaly could enter a mission-oriented research project and lead to revolutionary technological change at that level without significantly altering the overall system.[56]

The process of engineering design depicted in these accounts is commensurable with the model for technological change presented here and would seem to bound it from below.[57] Certainly these models of the design process are relevant to the turbojet revolution. Hierarchically above the turbojet is the total aircraft system; hierarchically below the turbojet are the various engine subsystems: compressors, combustors, turbines, and so on. Actual design of turbojets required both integration into higher-level systems (aircraft) and solution of esoteric subproblems, some of which posed major challenges to other communities of technological practitioners—high-temperature alloys for metallurgists, for example. To explore the full ramifications of the complete design process, however, is beyond the scope of this study.

Economic Factors and Patents

Economics permeates technological change. Economics acts as a direct selector both for entire macrosystems and for individual subsystems. Perceived economic factors operate as vicarious selectors during the process of technological change, and may serve in the first place to inspire efforts towards such change. Yet the impact of economic factors is not homogeneous. Rather, the relative importance of economic aspects varies both with respect to time and with respect to hierarchical level within technological practice: at different times or stages of development at each level, the decisiveness of economic factors differs.[58]

At a given hierarchical level, the stages or phases of technological change might be typified as motivation to invent, creation or invention, development and innovation, and adoption or diffusion.[59] Motivation, because it represents an internal state rather than an overt behavior, is always elusive. Professed motivation may be, and usually is, economic, although admitted desire for personal gain is frequently hedged about with invocations of other values: advancement of the state of the art, social progress, and so on. Religious and cultural values, as well as a perpetual fascination with "pure technology,"[60] may underlie, if not supercede, much ostensibly economic motivation.

Individual cases vary widely: Edison, Sperry, and Eastman, for example, all seem to conform fairly well to the conventional want-driven, economically motivated model for inventive activity.[61] Other inventors, notably Frank Whittle, appear to have been much less responsive to perceived need, and seem to have invented first and sought a market later. Yet however loosely a particular inventor's endeavor may be coupled to market demand, all inventors do share a common utilitarian ideology: all believe, often fanatically, in the ultimate usefulness and value of the devices they create.

Even in those cases where economic motivation is most preeminent, choices of specific areas of invention represent an interaction between perceived need or market opportunity and an individual's particular interests and expertise.[62] Few inventors seem to be capable of entering totally alien areas. These limitations probably result both from any given individual's limited information field and from his marginal inefficiency in entering many new intellectual domains. In this sense, inventors' responses are satisficing rather than optimizing.

The actual process of creation or invention likewise may or may not be closely coupled to economic factors. Edison, for example, in his quest for an electric lighting system, structured every aspect of his inventive effort by strict economic criteria.[63] The creators of modern water turbines and the creators of steam turbines alike sought high technical efficiency in the certainty that technical efficiency meant economic efficiency. Yet such a conflation of technological and economic factors can be dysfunctional: those men who tried to design gas turbine aircraft engines to meet the economic criteria defined by conventional piston technology were unsuccessful, while those men whose inventive processes were much more closely coupled to scientific insight than to overt economic considerations created the turbojet revolution.

It is more often after a new system has taken shape in its creator's mind, when he begins the long process of converting the relevant community, that economic factors become more consistently and directly determinant. First, the new system must be developed. If the resources for development are not forthcoming, the new system usually cannot be effectively presented to the community. The decision to allocate development resources is presumed to be economic: it is based at least ostensibly on objective evaluations of risk, possibilities for profit, and costs, which are, in turn, partly a function of perceived general economic conditions and of the perceived urgency of a particular problem.[64] Unfortunately, *ex ante* cost estimates for a new system are inherently uncertain. Since the old tradition's cost structure is inapplicable, and the new system's cost structure has not yet been defined by experience, cost estimates for the new system are based more on faith than on concrete evidence.[65] In general, all of the evaluations involved in development and in innovation, or the initial decision to adopt, depend on the new system's attaining its alleged performance. Those evaluations are intimately related to the aesthetic appeal of a new proposal and to the persuasiveness of its proponents' arguments. Until a technological revolution is well under way, the perception of estimated costs that determine success remains highly subjective.[66]

Finally, at the stage of community-wide conversion, or more conventionally, in the process of diffusion or firm-by-firm adoption, economic factors can perform their more conventional, determinant role. Yet that role depends upon new performance and cost criteria being already established by prior development and utilization of the new technological system. In this stage economic considerations not only determine whether a whole community adopts the new technology, but also, if such change does occur, determine the timing of firm conversion.[67] Likewise, the conversion of national subcommunities can be expedited or retarded by economic considerations. In the end, economic factors, expressed in relative costs, are the ultimate criteria by which a conventional technology is "falsified" and abandoned and a new tradition "corroborated" and adopted community-wide. Such historical choices do, however, always leave hanging the counterfactual question, If equal resources had been devoted to further developing the old system, could it not have proven superior to the new?[68] Economic analyses, like innovative processes, are always prisoners of the past.

In addition to marked variation with regard to specific inventors and inventions, and somewhat more consistent variation through the phases of technological change at a given hierarchical level, the impact of economic factors also varies with hierarchical level at any given point in time. For example, whether a device represents a normal tradition or a technological revolution, whether costs for the entire macrosystem are or are not well defined, selection of a bolt or other lower-level component still ordinarily will be made according to strictly economic criteria—unless some higher-level design constraint requires an especially critical (and expensive) capability at that lower level. Similarly, within well-developed middle-level technological traditions, choices are ostensibly purely economic. Assuming complete and accurate information derived from prior practice, choices, say, of engine model or type, are subject to rigorous discounted-returns calculations. Even at the very highest levels, choices, for example, between canal and railroad transportation systems, hypothetically can be cast in purely economic terms.

Yet these ascriptions are overly simple. Certainly consideration of economic factors can shape the inventive process at every level. As noted, Thomas Hughes has shown that Edison carefully computed the economic constraints on his electric lighting systems at all levels, from the very highest—the market price set by gas light, which he had to beat—to the very lowest—the price of copper wire, coal, and casual labor—and that those computations directed his experimental investiga-

A Model for Technological Change

tions. Hugh G. Aitken, also as noted above, proposes an information model in which science, technology, and economics continually interact in the inventive process. Meanwhile, other students of technological change have argued that activities promoting such change continually shift back and forth along a continuum ranging from exploratory research and invention through development to innovation and entrepreneurship as various problems are encountered.

Clearly much of this shifting about is not horizontal but rather is vertical, through the different hierarchical levels represented in complex systems. As economic or technical constraints are encountered at one level, effects ripple through the entire hierarchical system, much in the fashion suggested by Simon, Marples, and Booker. Economics, then, as it presents constraints and incentives at multiple levels within a hierarchical technological system, is not so much simplistically determinant as evocative of continuous technological, and sometimes scientific, stresses up and down the structure. Add to this situation the changing nature of technological traditions through time, and the continuous, multilevel interaction of economic and technical factors becomes apparent.

Corporate and national research and development (R&D) policies, to the extent that they represent an attempt to deliberately harness science and technology to economic and social objectives, add further organizational levels to these complexities. The devotion of corporate resources to a certain area of research, for example, usually will hinge on *ex ante* estimates of the relevance of what is likely to come out of that research to the corporation's specific capabilities, proprietary patent position, business areas, market expertise, and technological momentum.[69] Clearly, whether or not resources are directed into a basic research area shapes the probabilities that discoveries will be made in the area. These same internal corporate factors, as well as the perceived promise of a new system, also will help determine whether a corporation supports technological development of a specific new system. As John Kenneth Galbraith and others have demonstrated, the complexity of modern technological systems requires long lead times, massive capitalization, and extensive planning: virtually all product development and most production are done in anticipation of demand rather than in response to it.[70] Corporate R&D policy, then, will act as a more remote vicarious selector for whole areas of endeavor, and may attempt to apply economic decision criteria at that remote level. Similar comments apply to national R&D policies at even higher levels.

These considerations raise a fundamental question as to where the major cognitive content of technological practice is lodged. The protagonists of the major revolutions of interest here were generally outside the relevant communities and therefore outside the relevant corporate structures also. The one case in which corporate and community boundaries were not congruent—that of the Junkers aircraft and aero-engine companies—suggests, however, that commitment to community membership external to the firm may be in some circumstances preeminent and that such commitment is fully capable of generating intrafirm quarrels no less violent than those which cross firm lines. Whether more modest levels of technological change would engage commitments with a different locus and be impinged upon by fundamentally different factors remains at this point an open question.

Patents, because they represent a proprietary economic interest, can affect the process of technological change at every level and in every phase: essentially, all that has been said about economic factors in general applies to patents in particular. Like other economic factors, patent considerations interact with technological and scientific factors in the direction of technological change. Yet patents are not in themselves always determinant. For the technologies of interest here, numbers of patents do not provide meaningful measures of the process of change. More strikingly, while patent considerations did in some instances sharply affect the direction of inventive effort, in general they provided neither an effective information channel for prior work nor an effective barrier to further work. Many of the protagonists of technological revolution portrayed here developed their ideas in ignorance both of prior patents and of each other, and were successful despite what turned out to be conflicting patent claims.

Because this study seeks to examine a horizontal slice through community practice at a specific level and over a specific time period, economic factors and patents are of less consequence than they probably would be in a differently structured study.[71] Economic factors most directly impinged at levels above (aircraft procurement policies, for example) and below (high-temperature alloy costs) the level considered here. Similarly, patents had little effect on the process of technological change at this level. A different focus would yield a different picture, as Frank Whittle's exhaustive treatment of the role of patents and finances in the formation of Power Jets, Ltd., indicates.[72]

Sociocultural Factors

Values, like economics, permeate technological change. Indeed, sociocultural values, especially their institutional expression, underlie derivation of economic utilities themselves. Values, however, are broader in expression and in impact. All of the activities considered in this study depended upon sociocultural sanctioning of science and technology, and on their specific institutional structures and reward systems.[73] Specific culture dispositions affect technological change: all of the major achievements considered here occurred in cultures in which the dominant elements at least were positively disposed toward large, high-performance, high-efficiency technological systems. Yet cultural differences existed and were important: different styles of science (theoretical versus empirical), different expressions of technological excellence (performance versus utility), have real effects. While a detailed examination of sociocultural values in technology is far beyond the scope of this study, patterns of discovery even at the middle levels of technological practice of interest here are strongly influenced by these broad cultural factors.

Because means are limited, the principal focus of this study must be a series of major technological innovations culminating in the turbojet. Over a strictly limited domain, the primary concern will be the relationship of those innovations to prior practice, to scientific advance, to traditions of technological testability, to the nature of technological design, and to more general societal values. The model for technological revolution presented here, it is hoped, comprehends the essential elements of radical change at the level of analysis chosen. The formal unit of analysis—the community of practitioners—should permit at least tentative substantiation of middle-level generalizations about technological change. Most of all, the model should render intelligible the complex interplays and interrelations that constitute the turbojet revolution.

Still, it is men, not abstractions or "forces," who created the turbojet and who in general cause technological revolution. It is men, often alone and not infrequently ridiculed, who recognize anomaly, formulate new systems, and provoke change. It is men, not institutions or governments, who ramrod the development of what to others may seem an unlikely device to the point of community acceptance. More often than not, it is individuals external to the prevailing normal technology, outsiders, who perceive presumptive anomaly and create the crucial variety of technological revolution founded upon it. It is the faith of those men in themselves and in their works, a faith that can only be described as fanaticism, that provides the critical motive force for technological revolution.[74] It is men in the tradition of an older, more heroic history who are the protagonists of that revolution.

The Origins of the Turbojet Revolution

2

Structural Antecedents: Water Turbines and Turbine Pumps

The previous chapter offered a general if idealized description of how communities of technological practitioners operate and of how traditions of technological practice are begun and developed. The present chapter turns to the specific technological heritage of the turbojet, to its remote structural antecedents. Those antecedents, which share basic structural principles with components of the turbojet, include water turbines and turbine water pumps. Subsequent chapters will examine structural antecedents more closely related to the turbojet: steam turbines, internal combustion gas turbines, rotary air compressors, and piston engine superchargers and turbosuperchargers.

The reasons for examining these prior technologies in some detail are twofold. First, it is important to define precisely what the structural antecedents of the turbojet were and were not, to specify similarities, and, more critically, to discover essential differences. Second, the historical development of the turbojet's antecedents will give some depth and reality to the concept of traditions of technological practice and will illustrate some of the important elements in the creation and sustentation of such traditions.

WATER TURBINES

Water turbines represent the first fully developed turbine technology and thus are the direct mechanical antecedents of all succeeding turbine systems. The invention and development of water turbines exhibits the dynamic interaction of what have been suggested above as critical elements in modern technology: scientific theory and analysis, technical systems conjecture, quantitative systems testing, and the emergence of well-defined communities of practitioners.

The word *turbine* itself, first used to describe a water-powered prime mover by Claude Burdin in 1822, was and remains ambiguous in usage. As Norman Smith ably argues, "the" water turbine is not adequately defined by such criteria as reaction (moved by the reaction of water issuing from the blades) versus impulse (moved by the impact of water striking the blades) or vertical versus horizontal rotational axes.[1] As Smith notes, the essential connotation of *turbine* initially was that of a relatively high-speed prime mover, although later evolution of the term implied also high efficiency and, usually but not always, at least partial dependence on reaction.[2]

What became in the nineteenth century a well-defined water turbine tradition emerged, in Smith's felicitous phrase, from a rich and diverse "repertoire" of pre-nineteenth-century water wheel designs. Those products of essentially craft traditions included "tub wheels," horizontal impulse wheels, and Barker's mill reaction wheels (lawn-sprinkler-type rotary devices apparently inspired originally by Hero of Alexandria's steam-powered aeolopile).[3] Before 1800, however, none of these devices developed efficiencies above 25 or 30 percent.[4] Indigenous water-power craft traditions emerged relatively independently and approximately concurrently in England, Europe, and the United States; and while significant variations based on local conditions did occur, the basic national repertoires were generally similar.[5]

Yet the water turbine as it emerged in the nineteenth century differed markedly from its craft antecedents. The new water turbines were based both upon a well-articulated if incomplete tradition in fluid mechanics and major advances in the comprehension of mechanical work, as well as upon detailed empirical investigations of water wheels. The first true, widely adopted water turbine, that of Benoit Fourneyron (1824), was carefully designed to meet the prescriptions of known theory. More importantly, Fourneyron was the first person to successfully employ the Prony brake, or dynamometer, which he used to accurately test the output of his turbine. Fourneyron thus not only designed and built a highly efficient, if ultimately superseded, water turbine, but also established a precise, replicable, quantitative means of ascertaining and comparing the output of all water turbines and wheels—indeed, of all primary movers. From Fourneyron's work very rapidly developed a coherent tradition of water turbine practice, culminating in the water-power systems engineering of James B. Francis and his successors, as well as an independent, hierarchically more general tradition of technological testability.

Prior relevant science for the water turbine includes, at least indirectly, most of the major achievements in hydraulics and fluid mechanics from Galileo to Euler, as well as the evolution of the related concepts of potential and kinetic energy and of work. In addition, theoretical and empirical analyses directed specifically toward water power problems formed a foundation, albeit incomplete, for the eventual perfection of the water turbine and, moreover, were related to the crucial transformation of the physical concepts just mentioned.[6]

Even by Galileo's time, practitioners had accumulated considerable empirical knowledge of hydraulic phenomena, such as rates and patterns of flow in channels and rates of discharge from orifices. Galileo's disciple Evangelista Torricelli, in a work published in 1644, first noticed that rates of discharge from an orifice in a tank were proportional to the fluid level in the tank and "that a jet from an orifice directed vertically upward would rise almost to the level of the free surface in the tank."[7] The French experimentalist Edme Mariotte was the first to enunciate, in 1686, the principle that the force exerted by a jet of water is proportional to the square of its velocity.[8] Blaise Pascal, meanwhile (in 1663), had succeeded in uniting hydrostatics and aerostatics and in establishing unequivocally the principle that fluid pressure is exerted equally in all directions.[9]

A number of eighteenth-century technologically inclined investigators—among them, Antoine Parent, Bernard Forest de Belidor, J. T. Desaguliers, and Colin MacLaurin—made direct contributions to understanding the operation of water wheels.[10] C. A. Deparcieux, in 1752, introduced the crucial idea of reversibility, arguing that if an overshot wheel were attached to an identical bucket wheel running in reverse, it should be able to lift back to the original height almost as much water as was used to turn the wheel. Deparcieux thus contended that the theoretical efficiency of an overshot wheel should approach 1. Johann Albrecht Euler (Leonhard's son), reasoning purely from hydrodynamic theory, reached the same conclusion in 1754.[11]

Perhaps the most thorough water power experiments prior to those of Fourneyron, however, were conducted by John Smeaton in England. Smeaton's work is especially significant here because it probably represents the most potent development of purely empirical approaches, only modestly informed by sound scientific theory, to water wheel analysis. Smeaton, a former instrument maker (like James Watt), was a consummate but unlucky experimentalist: his work on water wheels was rapidly eclipsed by the more theoretically oriented French, and he later devoted himself wholeheartedly to the incremental (and successful) improvement of the Newcomen steam engine at the same time Watt was independently discovering the latent heat of steam and conceiving his separate condensing engine. Be that as it may, Smeaton's work on water wheels is the epitome of Baconian method.

Smeaton began experiments in 1752 but did not publish results until 1759, in order that he should have "an opportunity of putting the deductions made therefrom in real practice."[12] Smeaton used a carefully constructed, small-scale test apparatus which permitted precise measurements of water used, head, and weight lifted. He was fully aware of the dangers of making full-scale inferences from model data:

> What I have to communicate on this subject was originally deduced from experients made on working models, which I look upon as the best means of obtaining the outlines in mechanical enquiries. But in this case it is very necessary to distinguish the circumstances in which a model differs from a machine in large; otherwise a model is more apt to lead us from the truth than towards it. Hence the common observation, that a thing may do very well in a model that will not answer in large. And, indeed, though the utmost circumspection be used in this way, the best structure of machines cannot be fully ascertained, but by making trials with them, when made of their proper size.[13]

Smeaton properly defined the "power" developed by a water wheel as "weight raised multiplied by the height to which it can be raised in a given time," provided the motion of the body raised be "slow and equable." He defined the theoretical maximum "power," or "effect," available from the water as the product of "the quantity or weight of water, really expended in a given time, by the height of a head so obtained."[14] Realizing that power actually developed was the net of frictional losses in the machinery, he made measurements accordingly. To assure nonaccelerated lifting of the measuring weight, he provided a button-operated clutch mechanism such that the weights could be set in motion before engaging the water wheel. He also provided a separate hand-operated pump to keep the head constant during each experimental run.

Smeaton securely established the maximum efficiency obtainable in practice from conventional plane-float undershot wheels to be 1/3 and that of overshot wheels to be 2/3. Smeaton thought that the difference between undershot and overshot wheels resulted from the inelastic impact of the stream on the undershot floats.[15] He argued, therefore, that the theoretical maximum efficiency of an overshot wheel was 1, of an undershot wheel, 1/2. Smeaton's observations of various full-size installations corroborated, although crudely, his experimental analysis. He did not develop the kinetic energy considerations which would have fully explained the observed differences in wheel efficiencies.[16]

By the time Smeaton's work was published in England, stylistically and intellectually contrasting classical hydrodynamic theory was reaching maturity in the hands of continental scholars: the family Bernoulli,

John Smeaton's scale model water wheel test apparatus, 1759. (From John Smeaton, "Experimental enquiry concerning the natural powers of wind and water..." [London, 1794].)

Jean d'Alembert, and Leonhard Euler. Euler applied his immense intellectual gifts directly to the theoretical problem of reaction water turbines. Before 1750, a professor of mathematics and physics at Göttingen, Johan Andreas von Segner, had proposed a water turbine not unlike the "Aeolipile" of Hero of Alexander. Euler, then at the Berlin Academy of Sciences, became interested and, in 1750, offered a memoir containing a hydraulic analysis of the proposed machine. He correctly deduced that for efficient operation such a turbine would have to attain very high velocities. Euler became so interested that he presented two more papers on turbines of his own design. Using fixed, angled conducting tubes to adjust the velocity of the incoming water to that of the wheel, he showed his machine to have a theoretical efficiency of 100 percent at designed rpm.[17] In his 1754 memoir to the Berlin Academy, Euler "first expressed the basic relationship of reaction turbines, by equating the torque to the change in the moment of momentum of the fluid as it passed through the rotating part."[18]

Although in 1944 Jakob Ackeret, at Zurich Technical University, had a small test machine built to Euler's specifications which gave an efficiency of 67 percent,[19] Euler's design was not adopted in his own time for several reasons. First, Euler's esoteric mathematical analysis was inaccessible to virtually all contemporary practitioners. Second, and perhaps more critically, that theoretical analysis depended upon the properties of an ideal fluid, inviscidity and irrotationality. While design inferences drawn from such abstract scientific theory could and ultimately did serve as an appropriate and compelling outer boundary or goal for practical designers to attempt to attain, such theory alone could not suffice as a foundation for practice. Means of testing and evaluating real, full-scale machines in light of theory, but not only by theory, were imperative. Those means did not exist in Euler's time, and, indeed, did not exist until Fourneyron applied the Prony brake to his own turbine. Finally, Euler's turbine required a precision in forming and shaping parts and a perfection in bearing manufacture unattainable until after 1850.

Two French scientist-engineers complete the scientific and analytical background to creation of the modern water turbine tradition. Jean Charles Borda in his 1767 analysis of ideal water wheels enunciated the design desiderata for all such machines: that water should enter the wheel without shock and leave it without (relative) velocity. Borda also introduced the concept of "stream tubes" parallel to the direction of flow, rather than perpendicular "slices" or fluid elements, for analysis of hydrodynamic flow in pipes, in passages, or through orifices.[20] Borda's design precepts were endorsed by Lazare Carnot, and by virtually every succeeding authority on water turbines. Finally, J.H.P. Hachette, a professor of mathematics at the Ecole Polytechnique, published a treatise on machines in 1811 in which he formalized previous notions of "duty," "power," "effect," "*force viva*," and head as the modern relationships between potential and kinetic energy and work.[21]

By the end of the French Revolution, then, there existed basic empirical understanding of water wheels plus relatively complete theoretical understanding both of ideal fluid turbines and of potential and kinetic energy and work relationships. The critical variables to consider in turbine design (energy, shock, flow velocity) were well known. What did not exist was either a practicable, efficient water turbine or a full-scale means of evaluating one if it had existed. Although Watt had furtively adopted the indicator diagram in lieu of gross coal consumption to measure output of piston steam engines, no general technique for comparing the output of prime movers had been developed. Hydraulic machinery was still compared in terms of water used (potential energy) divided into work actually done, which could only be defined for full-scale machines, not models such as Smeaton's, as bushels of grain ground, blows of a hammer struck, or volume of water pumped from a mine.[22] The incommensurability and susceptibility to undefined variables of such measures is obvious. The best theoretical analyses could not then be connected with the best empirical investigations, nor could either of them be linked to actual, full-scale water wheels or turbines. As a result, a cumulative, progressive water turbine tradition was also lacking.

The early years of the nineteenth century did bring several proposals for relatively sophisticated hydraulic machines, as well as the usual spate of poorly designed contrivances.[23] J. V. Poncelet in 1825 proposed an undershot water wheel with curved blades which, when tested in later years, gave efficiencies of 60 to 75 percent.[24] In Poncelet's design, water ran up the blades, stopped, and fell back under the pull of gravity, thus fulfilling Borda's maxim.[25] Poncelet also proposed, but did not construct, an inward-flow turbine the next year, 1826. Claude Burdin, a professor at the Ecole des Mines at Saint Etienne designed an outward-flow turbine in 1822. Although generally similar in configuration to that of Fourneyron, the execution of Burdin's turbine was inadequate: "the relatively flat surfaces of the mobile vanes would fail to secure the full effects of impulse and reaction essential to the significant operation of the wheel."[26] Realizing this deficiency himself, Burdin reverted to the Euler form in an 1828 design, although he may not have known of Euler's prior work.[27] He afterward abandoned turbines altogether to concentrate on locomotives.[28]

Although he thus had numerous antecedents, Benoit Fourneyron designed and built the first modern, high-efficiency water turbine. More importantly, he established the efficacy of his design by objective, replicable tests. Fourneyron was the son of a land surveyor, and had studied at Saint Etienne under Burdin. After leaving school, Fourneyron undertook a number of projects in connection with the quickening industrialization of the Loire region: surveying iron carbonate deposits and railroad routes, managing coal mines, erecting rolling mills, and so on. His work with rolling mills apparently led him to an interest in hydraulic machinery.[29] Fourneyron began work on his turbine in 1824, but did not announce his results until 1827, when he submitted a memoir in a competition sponsored by the Société Industrielle de Mulhausen.[30]

Fourneyron's turbine was of the radial outward-flow type. Inlet guide vanes and the turbine blades themselves were both carefully constructed to insure smooth, shockless entry and zero relative velocity outflow.

Structural Antecedents

Scale of 25 millimetres to a metre.
The Fourneyron outward-flow water turbine. (From Morin, 1843.)

The turbine was designed to run submerged, but ran almost as well above the surface of the water. It was not, however, efficient at partial gate. In 1836, at the request of the French Academy of Sciences, Arthur Morin, of the School of Artillery, evaluated five different Fourneyron turbines, also using a Prony brake. Morin found that the Fourneyron turbines developed efficiencies of 70 to 78 percent, were able to run under small or great heads, could operate submerged or above the tailrace water (both critical capabilities in regions plagued with frequent high backwater, as were most areas of France), and presented considerable constructional advantages over conventional water wheels.[31]

The Fourneyron type nevertheless was ultimately supplanted. Entry was not perfectly smooth, and because of the passage shape in the turbine, flow separation, leading to instability, occurred. Construction was

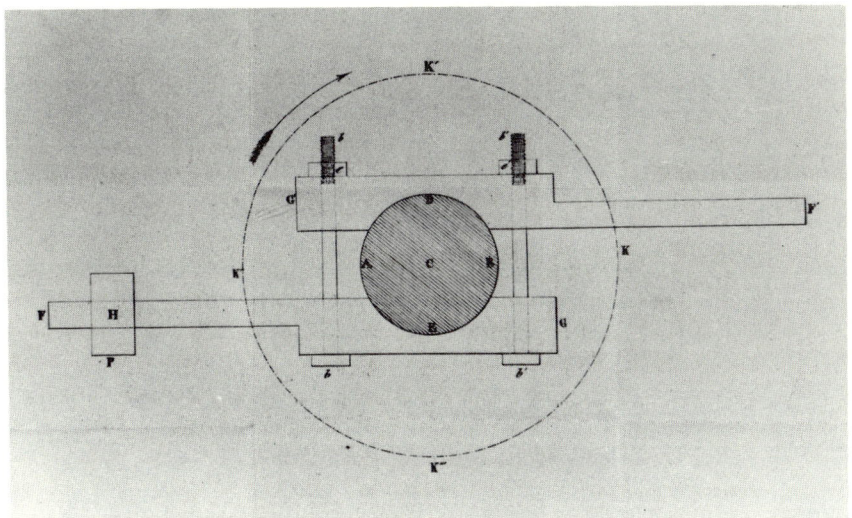

Prony's original sketch of the friction dynamometer, or Prony brake. (From M. Prony, "Rapport: Machines à Vapeur du Gros-Caillou; Deuxième Note: Sur un moyen de mesurer l'effet dynamique des machines de rotation," *Annales Des Mines*, 1st series, 12 [1826]: 91–99.)

complex and therefore expensive. Fourneyron did install over a hundred of his turbines, and, in 1855, solved the flow separation problem by designing and patenting a radial-conical diffuser.

Perhaps more critical to the emergence of a tradition of technological practice than the turbine itself, however, was Fourneyron's utilization of the Prony brake, or dynamometer, to determine net output. Fourneyron was the first person to actually use the Prony brake. Baron Riche de Prony, military engineer and later director of the Ecole des Ponts et Chaussées, had first proposed his dynamometer in a somewhat obscure memoir published in 1822.[32] He offered a clearer explanation, together with a partial diagram of his device, in 1826 in a note attached to an evaluation of a steam engine.[33] The idea of the Prony brake was quite simple, however, and was nicely summarized by Ellwood Morris in his 1843 memoir on the subject in the *Journal of the Franklin Institute*. Morris's memoir also beautifully illustrates the difficulty of relating abstruse scientific concepts to the experiences of ordinary technological practitioners:

> It is well known that *friction*, in all machines, consumes or wastes, more or less of the motive power, and that it may, at will, be so augmented, as to consume the whole power of a prime mover; thus by gearing to any steam engine, or other motor, a train of wheel work of sufficient dimensions, and

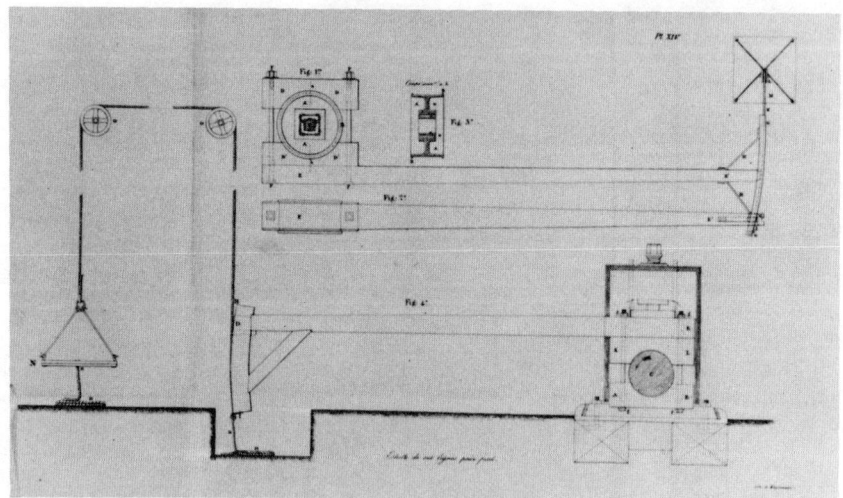

Fourneyron's development of the Prony brake, 1829. (From M. Fourneyron, "Sur l'emploi du frein M. de Prony, pour mesurer l'effet dynamique d'un moteur," *Bulletin de la Société Industrielle de Mulhausen* 2, No. 6 [1829]: 14–37.)

extent, it would be possible, so to multiply the friction of the moving parts, as to use all the power developed by the motor, and leave none for application to useful purposes.

This must be so evident upon reflection, that further illustration seems unnecessary, and it follows, consequently that in any case where the friction produced, consumes the whole power of the machine; if the amount of that friction could be, by any process, correctly ascertained, it would furnish an accurate measure of the power of that machine; *for the retarding force, which consumes the power of a motor, must be an accurate meter of that power.*[34]

The Prony brake consisted of a pulley attached to the output shaft of the water wheel around which was bolted a wooden friction collar. The collar was fastened to a beam to which weights were attached. As the pulley turned, the collar was tightened until the speed of the machine was stabilized. A formula taking the diameter of the pulley and the length of the beam multiplied by the weight suspended from the beam times the rpm of the water wheel (measured by a counter which was initially just a bell device) yielded directly force times distance per unit time, or power.

Like any other device, however, the Prony brake required considerable development before it would provide consistent, reliable results, a development which Fourneyron undertook concurrently with work on his turbine.[35] For the first time, in place of heterogenous bushels of grain, engineers had in the Prony brake a precise, replicable, objective measure of motor output. Different prime mover designs, different developmental modifications to existing designs, could for the first time be rigorously evaluated and compared. Indeed, virtually every innovation in the emerging water turbine tradition was evaluated by the Prony technique. Later, hydraulic and electric dynamometers supplemented the Prony brake for engines of higher speed and greater power.

Fourneyron published the first description of his use of the Prony brake in 1829 and the first detailed description of his turbine, together with a theoretical analysis, in 1834.[36] Significantly, both the order in which Fourneyron published his results and his decision to use the Prony brake in the first place may have been influenced by the heavily academic orientation of the dominant institutions of French technology at the time. The importance of industrial-society prizes in the highly centralized French system and the dominance of scientific ideology—an emphasis on formal, mathematical expression, in contrast to more empirical or utilitarian approaches—within those societies may have shaped the way Fourneyron presented his turbine.[37] Yet that that causation was one way, from scientific to technological practice, is not at all clear. Rather, norms of testability, replicability, and full disclosure to a relevant community of practitioners may have emerged more or less contemporaneously in both science and technology.

Practice of science in differentiated "communities without propinquity" has been common at least since the fifteenth century and probably since the thirteenth. Yet such communal practice does not imply homogeneous or unchanging norms governing, specifically, rigor in experiment or candor in reporting. Some scientists—notably Newton on optics, and Priestly, Cavendish, Lavoisier, and LaPlace in chemistry—gave relatively complete reports of the experimental apparatus used in some aspects of their work early on. Certainly their experimental reports are very different from Galileo's still controversial "thought experiments."

Thomas Kuhn argues that mathematical rigor in theory, quantitative precision in measurement, and candor in describing experimental apparatus became characteristic of different scientific specialties at different times. Kuhn suggests that these dimensions, first apparent during the scientific revolution, distinguished the "Baconian" sciences (chemistry and biology) from the "classical" sciences (astronomy and mechanics) until well into the nineteenth century. Kuhn, moreover, notes specific national patterns in the pursuit of science. He points out the extreme mathematical bias which imbued the French Academy of Sciences until long after 1815 and relegated the "Baconian" sciences or any nontheoretical work on technological practice to limbo. Clearly this circumstance is central to Fourneyron's course and, ultimately, to his fate. Kuhn, following Robert Merton, suggests that the opposite bias permeated

British science through the eighteenth and early nineteenth centuries, and that characteristic national patterns have continued to typify science in various countries since.[38]

Donald Cardwell, in contrast, notes an apparently independent but parallel trend towards greater quantitative accuracy and greater experimental candor in British technology. Cardwell examines the critical role which publication of detailed data about the performance of the Cornish steam engine played in promoting the rapid, incremental improvement of those engines after 1811. These reports were regularly republished in France after 1816.[39] Cardwell also relates the new experimental and reporting norms to the rise of professionalism in science. In his discussion of Gay-Lussac's experiments leading to the formulation of the gas law (1802), he points out that Gay-Lussac "had the new professional touch; he gave full details of his apparatus and the experiments he carried out."[40]

Complete description of experimental apparatus, then, seems to have become *de rigeur* in science, to have become a binding norm for all would-be practitioners, only in the first years of the nineteenth century. In this evolution of fundamental values for both science and technology, the two enterprises may have been very closely related. Certainly the role of norms governing experiment and reporting, and especially the existence of distinctive national styles or patterns in the pursuit of science, will prove crucial to historical understanding of the turbojet revolution.

Significantly, Fourneyron's work did not receive wide currency until the publication of Morin's test results in 1838.[41] The following ten years, however, saw the rapid diffusion of the Fourneyron turbine and the proliferation of other designs. The turbine was introduced into the United States by Elwood Morris, a classic "linker,"[42] a man attuned, probably because of language skills, to European developments. Morris built his first turbine in the United States in 1843. His *Journal of the Franklin Institute* account of his endeavor emphasizes both the importance of communal corroboration of turbine efficiency provided by the use of the Prony dynamometer and the usual travails of any protagonist of technological revolution:

> The above series of experiments on turbines, made by Morin, corroborated by other experimenters, and finally sanctioned by the approbation of the Academy of France—after undergoing the scrutiny of a special committee of their body deputed for that purpose—are beyond the reach of cavil, and must carry a conviction of the value of these hydraulic motors to the minds of all who are capable of appreciating subjects of this nature.
> Nevertheless, when the writer, some time ago, formed the resolution to

attempt the introduction of these valuable water-wheels into use in our country, and entered into a business arrangement with Merrick & Towne, the well known machinists of Philadelphia, for that purpose, they were met at the threshold of their enterprize by an unusual degree of scepticism, which, unfortunately, received here, as in France, a slender resting point from the abortive efforts to construct successful turbines, which were made by some not sufficiently acquainted with the principles of these motors, whose attention had been drawn to them by the articles upon turbines, published, by the writer, in this journal, during the autumn of last year.[43]

In Europe, other designs were quickly introduced. Several constituted major improvements. In 1841 Jonval introduced an axial-flow reaction turbine which overcame some of the constructional disadvantages of the Fourneyron design. Jonval was also among the first to employ the draft tube, which permitted the turbine to be mounted above the penstock and tailrace and therefore to be easily accessible for inspection and maintenance.[44] Girard, about 1850, designed an axial-flow partial-admission impulse turbine which gave good efficiency at partial gate. In the 1860s in England, James Thomson (William Thomson's brother) developed a highly efficient inward-flow radial turbine with movable entry guide vanes.[45] Other variants and detailed development continued the tradition.[46]

In the United States, although Morris had introduced the Fourneyron turbine, the most significant and far-reaching water turbine work was done by James B. Francis and Uriah A. Boyden. Francis was born in England, the son of a railroad superintendent. He worked on a canal project in England before emigrating to the United States to work on construction of the New York, Providence and Boston Railroad. In 1834, he entered the employ of the Proprietors of Locks and Canals on the Merrimac River, Lowell, Massachusetts, becoming superintendent of Proprietors in 1837.[47] The Proprietors was a corporation chartered by the state, originally in 1792, to control and improve navigation, and, incidentally, to sell power rights on the Merrimac River. By the early 1840s, the Merrimac represented perhaps the world's largest and most intensively developed water power system.[48] Francis's duties included canal and lock construction and supervision, but were extended to include even cotton mill construction. He was charged with making optimal and profitable use of the whole system's power and navigational resources. It was pursuant to his desire to maximize the power obtainable at the Lowell works that Francis undertook the comprehensive experiments on water turbines and on flow in canals and over weirs known collectively as the Lowell hydraulic experiments.

Francis first learned of water turbines through articles published in the late 1830s and early 1840s in the *Journal of the Franklin Institute*, especially those either written or translated by Morris. Francis did not begin his work on turbines, however, until 1849, when the Proprietors group bought rights to the turbine designs of Uriah A. Boyden. Boyden, a modestly educated but talented and widely experienced civil and hydraulic engineer, had constructed his first turbine in 1844 for one of the mills to whom the Proprietors group had sold power rights. Boyden installed additional, improved versions of his turbine for the same company beginning in 1846.[49] The Boyden turbine was essentially similar to that of Fourneyron, but Boyden had added a submerged conical diffuser to reduce instability and improve efficiency. Boyden's own tests of his first turbines, using a Prony brake, showed efficiencies of 78 to 88 percent.[50]

Francis and Boyden worked together in building and testing the Proprietors' Lowell turbines. A large-scale test facility was constructed on one of the Lowell canals. Extensive tests began in 1851. Virtually every significant variable in the turbine system was carefully controlled and measured. A large Prony dynamometer was employed to measure output. At Boyden's suggestion, a "hydraulic regulator," or damper, was added to damp out the inevitable irregularities, or "jolts," which resulted from uneven friction on the brake. Various lubricants were tried on the brake—water, resin oil, and mixtures—to get optimum smoothness and consistency. A revolution counter consisting of a bell struck mechanically every fifty revolutions was used. A vane, complete with its own damper, was installed in the wheel pit to indicate the direction of water outflow from the turbine. A straightening grate, which anticipated later wind-tunnel practice, was installed in the outflow channel to secure smooth, even flow over the weir where the measurements of water consumption were made. The variation in the weir itself "did not exceed 0.01 inch," and a "hook gauge," also of Boyden's invention, was used to measure the depth of water flowing over the weir. The hook gauge was claimed to be accurate to ".001 feet," or, with special lighting, to ".0001 feet." Francis also conducted equally classic experiments "on the flow of water over weirs and in short rectangular canals."[51] He was inspired to do so by the gross "discordances" in previous work.

Although the Boyden-type wheel that Francis tested gave excellent efficiency—75 to 80 percent at optimum gate—he was still dissatisfied. Francis analyzed the flow within the Boyden turbine according to a simple stream-tube hypothesis and found the theoretical results to contradict the direction of outflow indicated by the directional vane. He could only conclude that either a sharp change in direction had occurred or, more likely, flow separation with consequent energy loss was occurring.[52] Francis therefore conceived an elegant inward-flow turbine with "guide vanes and runner blades ... carefully designed (on the basis of mean velocities in relative motion) to yield separation-free entrance to and minimum-velocity exit from the runner."[53] Although a similar wheel had been patented by Samuel B. Howd of Geneva, New York, in 1838, as well as suggested by Poncelet in 1826, inward-flow turbines previous to that of Francis had been crudely built and could make no pretense to the design sophistication or efficiency (about 80 percent) of the Francis turbine.[54] The Francis type, moreover, exhibited significant constructional advantages over the Fourneyron-Boyden types.

Both Francis and Boyden were aware of European theoretical analyses of water turbines, including Euler's, but found them of little use because of their necessary idealization on the one hand and their mathematical complexity on the other. Edwin Layton, in fact, argues that Francis's and Boyden's work represents the amalgamation of earlier, indigenous American millwright craft traditions, perhaps most highly developed in the work of the Parker brothers, and the more empirical elements of European traditions.[55] Clearly, the eminence in the Lowell hydraulic experiments of experimental technology and technique, centered on the Prony brake, and of the scientific insight such procedures and equipment embodied, sets Francis and Boyden apart from purely craft or empirical practice. The Lowell experiments, then, represent both the highly effective development of a dynamic tradition of technological practice—water turbines themselves—and the further articulation of a related but separate and hierarchically higher or more general tradition of technological testability. Francis's turbine was technically successful, although not widely adopted because of its extraordinarily expensive construction, while his experimental techniques became the model and ideal for two generations of turbine practitioners.[56]

Water turbine development in the United States continued in the directions of greater efficiency and, particularly, of larger size and power. In 1855, the year the first edition of *Lowell Hydraulic Experiments* was published, A. M. Swain, a mechanic who had worked in the Lowell shops on the Boyden and Francis turbines, introduced a turbine wheel which combined inward and downward flow. He eventually perfected a superior gate mechanism which overcame the inefficiency of the Francis turbine and most other turbines at partial gates. By 1875, the developed Swain wheel could attain efficiencies of 66 percent at half gate (versus 30 percent for the Francis) and of over 83 percent at full gate.[57] Other similar types, with many slight variations, were introduced by a large number of manufacturers. In general, those manufacturers offered a line of "stock," or off-the-shelf, wheels of different sizes, powers, and speeds, from which customers could choose.[58] Very rarely during this period were wheels custom-designed for specific installations. Stock wheels served the needs of rapid industrialization quite well. Cumulative

Structural Antecedents

progress did occur: by 1895, the advance from the original Fourneyron type provided "turbines of equal power in one-half the space and at one-fifth the cost, being single castings of iron or bronze, instead of built up of many parts.[59]

If water turbines in the United States became standardized, testing was both standardized and institutionalized. Virtually anyone doing serious work on water turbines employed what became known as the Francis technique—a Prony brake, a Francis-type weir, and careful measurement. From the mid-1880s on, most manufacturers of turbines had their wheels tested under standard conditions in the Holyoke testing flume of the Holyoke Water and Power Company. The Holyoke flume had been built originally in 1872 by James Emerson, who had begun comparative testing of water turbines in 1869.[60] Emerson's flume was used until 1880, when a new, much improved flume was designed and built under the direction of Clemens Herschel. Between the opening of Herschel's flume in 1882 and 900, over 1200 different wheels were tested.[61] Tests were done on a commercial basis and the results communicated confidentially to the manufacturers. The Holyoke flume employed a Prony brake and a Francis-type weir, as well as measuring instruments of Herschel's design.[62] Although test results were confidential, the repute of the Holyoke tests was such in the United States that most important manufacturers submitted their wheels for testing: manufacturers needed the test data both to set their own contractual performance guarantees and to sell their wheels.

By the mid-1880s, at least in the United States, a clear bifurcation between turbine technology, as development or design, and a tradition of testability seems to have occurred. Turbine design remained confidential, proprietary, patentable, and usually empirical. Turbine testing became communal and consensual, the common property of a well-defined community of practitioners. Thus a tradition of testability, comprising both techniques of testing (hardware and its utilization) and a set of behavioral norms, had evolved quite distinct from water turbine practice itself. To be a serious water power engineer was not only to design water turbines, but also to hew hard and fast to the communally sanctioned norms of careful testing, precise measurement, rigorous experiment, and "scientific" objectivity.

This stable tradition of technological testability and this orderly development of water turbines were rudely disrupted by a purely exogenous if not alien force: electricity. Where 100 or 200 h.p. was about as much as most people would expect to distribute by mechanical means out of one water turbine, now 5,000 or 10,000 or 20,000 h.p. could be absorbed by one dynamo and the resulting power distributed over half a state. Suddenly minor fluctuations in rpm or in output held disastrous potential for electrical distribution systems. Suddenly slight reliability problems that had been nuisances became lurking catastrophe. Suddenly the stock wheel, the Holyoke flume, the Prony brake itself, were all painfully obsolescent.

The driving of electric generators required increased turbine power, higher turbine speeds, and close speed regulation. At first, multiple reaction wheels were placed on a single shaft for more power, and improved forms of gates with automatic governors were designed to secure speed regulation—all of which resulted in increasingly complex and therefore failure-prone mechanisms. Designers therefore adopted volute casings with draft tubes mounted above the penstocks to at least insure accessibility if something did go wrong. Nevertheless, the consequences of losing a high-horsepower unit remained unacceptable. As a result, engineers returned to large, heavy, but simple reaction turbines which furnished ample power and controllability but which, because of their size, turned quite slowly.[63] Efficient electrical generation demanded a large, simple, powerful, yet high-speed turbine.

In the years after 1910, two men, one American, one European, independently developed a solution. In the United States, an engineer, Forrest Nagler, realized that the limitation on turbine speed was in large measure the result of frictional resistance—that for higher speeds, the wetted surface area of the turbine had to be reduced. Nagler first eliminated the stiffening band around the circumference of the runners, since it produced no thrust and was subject to significant skin friction. Next, he made the runner vanes radial (resulting in an axial flow turbine) "because this required the least exposed surface." Nagler's experiments "led to what is now called a propeller runner, which, when associated with guide vanes and a whirl chamber, provided a new type of hydraulic turbine giving nearly double the speed which had been before attained by the use of the Francis runners for the same conditions of power and head."[64] Nagler's first commercial unit was installed in 1916.

Meanwhile, Dr. Victor Kaplan of Brunn, Czechoslovakia, created an even more advanced turbine. In 1912, Kaplan proposed a propeller-type turbine with blades mechanically adjustable in pitch. Because of the reduction of skin friction, such a turbine would offer the same efficiency advantage at full load as the Nagler turbine. But in addition, since the blades would be adjustable, the Kaplan design also could run efficiently at part load and could maintain a constant speed over a wide range of load and flow conditions. In the course of his experiments, Kaplan "discovered the marked advantages to be obtained by adjusting the blades and the wicket gate simultaneously,"[65] and such mechanical inter-

Structural Antecedents

connection became, with the more basic concept of controllable pitch blades, the distinguishing feature of the Kaplan turbine.

Although Kaplan applied for European patents in 1913 and for U.S. patents in 1914, the World War delayed commercial construction of a Kaplan turbine until after 1920. Once the Kaplan turbine was shown to be a practical proposition, it was widely adopted. An automatic governor was introduced in 1928, and various constructional advances were made continuously. From researches begun in 1928, an American, R. V. Terry, introduced just before the Second World War a Kaplan-type turbine which obviated the need for mechanical interconnection between the gate and the turbine blades. The blades automatically adjusted themselves to optimal pitch as a result of the movement of the center of hydrodynamic pressure on the blades' surfaces as water flow was changed by regulation of the inlet gate.[66]

After 1890, the direction of water turbine development was clearly very closely related to the demands of electrical power generation and distribution. Simultaneously, design of dynamos, switching apparatus, and electrical grids had to be matched to the power water turbines could provide. This systems interaction, in which each technology defined the primary selective criteria (or the selective pressure) for the other, exemplifies technological co-evolution.

Testing techniques and turbine design approaches, no less than water turbines themselves, were revolutionized by the coming of electricity. No known friction brake could absorb the power of a really large turbine. In other applications, hydraulic dynamometers had been developed. But since virtually all large water turbines or wheels were used to generate electrical power, clever engineers soon devised means, using power generated in full-scale installations, to infer the output of the colossus-sized turbines.[67] Yet such *ex post* testing was hardly adequate: it was one thing to have a single wheel do poorly at Holyoke, quite another to have an entire, multi-million-dollar hydropower installation fail.

Fortunately, the closing years of the nineteenth century saw secure establishment of flow similarity relationships which would permit effective and relatively predictable application of scale model results to full-scale turbines. The work of Osbourne Reynolds on flow similarity and on the transition between laminar (smooth) and turbulent flow (1883),[68] together with the work of Lord Rayleigh in developing dimensional analysis (1892),[69] enabled engineers for the first time to cope confidently with scale effects and to thereby use inexpensive and easily observed and manipulated turbine and installation models in design. This capability, together with a generation of more mathematically adept practitioners, also inspired the reintroduction of formal scientific theory, essentially Euler's, into the study and design of turbine systems. These trends reached fruition within water turbine practice with the introduction of the concept of specific speed by R. Camerer in 1903.[70]

Thus, within fifteen years of the coming of hydroelectricity, practitioners could deploy revolutionary and well-integrated new theoretical, experimental, and design resources in turbine development. Moreover, hydroelectric engineers quickly realized, very much as James B. Francis had half a century earlier, how crucial total-systems engineering was. Their problem was not merely to design a high-performance turbine, but to design a turbine, an intake or penstock system, surge tanks, casing, gates, draft tubes, tail-races, shafting system, access galleries, heavy crane installations, and so on.[71] All had to be internally integrated and had to fit within the hydrological, legal, and electrical power constraints of a certain system.[72] To find the power of a turbine, that which had been impossible until Benoit Fourneyron first applied a Prony brake to his water turbine, now was perhaps the least of an engineer's worries.

Yet the radical transmutation water turbine practice and water turbine testing underwent with the coming of electrical power should not obscure the generic importance of the nineteenth-century experience. Until Fourneyron's use of the Prony brake, no means existed for clearly relating scientific theory and careful empirical study to full-scale, real-world water power practice. The epoch-making creation of a tradition of technological testability for the first time permitted practitioners to know which designs and which modifications represented real progress and to know precisely how close they were approaching the theoretical ideal.

One other tradition of water power practice, that relating to what are commonly called Pelton wheels, is worthy of mention, for three reasons. First, Pelton wheels have as their central distinguishing characteristic a uniquely shaped split bucket (called the Pelton bucket) which some steam and gas turbine designers later attempted to use. Second, the creation of the first Pelton water wheels yields further insight into the relationship among traditions of technological practice and their relationship to hierarchically higher traditions of testability. Third, the development of Pelton wheels again exemplifies technological co-evolution.

The Pelton is a pure impulse wheel. It exploits full conversion of potential (head or pressure) energy into kinetic (or velocity) energy in an efficient nozzle. The resulting high-speed water jet is directed into "split buckets" fastened around the rim of the wheel. The bucket shape provides optimum conversion of the kinetic energy of the water jet into motion of the wheel. Ideally, the water should fall away from the wheel with no residual (relative) velocity and the wheel should run at a peripheral velocity one-half that of the water jet. The Pelton wheel is

Structural Antecedents

especially well suited to high-head (high-pressure) but relatively low-volume water power resources. Those water power conditions were prevalent in the late-nineteenth-century mining districts of northern California, and that locale marked the origin of the Pelton wheel.[73]

Prior to introduction of the Pelton type, indigenous, itinerant wheelwrights commonly erected water wheels known locally as the "hurdy-gurdy" type: "The buckets were shaped like saw-teeth, and wooden flanges covered the sides of the buckets, to confine the water; a round nozzle was used; and the general results were considered at that time highly satisfactory."[74] Yet as mines became deeper and ore qualities poorer, larger pumps for greater lifts and larger ore crushers and other machinery demanded more powerful and more efficient prime movers. It was this need which the Pelton water wheel initially would meet.

The development of the Pelton-type wheel comprises the work of at least four men. About 1872 S. N. Knight of Sutter Creek, California, introduced curved iron buckets in place of wooden boxes and a rectangular nozzle in place of a round one. Nicholas J. Coleman, Railroad Flat, California, in 1873 patented a hurdy-gurdy type of wheel with split buckets, although it does not appear that any wheels were actually constructed to his design.

In 1874, or possibly several years earlier, Joseph Moore of the Risdon Iron Works, San Francisco, consulted Professor F. G. Hesse of the University of California at Berkeley concerning the design of a water wheel "answering the following conditions: High head, good efficiency, and such construction as to admit of its being built of wood at the mill, except flanges, shaft, and such light castings as could be readily transported on pack animals."[75] Hesse realized that under those conditions only wheels "in which the energy of the water to be converted into work is received by the wheel in the form of kinetic energy" need be considered. For Hesse, a wheel of the Jonval type fulfilled the essential desiderata. But the classic Jonval type, if operated on a horizontal axle, produced an end thrust on one bearing, which, given the sizes and construction techniques prevalent in the mining districts, was unacceptable. Hesse knew that two Jonval wheels on the same shaft had been used before to balance end thrust. It then occurred to Hesse

> that two such wheels might be placed together, so as to form one wheel, and one bucket out of every pair of buckets, reducing thus the entrance angle to 0, causing an increase of efficiency. The jet entering in a direction tangential to the wheel is divided and discharges in two streams at the opposite sides of the wheel. Another advantage is to be found in the increased passage-way of the discharge-water, one on each side of the bucket, a fact which greatly lessens its weight and facilitates its free discharge.

For lack of time, Hesse performed no tests. He did furnish drawings of his bucket design to Risdon and "was afterwards informed ... that castings were made from this design, and were sent to a mine to be bolted to the rim of a wooden wheel." Most importantly, Hesse further notes, "having never contemplated taking out a patent for what I considered so obvious an improvement, I lost sight of the matter from that time."[76] Hesse nevertheless later did take out a patent covering both his bucket design and an automatic governor.[77]

Finally, Lester G. Pelton invented the Pelton water wheel:

> I crossed the plains from Ohio in 1850, and engaged in mining almost continuously until 1864, when I took up millwrighting, in connection with mining, at Camptonville, Yuba County, and other places north of that town, in which business I was employed until 1878; and during this period I constructed a number of waterwheels, of the type commonly known as hurdy-gurdy wheels, having an efficiency of 40 per cent, and upwards, according to the style of buckets used. Here, I conceived, was a chance for improvement; and in 1878 I procured the necessary appliances for testing the efficiency of buckets for pressure- or jet-wheels, and devoted most of the time for two years following to designing a bucket which would give a higher efficiency. I tested between thirty and forty different shapes of buckets, and finally noticed that a curved bucket having a jet-strike on the side, ... instead of in its center ..., gave a marked increase in the efficiency of the wheel, but caused an end-thrust against one bearing. To avoid this, I experimented with placing the buckets alternately, ... when it was but a step to combining the two curved buckets and splitting the stream.... This bucket, when tested, gave such astonishing results that I immediately took steps to secure my invention.
>
> I introduced my wheel to the public, after obtaining a patent, in October, 1880, and claim to have invented what is known as the "Pelton Water-wheel" independently, and without any knowledge whatever or aid from the efforts of others in that line.[78]

Pelton conveys an exaggerated impression of his isolation. According to Hamilton Smith, in a paper presented to the American Society of Civil Engineers in 1884, Pelton had procured a copy of Francis's *Lowell Hydraulic Experiments*. Pelton constructed and utilized both a Prony brake and a Francis-type weir in developing his wheel.[79] Thus, although credit is certainly due Pelton for invention of the distinctive buckets that bear his name, he was no isolated, ignorant "ingenious mechanic." His method was not that of crude tinkering. Rather, he was a skilled millwright with access through the medium of Francis's work to the best previous water turbine practice, to the best experimental procedure, and to the scientific information embodied in that procedure.

There were other alleged inventors of Pelton-type wheels, although less is known of them. In any event, the other inventors and their financial backers, together with the men already mentioned, engaged in running court battles over patents for most of the last two decades of the nineteenth century. The details and outcomes of those squabbles are not germane here. What is important here is simply this: the Pelton water wheel does in fact represent a genuine instance of multiple invention.[80] The essential feature of the Pelton design—the split bucket—was independently, identically, and multiply invented by several individuals. The locations were identical: northern California. Purposes were identical: to make use of the special water power resources of that area. Time proximity is very close: at the most ten or twelve years, more likely three or four years.

What is especially striking about the Pelton wheel is that it could represent either the extension of European practice to the novel conditions of California (Hesse's alleged adaptation of the Jonval design) or the radical adaptation of California practice to new and more demanding conditions (Pelton's transmutation of the hurdy-gurdy), or some combination of the two processes. More importantly, however, the Pelton wheel is clearly also the product of a more general, higher-level tradition of technological testability comprising both hardware (Prony brake, Francis weir) and a normative commitment to rigorous, "scientific" engineering standards. Pelton as much as Hesse is a disciple of this general engineering tradition. Thus, at one level the Pelton wheel may be viewed as the product of prior water-wheel or turbine traditions or, at another level, as the specific outcome of a general tradition of technological testability. That such diverse inventive processes should lead to identical results may indicate the existence of a clearly delineated but previously unoccupied niche in a technological ecology. If that is the case, the sudden, equifinal occupation of the niche from several adjacent niches is to be expected.[81]

Finally, the ultimate Pelton wheel—the 90 percent efficient, 10,000 to 20,000 h.p. units with precisely shaped, hydrodynamically correct ellipsoidal buckets, into which the original Pelton type developed by the early twentieth century—was the product of technological co-evolution. Ironically, much of that intensive development, especially the work on the ellipsoidal buckets and on needle-valve nozzle control systems, was done by William A. Doble at the rival Doble company (the Pelton and Doble companies merged under the former name in 1912).[82] As Doble himself noted in 1915,

> The progress in the art from that time (1890) has gone hand in hand with a similar development in the co-related work of the electrical engineer. Thus the hydro-electric prime-mover has brought about a great and rapid development in the water wheel, with its accessories, as it has in the development of the dynamo. A similar development has also taken place in steam-electric

· 55 ·

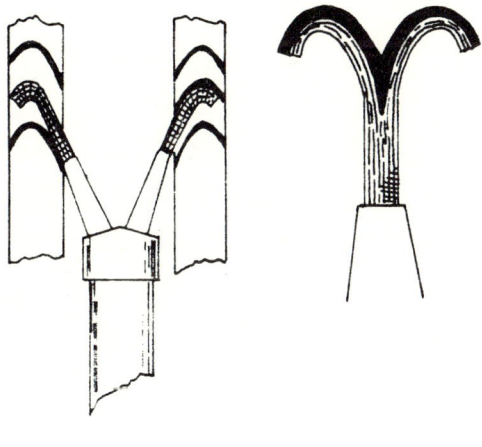

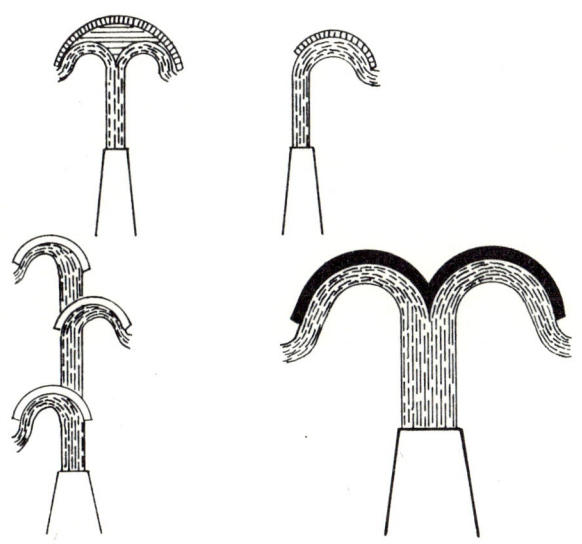

Hesse's (top) and Pelton's sketches.

prime-movers. New problems in the designing of both types of prime-movers—particularly those connected with safety, reliability and accurate regulation—have thus been brought about by the special requirements of the electric dynamo, with the co-related problems introduced by the requirements of long-distance transmission, continuous service, and speed regulation of a character heretofore unknown, in combination with the demand for prime-movers of much greater power output, high rotative speed, and to operate under extremely high heads and therefore high water pressures.[83]

Structural Antecedents

Single wheel of a 25,000 h.p. Pelton double-overhung unit. (From William F. Durnad, "The Pelton Water Wheel," *Mechanical Engineering* 61 [1939].)

The Pelton water wheel, like the water turbine and the steam turbine, co-evolved with electrical power systems.

Three elements, then, are central to the development both of water turbines and of Pelton water wheels: scientific conceptions of energy, work, and efficiency; technological testability; and technological co-evolution. Technological testability is clearly critical, for without it, idealized scientific theory, however correct, could have only limited impact on the design of real-world systems. Traditions of technological testability, moreover, are a highly effective medium by which to transform and communicate scientific information beyond the immediate scientific community to those not scientifically adept. Esoteric scientific knowledge as well as the essence of scientific methodology is materially vested in the hardware of the Prony brake and the Francis weir and is behaviorally manifest in the procedure and communal sanctions attached to that hardware's utilization. The act of testing a water turbine becomes a "behavioral template" containing a significant subset of existent scientific and technological information. Lester Pelton was not a scientist and questionably an engineer, yet he was able to possess fully the knowledge embodied in Francis's work and hardware; he was able to use that information effectively to advance technological practice, indeed to create a new water power system; he was able to join a international community of technological practitioners not merely at a systemic level but at a hierarchically higher level of more general engineering practice. This potent tradition of technological testability as it emerged in water power development from Fourneyron to Francis was perhaps not totally without precedent. Galileo in his experiments on beams and columns, Watt in his experiments on steam engines, and numerous early chemists in their experiments to some greater or lesser degree anticipated the later pattern. Nevertheless, the creation of a theoretically and mechanically sophisticated technology, dynamometers, the sole purpose of which was to provide precise, replicable tests of another technology, prime movers, appears to be a genuine departure from previous practice. Furthermore, communal ownership (in Merton's sense)[84] of test apparatus and of test results would appear to have been a unique concomitant of deployment of this particular testing technology. Certainly Watt's attempt to keep the indicator diagram secret[85] is radically different from Fourneyron's full and enthusiastic disclosure of his construction and use of the Prony brake. The antithetical responses of Watt and Fourneyron are perhaps illustrative of a profound shift both in technological practice and in technological community norms during the first two generations of the industrial revolution.

TURBINE PUMPS

Modern high-efficiency, high-pressure turbine pumps, like water turbines, emerged from the amalgamation of a rich craft repertoire of water-lifting devices and ideas with careful empirical investigations and abstruse theoretical analysis.

Rotary devices for lifting water—the ingenious Egyptian tympanum and the Archimedean screw, for example[86]—are ancient in origin. All such early contrivances apply the simple mechanical principle of the inclined plane. A second class of rotary pumping devices acts by positive displacement. That is, water in the pump is forced out by having the volume the water occupies physically constricted by the parts of the pump, in the same manner as a reciprocating piston pump. In the seventeenth century, Serviere in France suggested a positive displacement rotary pump which essentially was reinvented later as the Roots blower or air compressor. Other rotary pumps, operating at least partially on the displacement principle, were proposed in the sixteenth century by Ramelli and in the eighteenth century by James Watt.[87] Such early rotary pump designs are relevant here not because they were esspecially successful (most were probably never built), but because their basic mechanical configurations were repeatedly copied by later designers of rotary air compressors.

A centrifugal pump, as distinct from other rotary types, employs a rapidly turning impeller to impart kinetic energy (velocity) to the water to be pumped. Water enters the pump near the pump axis and is discharged at the circumference; the action of the impeller "produces a forced vortex in the contained water, with a consequent increase of pressure in an outward radial direction and a tendency to outward flow."[88] No attempt, however, is made "to convert this kinetic energy into potential energy of head, the impeller being simply surrounded with a collecting chamber, which received the liquid from the impeller and conducted it to the delivery pipe."[89] Denis Papin apparently originated the centrifugal pump in the late seventeenth century, and published full descriptions of both a centrifugal pump and a centrifugal blower in 1705.[90] The irrepressible Euler presented an idealized theoretical analysis of the action of any centrifugal pump (based on a conceptualization of his tubular turbine run backwards) in his 1754 memoir, while other eighteenth-century investigators made less original and less sanguine suggestions.[91]

The first half of the nineteenth century witnessed the proliferation of apparently independently derived, craft-originated centrifugal designs. A simply constructed centrifugal pump was in use in the United States as early as 1818, while an improved version was installed at the New York dockyards in 1830.[92] Blake (1831) employed a "disc," or single-sided, impeller in place of free-standing blades, while W. D. Andrews introduced a scroll, or volute, discharge chamber (1839) and a double-sided impeller (1846).[93] In England, John Appold conducted an exhaustive series of empirically directed experiments which culminated in his discovery of the efficiency enhancing qualities of curved rather than straight impeller blades: his pump, when tested at the Crystal Palace Exhibition of 1851 showed an efficiency of 68 percent, more than three times better than any other pump tested.[94] Meanwhile, W. H. Johnston (United States, 1846), and John Gwynne (England, 1851) proposed multistage pumps (several pumps joined in series to achieve higher heads).[95]

None of the early pumps, however, employed a diffuser, and all were used for comparatively low heads (4½ to 15 feet). Most developed efficiencies of 46 to 60 percent and were driven directly by reciprocating steam engines.[96] Before the late nineteenth century, use of centrifugal pumps was limited. Writing in 1915, R. L. Daugherty explained:

> The centrifugal pump is a relatively high-speed machine and until recent years there was no form of motive power well suited to it. In the days of the slow-speed steam engine the reciprocating pump was better adapted to the conditions. But with the introduction of the steam turbine and the electric motor the conditions were reversed. For such sources of motive power the reciprocating pump is not as well adapted as the centrifugal pump.[97]

Adoption and development of centrifugal pumps ultimately would depend upon co-evolution with prime movers.

Turbine pumps, a subset of centrifugal pumps, are more directly relevant to the turbojet revolution and represent, as does Fourneyron's turbine, the merger of empirical craft and formally scientific approaches to pump design. A turbine pump was distinguished by having the impeller "surrounded by a series of guidevanes, enclosing expanding passages, which received the liquid issuing from the impeller and converted its kinetic energy or a part of it into potential energy of head as its velocity diminished in passing through the expanding passages."[98]

The first true turbine pump was invented by Osborne Reynolds. Reynolds was born in 1842, the son of a Belfast minister. Like his father, Osborne Reynolds took the mathematics tripos at Cambridge. After a brief apprenticeship to a civil engineer, he was elected, in 1868, to a newly created chair of engineering at Owens College, later Manchester University.[99] Reynolds was a first-rate scientist and, among other achievements, formulated the law for hydrodynamic (and, by extension, aerodynamic) flow similarity. That measure of similarity, derived, as noted, in two papers of 1883 and 1884, now bears his name: Reynolds' Number.

Reynolds' contribution to turbine pump development was contained in a much broader 1875 British patent covering turbines and pumps operating in liquid or gaseous media. The patent included both the idea of multistage pumps and turbines and the use of movable guide vanes with divergent passages in pumps. Reynolds thus not only invented the turbine pump but also to some extent anticipated the steam turbine designs of Parsons and Ljungstrom.[100] Reynolds' turbine pump was very nearly identical in configuration to the Thomson water turbine, which also used moveable guide vanes.[101] The source of novelty and of efficiency in Reynold's design, however, was his complete scientific understanding of the value of divergent diffuser passages.

In his turbine pump design work, Reynolds was probably not doing completely original science. He was essentially applying old science:

Structural Antecedents

Impeller and diffuser, Reynolds designed Mather-Reynolds centrifugal water pump, 1895. (From Edward Hopkinson and Alan E. L. Chorlton, "The Evolution and Present Development of the Turbine Pump," *Engineering* 93 [1912]: 112.)

Bernoulli's theorem (1738), which held that fluid pressure in a pipe is inversely proportional to flow velocity. Yet Reynolds' application of theory was entirely novel and may have involved fundamental theoretical development on his part.[102] For technology, what is important is that Reynolds' turbine pump, designed according to his scientific precepts, was substantially more efficient than previous centrifugal pumps, especially at high heads.

Two other aspects of Reynolds' work relating to the 1875 patent are of interest here. First, soon after 1875, Reynolds "constructed and successfully operated a small experimental multi-stage axial steam turbine, essentially the same in principle as the Parsons turbine." Reynolds' turbine ran at 12,000 rpm.[103] Clearance leakage losses between the blades and casing were high, however, and Reynolds concluded that his turbine could not be developed into an effective competitor to the steam engine. He therefore discontinued work on steam turbines. It is ironic that a scientist who within ten years would formulate the fundamental law of flow similarity and who would make basic contributions to the physics of gases did not realize either that clearances would be proportionately smaller the larger the turbine and that clearance losses would thus be proportionately less, or that the steam turbine could expand steam more fully than a piston engine and therefore utilize more of its energy. Second, also ironically, Reynolds in 1885 mathematically described the convergent-divergent nozzle,[104] the efficiency-enhancing properties of which were soon to be independently and empirically discovered by De Laval and exploited by him in his steam turbine.

The first Reynolds turbine water pump actually was built in 1887 by the firm of Mather and Platt for the engineering laboratories at Owens College. That pump used fixed, not movable, guide vanes. The first commercial high-head Reynolds pump was manufactured by Mather and Platt in 1893 and marks the emergence of a community tradition of turbine pump practice. Turbine pumps were intensively developed following the commercial success of Mather and Platt's Reynolds-designed model. The first successful imitator was Sulzer Brothers, a Swiss engineering firm, which had built low-lift, diffuserless centrifugal pumps since the early 1860s,[105] and had introduced a not-particularly-successful three-stage diffuserless pump in 1894. In 1896, Sulzer introduced a four-stage turbine pump with diffusers and in 1898 designed and installed a remarkable system of high-head turbine pumps in the silver mines at Horcajo, Spain.[106] In 1900, Sulzer entered into a patent and information-pooling agreement with Mather and Platt in which both parties fully acknowledged the priority of Osborne Reynolds' work.

As a result of the success of the Mather and Platt and the Sulzer turbine pumps, a number of other firms entered the market. Among the earliest entrants were a French company headed by Auguste Rateau, and the John Richards and Byron Jackson companies, both in California. Soon after the turn of the twentieth century, the De Laval Steam Turbine Company, Allis-Chalmers, and Worthington began manufacture of multistage turbine pumps. Detailed, incremental development produced the significant improvements characteristic of a vital normal technology: in this case, shrouded impellers, improved casting techniques, new methods of balancing the end thrust of the impellers (balancing valves or pistons, as in steam turbines, or, notably, the use of double-sided impellers, which obviated the problem entirely), special bearings and lubrication systems, and seals. Some very large, relatively low-head turbine pumps were built without diffuser guide vanes. Instead, those pumps employed a volute diffuser chamber "in which the rotating liquid formed a vortex, which also converted the kinetic energy or a part of it into potential energy of head." As a result of such developmental efforts, capacity, pressure, and efficiency of turbine pumps continually increased. Multistage turbine pumps by the first decade of the twentieth century could lift large volumes of water well over a thousand feet at efficiencies of 60 to 85 percent. Furthermore, turbine

Structural Antecedents

pump units frequently cost one-fourth to one-half as much as comparable reciprocating units.[107] Turbine pumps were adopted for uses as diverse as deep-mine drainage, cross-country pipeline pumping, and, much later, rocket engine fuel pumps.

Both water turbines and turbine water pumps, then, evolved from rich, preexisting craft repertoires. Both were early subjected to the most elegant if idealized scientific analysis, the usefulness of which was long delayed. Both depended upon careful empirical experimentation in the development of actual practice, and both ultimately benefited from the amalgamation of high scientific and more empirical traditions. Development of water turbines especially also encompassed the evolution of hierarchically higher-level traditions of testability comprising both hardware and normative dimensions. Finally, all the basic configurations (but certainly not designs) of turbojet air compressors and gas turbines were somewhere anticipated in prior water turbine or turbine pump practice.

3

Prior Technology: Steam Turbines, Turbo-Air Compressors, and the First Internal Combustion Gas Turbines

Steam turbines and turbo-air compressors are direct structural antecedents of the turbojet. The turbojet is an internal combustion gas turbine, with power extracted in the form of trust rather than as shaft output. Yet the turbojet is a strict, linear extrapolation of none of these prior technologies. Rather, steam turbines, turbo-air compressors, and early experience with internal combustion gas turbines constituted the most salient aspects of the technical milieu within which the turbojet was created. Moreover, beyond this purely technical role, each of these prior technologies offered a sharply contrasting inspirational message to would-be turbojet creators: steam turbines were spectacularly successful, indeed, together with the dynamos which they powered, were the very apotheosis of late-nineteenth- and early-twentieth-century technological progress; turbo-air compressors were more modestly, and more controversially, successful; while the first internal combustion gas turbines were abject failures.

STEAM TURBINES

The idea of a steam-turbine-like device, which operates in a manner similar to a reaction lawn sprinkler, is ancient and is usually traced to Hero of Alexandria.[1] Excluding the medieval "smokejack," technically a form of gas turbine, the first recognized proposal for a steam turbine proper, a simple impulse design, was that of Giovanni Branca in 1629.[2] James Watt considered a steam turbine of the Hero type, but rejected it because he thought the high rotational speeds necessary for efficient operation unobtainable with then-available materials and fabrication techniques. Of the steam turbine Watt wrote to Matthew Boulton in 1784: "In short, without God makes it possible for things to move them one thousand feet per second, it cannot do us much harm."[3]

At least one "turbine" was produced commercially during the 1830s. William Avery in the United States built about fifty Hero, or "lawn sprinkler," turbines for use in sawmills, in woodworking shops, and, reportedly, in a locomotive. Although Avery claimed for his turbine efficiencies comparable to those of contemporary piston engines, his design proved noisy, dangerous, and unreliable, and hence was abandoned.[4] The first half of the nineteenth century saw numerous other steam turbine proposals and patents, at least one of which—that of Robert Wilson (1848)—went far in anticipating the principles of the Parsons design.[5] Yet, as Charles Parsons himself noted, all the alleged anticipatory designs "showed a want of knowledge of the properties of steam and materials, and could not have given a satisfactory performance."[6] Less charitably, the steam turbine authority Lester French observed in 1907 that a painstaking survey of early steam turbine patents (of which there were legion) reveals "a great many more features... which would contribute to an *unsuccessful* turbine than to a *successful* one."[7]

One of the earlier steam turbine proposals, nevertheless, does demand attention, for it prescribed virtually all the particulars of later steam and even internal combustion gas turbine practice. In 1853, Tournaire, a mining engineer trained at St. Etienne, read a paper before the French Academy of Sciences. He proposed a turbine to be driven by "elastic fluids," by which he meant not only steam but also externally heated air or the products of internal combustion. Tournaire drew inspiration from two sources. First, Burdin, who had also been Fourneyron's professor, in 1847 had proposed a "hot air" turbine which was to employ a multistage blower. Burdin had discussed his ideas with Tournaire. Second, Tournaire was fully aware of the cumulative insights into turbine operation developed in water turbine practice.

Tournaire proposed a multistage reaction turbine to be run either by steam or by some other expansive gas.[8] He realized that a single-stage "vapor" turbine would have "a rotative motion of extraordinary rapidity." To avoid that difficulty, Tournaire would provide for the gradual expansion of the vapor through a large number of turbine stages. From water turbine practice, Tournaire was cognizant of likely sources of loss: overly large clearances, "shocks and eddies," and skin friction. He recognized the advantages for a turbine of small size and light construction, and commented on the possible benefits of internal combustion.

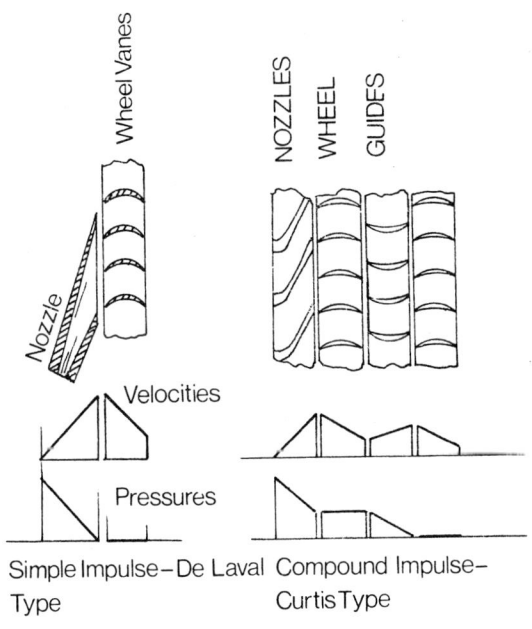
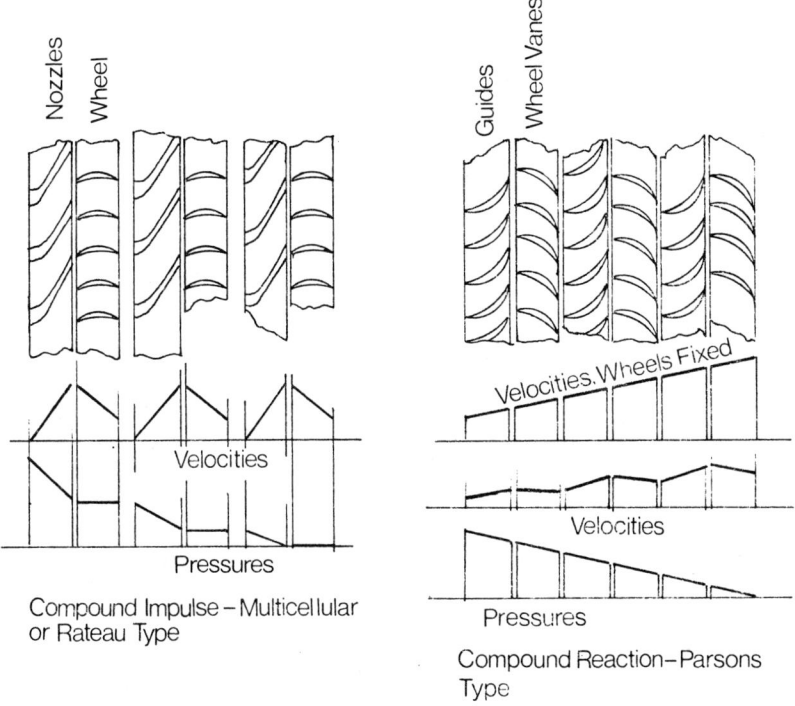

Steam turbine types.

Even though the committee of the Academy which reviewed Tournaire's paper included the outstanding water turbine specialists Morin and Poncelet, Tournaire's proposal received little notice until the twentieth century, when it was unearthed in connection with renewed interest in internal combustion gas turbines. At that time, Henry Harrison Supplee, an English propagandist for the internal combustion turbine, attributed the failure of Tournaire's ideas to the "constructional difficulties" which he had foreseen but could not overcome.

Constructional obstacles, however, are perhaps only a partial explanation for the failure of Tournaire's views to win contemporary converts. Tournaire's grasp of the basic rationale of the steam turbine was both modern and correct. Yet his presentation is purely qualitative. In 1853, the quantitative analysis of heat and of work in general, and of steam engines in particular, was not advanced enough for Tournaire to be able to specify how great the gains from his turbine should be. Unlike the situation when Fourneyron developed his water turbine, for Tournaire there was no theoretical ideal against which to measure his proposal. Tournaire knew, on the basis of water turbine criterion, why his turbine should be efficient: the last turbine stage would "discharge the fluid with a very low velocity," which indicated in water turbine practice full utilization of available energy. But nowhere does Tournaire discuss the extent to which the reciprocating steam engine failed to make full use of the expansive power of steam. Tournaire had access to no theory of energy conservation or of the mechanical equivalent of heat.

With no theoretical ideal specified, with no conception of how nearly any heat engine, piston or turbine, approached the best it was theoretically possible to do, choice between alternative systems could be only crudely empirical. A steam turbine might perform better than a piston engine, but without a well-articulated theory of heat engines, no rational expectation of the performance improvement to be obtained from incremental developmental effort on either system could be formulated. Well-founded theory not only specifies what a new system might be able to do, but also specifies what each system cannot do and indicates how closely each approaches its maximum possible performance. Furthermore, in 1853, there was virtually no technological use for high-speed rotary engines. Any turbine would require some speed-reduction mechanism.

Decisions by practitioners to continue development of piston engines rather than to begin work on turbines were thus quite reasonable. No quantitative advantage for the alternative system could be shown. The turbine might perform better, but practitioners were being asked to incur certain problems associated with high rotative speed and speed reduction in exchange for highly uncertain advantages. In essence, the absence of a well-defined theory of the heat engine to buttress a development program based on early- and mid-nineteenth-century steam turbine proposals provides, by analogy, a counterfactual test of the role attributed to scientific theory in the development of water turbines.[9] While Tournaire's is a remarkably, even brilliantly prescient memoir, it stands as clear evidence of the folly of indiscriminately considering as full technological antecedents of later systems proposals built upon incomplete, qualitative, or nontestable scientific foundations.

Steam turbines as actually developed during the last quarter of the nineteenth century drew upon intervening fundamental progress in several areas. As noted, water turbines underwent rapid and successful development following Fourneyron's work. With the publication in 1855 of the first edition of Francis's *Lowell Hydraulic Experiments*, water turbine practice had become very well articulated indeed.

During approximately the same period, piston steam engines were radically transformed, from large, slow-speed, low-pressure beam engines of the original Watt type to relatively high-speed, high-pressure, double-acting engines. The new, lighter engines proved applicable to steamboats and locomotives. More importantly, the rise of steam power led directly to the creation of classical thermodynamic theory. In 1824, Sadi Carnot, caloricist though he was, had introduced the basic ideas of a thermodynamic cycle, its reversibility, and "the principle that no cyclic device could produce more work than a reversible cyclic device for a given amount of heat and for given reservoirs."[10] Carnot's work, however, was little known and had no effect on practice—Kelvin could not even find a copy of Carnot's book in Paris when he went looking for one in the late 1840s. More directly, by the late 1850s, the work of Joule, Mayer, Helmholtz, Clausius, and Kelvin had established the essential principles of thermodynamics: the conservation of energy and the mechanical equivalent of heat.[11] With the publication of W.J.M. Rankine's *Manual of the Steam Engine and Other Prime Movers* (1859) and his *Manual of Applied Mechanics* (1864), the implications of the new theoretical insights of thermodynamics for heat engine design were readily available to technological practitioners. Those practitioners responded by intensively developing the piston steam engine. Higher speeds, pressures, and temperatures were introduced, and triple and even quadruple expansion designs manufactured.

After 1865, then, any moderately well-educated aspirant to steam engine practice could partake of well-developed if increasingly esoteric scientific understanding of heat engines, and of vigorous traditions of practice in piston steam engine design. Should he be of a mind to, he could have access to equally well-developed water turbine practice. Yet that these "cultural elements" existed does not imply that individual inventors of the steam turbine were exposed to them homogenously, responded to them identically, or were limited in their interests or motivations only to them. Four men are commonly portrayed as the early prioneers in the invention of the steam turbine: Carl Gustav Patrick De Laval (1883), Charles A. Parsons (1884), Auguste C. E. Rateau (1894), and Charles G. Curtis (1897).[12] Each invented, patented (in the years indicated), and developed a steam turbine. Yet examination of each pattern of invention reveals notable diversity even among these four.

C.G.P. De Laval was born in 1845, the son of a Swedish army captain. He attended the University of Upsala, then the Technical University at Stockholm, from which he graduated in 1866. He returned to Upsala, took courses in physics and mathematics, and received his doctorate in 1872. While working for the Kloster Iron Works in Germany (1875), he began experiments with centrifugal machinery, apparently blowers for Bessemer converters.[13] That experience seems to have led him to design a novel and ultimately successful centrifugal cream separator.[14]

Even early versions of the separator turned at 7,000 to 9,000 rpm. No prime mover then available could provide direct rotary power at those speeds. Belt drives used to step up the speed of reciprocating steam engines proved unsatisfactory. Neither helical gearing nor usable electric motors then existed. De Laval had earlier (1870) done experiments on nozzle configurations for sand-blasting equipment, so he was well acquainted with the thrust or reaction obtainable from high-velocity nozzles.[15] In order to power his cream separator, he therefore turned to the steam turbine. He first tried a simple Hero-type reaction design which, although it ran at 43,000 rpm, proved highly inefficient because of poor nozzle shape.[16]

In 1887, De Laval adopted a convergent-divergent nozzle to more fully convert steam pressure or potential energy into velocity or kinetic energy. Although the convergent-divergent nozzle had been suggested earlier by others, De Laval discovered its efficiency-enhancing properties.[17] With the exhaust of such nozzles directed onto a simple impulse wheel in the manner of Branca, De Laval was quickly able to achieve efficiencies comparable to those of conventional steam engines while running his turbine at speeds of up to 30,000 rpm.[18] At such speeds, turbine wheel peripheral velocities reached 1,200 feet per second, "or about one-half the speed of the projectile from a [then] modern cannon." Steam exited the nozzles at 3,000 to 5,000 feet per second, according to Rankine's calculations.[19] Such awesome velocities required notable innovations in steel turbine-wheel fabrication and in elastic shaft design.[20]

In 1889, De Laval introduced a two-stage velocity-compounded impulse turbine which secured higher efficiency and lower wheel peripheral velocities. The De Laval turbines were quickly applied to centrifugal pumps and blowers and to electric power generation, which by 1890 was the primary source of demand for high-speed steam engines. Before 1897, De Laval perfected manufacture of double-helical reduction gearing to step down the speeds of his turbines to levels more appropriate for driving dynamos.[21] He also began using very high steam temperatures and pressures in a quest for greater efficiency.[22] Clearly, De Laval's development strategy after 1890 was directed toward exploiting the rapidly expanding electrical market rather than toward perfecting a prime mover to run cream separators. Nevertheless, the De Laval turbine remained better suited to smaller than to larger powers: most units produced between 5 and 400 h.p.

The original De Laval turbine, then, was a simple impulse design intended to run a centrifugal cream separator. Novel features of the original design included convergent-divergent nozzles and the turbine wheel and shaft construction. Later, in response to demand for prime movers for electrical generation, De Laval developed velocity-compounded turbines and helical reduction gearing and pioneered the use of high-temperature, high-pressure steam.

Charles A. Parsons was born in 1854, the sixth son of the third Earl of Rosse. His family was one of the most remarkable in England: his grandfather had been a vice-president of the Royal Society, his father its president. The family seat, Birr Castle, Ireland, was a center of mid-nineteenth-century science. There, Parsons' father, an avid astronomer and engineer, had constructed one of the largest telescopes in Great Britain and also maintained a fully equipped workshop. Charles Parsons, with his brothers, was tutored at home, then spent two years at Trinity College, Dublin. During 1873-77, he attended Cambridge, passing out eleventh Wrangler in the Mathematical Tripos. His peers generally contended that his rather poor showing did not truly reflect his abilities—he was better at problem-solving than at "analytics."[23] While at Cambridge, Parsons presumably attended lectures in mechanical engineering and thermodynamics, although no formal courses in those topics were then offered.[24]

In 1877, Parsons became an appretice engineer in the Elswick Works of W. G. Armstrong and Company, at the time the world's leading naval ordnance manufacturer. At Elswick, Parsons initiated two interrelated projects that were to prove relevant to his steam turbine work. First, apparently inspired by the Peter Brotherhood torpedo engine (1872) and his own experiments while still at Cambridge, Parsons designed and had constructed at his own expense in the Elswick shops a high-speed, compound, four-cylinder, opposed-piston rotary steam engine of about 10 h.p. to drive a small arc-lighting generator.[25] Second, Parsons worked on torpedo propulsion. He seems to have done experiments on shrouded marine propellers at this time. He also worked on rocket propulsion for torpedos, which caused the Armstrong board of directors some consternation, since Parsons' experiments "sometimes resulted in shattering explosions under the windows of their luncheon room."[26]

Whether or not as a result of the Armstrong directors' perceiving some deficiency in his experimental method, Parsons moved in 1881 to Messrs. Kitson and Company to continue his torpedo investigations. Kitson had earlier taken up manufacture of the Parsons rotary piston engine. The torpedo project proved unsuccessful, and in 1884 Parsons left Kitson to join Clarke, Chapman and Company. He was placed in charge of their electrical department and began work on small, high-speed generator sets for the burgeoning market in ship's lighting plants.[27] Clearly then, by 1884 Parsons had acquired, by virtue of his education and apprenticeship, a secure grasp of mechanical engineering principles and of applied thermodynamics plus an intimate familiarity with high-rotative-speed machinery, shrouded propellers, and reaction propulsion, specifically rockets.

Parsons designed and built his first steam turbine in 1884 specifically to drive a dynamo. The Parsons turbine was from the very beginning completely different from that of De Laval. Rather than a single-stage, simple impulse turbine, Parsons designed a multistage reaction turbine. His reasons for choosing that configuration were quite clear:

> It seemed to me that moderate surface velocities and speeds of rotation were essential if the turbine motor was to receive general acceptance as a prime mover. I therefore decided to split up the fall in pressure of the steam into small fractional expansions over a large number of turbines in series, so that the velocity of the steam nowhere should be great. This principle of compounding turbines in series is now universally used in all except very small engines, where economy in steam is of secondary importance. The arrangement of small falls in pressure at each turbine also appeared to me to be surer to give a high efficiency, because the steam flowed practically in a non-expansive manner through each individual turbine, and consequently in an analogous way to water in hydraulic turbines, whose high efficiency at that date had been proved by accurate tests.
>
> I was also anxious to avoid the well-known cutting action on metal of steam at high velocity.[28]

Parsons' efficiency expectations were derived directly from Francis's work on the "Tremont" turbine, one of the best designs reported in the *Lowell Hydraulic Experiments*. As Parsons phrased it:

> For small differences of pressure the velocity of air or steam passing orifices or short tubes is given by the same formula as that for water. The heads of pressure have to be calculated in each case in terms of the respective fluids, and for small differences of pressure the velocity of efflux is the same for the same head in all cases. This has been recognized as a fundamental principle; it has also been verified in some of the cases we are going to consider by actual measurements.
>
> On comparing this ratio of velocity of wheel to velocity of flow in the steam turbine with that of the Tremont water turbine we find a corresponding efficiency of a little over 72 per cent.[29]

Parsons also may have turned to Francis because he was less mathematically intimidating than most European sources.

Furthermore, although direct evidence is lacking, Parsons' determination to minimize expansion of steam through each stage of his turbine also probably derived from the English engineer Robert D. Napier's demonstration, published in 1866, that for free expansion of a compressible fluid (such as steam or air) through an orifice, flow rates did not increase above a pressure differential of about 2:1. This behavior (found later to be related to shock wave formation associated with attaining local sonic veolocity in the orifice) was in sharp distinction to that of an incompressible fluid, such as water, for which flow ordinarily increased linearly with pressure differential. Napier's argument implied that the efficiency to be obtained by rapid, full expansion of a compressible fluid in a turbine would be severely limited.[30] It is not clear, at least early on, whether or not Parsons realized that a convergent-divergent nozzle, independently discovered by De Laval and theoretically described by Reynolds, could fully convert the potential or pressure energy of steam into kinetic energy virtually without loss.

Although the blades of Parsons' first turbine were straight, he adopted curved blades (the shapes of which were empirically derived) for all succeeding turbines. In his earlier machines, he compensated for gradually increasing steam volume through the turbine stages by employing blades of different size and shape in each stage; in later turbines he used increasing drum diameter and multiply compounded turbines in series to the same purpose. The first Parsons turbine produced 10 h.p. at 18,000 rpm. The early machines used brass blades fastened to steel turbine discs. Parsons also contrived a very clever self-centering multiple washer bearing, later superseded by an equally clever concentric tube design.[31] Parsons realized that his turbine, running with a condenser, could make use of the expansive power of steam to a degree no

Straight blading of first Parsons steam turbine, 1884. (From Alexander Richardson, *The Evolution of the Parsons Steam Turbine* [London: Offices of "Engineering," 1911].)

piston engine ever could. The low-pressure stages of the Parsons turbine could extract usable power from very small pressure differentials; the size, mass, and friction of a piston engine cylinder designed for the same purpose would make it impractical. Thus the steam turbine would make possible levels of performance the old system could never hope to attain.

Over 300 marine turbogenerator sets designed by Parsons were installed before he dissolved his partnership with Clarke, Chapman in 1889, apparently in a dispute over funding large-scale central-station turbine development. There ensued a bitter dispute concerning the validity and value of Parsons' patents.[32] For five years, Parsons was barred from developing his reaction turbine. After founding his own firm, C. A. Parsons and Company, he turned instead to a compound radial-outflow design he had also patented.

Parsons had great difficulty getting his turbogenerator designs adopted for land use. Initially, central station builders simply refused to consider the steam turbine in lieu of conventional reciprocating engines. In desperation, Parsons and a group of his friends formed the Newcastle and District Electric Lighting Company to build the first central station using turbines (1890). But even success at Newcastle was insufficient to induce the user community to convert to the steam turbine: "Parsons

Early Parsons portable turbo-generator set, 1886. (From Rollo Appleyard, *Charles Parsons* [London: Constable, 1933].)

had to accept a financial risk in the businesses of companies formed to supply electricity to Cambridge in 1892 and to Scarborough in 1893 in order to give them sufficient confidence to install turbine machinery."[33]

In 1893, Parsons repurchased his original patents from his former partners (for £1500) and again took up the multistage reaction turbine.[34] The success of his earlier central station installations, together with the spectacle of his turbines saving the principal station of London's power company from being shut down altogether because of the noise nuisance of its large reciprocating engines, assured more favorable adoption decisions after 1894. Rapid developmental progress with the multistage axial-flow design produced ever greater size and efficiency: 1000 kW. units by 1900, 25,000 kW. by 1912 (the Fisk Street Power Station for the city of Chicago).[35] Parsons designed and built turbines and dynamos became the new normal technology for central stations, the standard against which all other systems and variants were measured. Liberal licensing policies promoted adoption of the basic Parsons type turbine worldwide. By 1923, Parsons had installed turbogenerator sets of 67,000 h.p. and 50,000 kW. output with thermal efficiencies approaching 30 percent.[36]

In addition to revolutionizing the technology of electrical power generation, Parsons also revolutionized marine propulsion. He had begun

Prior Technology

Parsons 67,000 h.p. turbo-generator set, 1923. (From Rollo Appleyard, *Charles Parsons* [London: Constable, 1933].)

work on a small turbine-powered demonstration vessel, the *Turbinia*, in 1893. The *Turbinia* displaced 44 tons and was 100 feet in length. As first built, she had a single radial-outflow turbine engine turning a single screw. As a result of what was later recognized as propeller cavitation, the initial trials were disappointing. After the recovery of his axial-flow turbine patents, Parsons re-engined the *Turbinia* with a three-stage axial-flow turbine, each stage of which drove a separate propeller shaft. Each of the three shafts, in turn, carried three propellers. With the revised engines and new propellers, *Turbinia* attained the phenomenal speed of 34 ½ knots from 2,300 h.p.[37] Parsons had done a great deal of work with model hulls to develop an appropriate shape for high speeds, and he had made fundamental contributions, both in knowledge and in experimental technique, to the study of propeller cavitation. Moreover, in order to separate the effects of hull shape, propeller cavitation, and variations in turbine performance, he had developed an air torsional dynamometer for his model experiments as well as a full-scale torsional dynamometer for use in *Turbinia*.[38]

Again, however, Parsons met resistance to adoption. The British Admiralty of the period was not noted for openness to innovation of any sort. Private ship constructors were likewise averse to the steam turbine. Parsons therefore staged what is perhaps the archetypal demonstration

Turbinia at 34½ knots, 1897. (From Alexander Richardson, *The Evolution of the Parsons Steam Turbine* [London: Offices of "Engineering," 1911].)

of technical feasibility. The year 1897 was Victoria's Diamond Jubilee. In celebration of that auspicious occasion, the Royal Navy arranged a huge review of its own and other nations' ships. It was a tradition uninterrupted since the time of Elizabeth. The review was to be held at Spithead in June 1897.[39]

Parsons sailed (or rather steamed) the *Turbinia* down from the Tyne, no mean feat in itself for a prototype vessel, and offered a demonstration of his own. He raced the *Turbinia* up and down the serried ranks of great ships of the line at something over 30 knots, making a mockery of the picket boats sent out to stop him and nearly ramming one of them as well as a French yacht which clumsily crossed his path.[40] The *Turbinia* was clearly the fastest thing afloat. A recalcitrant Admiralty finally had to contract for a turbine-powered destroyer, H.M.S. *Viper*. Even then the Admiralty required Parsons to deposit £100,000 "as a security, in case the vessel should not come up to expectations."[41]

Spectacular demonstrations of feasibility like the *Turbinia* episode seem to occur when a system which has already had considerable developmental effort devoted to it is adapted to a new use. Such demon-

Prior Technology

Ten years' progress in normal technology: *Turbinia*, 2300 h.p., and *Mauretania*, 70,000 h.p. (From Alexander Richardson, *The Evolution of the Parsons Steam Turbine* [London: Offices of "Engineering," 1911].)

strations are a strategy for persuading recalcitrant user groups, and depend for their effect on the efficacy of the demonstrated hardware. Such efficacy requires an advanced stage of development not characteristic of entirely novel systems. A parallel to the *Turbinia's* demonstration was the use, in the mid-1930s, of General Motors' Electromotive Division's first diesel road locomotives, pulling the Burlington Zephyr and the City of San Francisco (Union Pacific), to establish new average speed and nonstop distance records. G.M. was thereby able to convince at least part of an otherwise very conservative user community (railroads) of the suitability of an otherwise well-developed technology (diesel engines) for a new and demanding use.[42]

Following the construction of H.M.S. *Viper,* Parsons built a series of ever-larger destroyers, packet ships, and light cruisers, each in turn demonstrating new levels of marine turbine performance. Less than ten years after *Turbinia* had made a spectacle of herself, the Parsons Marine Turbine Company had contracted to build the turbine engines for H.M.S. *Dreadnought* and the 70,000 h.p. turbines of the fast auxiliary-cruisers-as-Cunard-liners *Mauretania* and *Lusitania*. The *Dreadnought* perhaps best exemplifies the advantages of the marine steam turbine:

Low-pressure turbine of the *Mauretania*. (From Alexander Richardson, *The Evolution of the Parsons Steam Turbine* [London: Offices of "Engineering," 1911].)

In stark statistical terms, the *Dreadnought's* predecessors of the *King Edward VII* class with a standard displacement of 16,350 tons could steam at 18.5 knots with 18,000 h.p. from their reciprocating engines; *Dreadnought*, of 17,900 tons, steamed 21.6 knots on her trials from 23,000 h.p. But that was only the beginning of the story. She could sustain high revolutions reliably and with far less vibration and noise. On her trials she steamed, with no evident strain, 7,000 miles at an average of 17 ½ knots, a figure far beyond the capacity of any warship afloat. To have reached the same *maximum* speed would have required an additional 1,000 tons in weight and £100,000 in cost. Finally, the conditions in the engine room at high speed were transformed from a sodden cacophonous hell to a paradise of quiet orderliness.[43]

Parsons marine turbines, like Parsons generator systems, became the epitome of the new normal technology.

Although De Laval and Parsons probably shared a similar basic theoretical knowledge of thermodynamics, there is little else in common between either their design processes or the designs themselves. The De Laval and Parsons turbines are mechanically dissimilar, in principle and in execution. De Laval drew upon earlier experience with nozzles; Parsons, on water turbine theory and practice. De Laval wanted to turn a cream separator; Parsons, to generate electricity. De Laval sought high rpm; Parsons, high efficiency. The De Laval design proved appropriate for small sizes; the Parsons, to the very largest and most complex land and marine installations. Both designs were ultimately shaped by the compelling demands of electrical power generation.

Prior Technology

The third of the original four steam turbine pioneers, Auguste Rateau, was born at Royon, France, in 1863, and attended the Ecole Polytechnique, Paris from 1881 to 1883. He then studied for three years at the Ecole Supérieure des Mines. From 1888 to 1898, he taught "industrial electricity" at the Ecole des Mines, St. Etienne, and after 1901, taught at the Ecole Supérieure des Mines, Paris.[44]

Rateau began steam turbine experiments in 1894 in cooperation with the firm of Sautter-Harle, Paris. He first tried a De Laval impulse wheel with Pelton-type buckets, but found it inefficient. By 1900, he had evolved the multicellular, pressure-staged impulse turbine that bears his name. Rateau formed his own company in 1903 to manufacture his turbine and other rotary machinery of his design. Rateau's description of his definitive turbine (1904) claimed numerous advantages:

> The most recent Rateau turbine is of the action type, that is to say, expansion of the steam is fully carried out in the distributor for each group consisting of a distributor and one moving wheel. The steam therefore acts by its velocity and not by its pressure. These turbines are moreover multicellular, that is to say, they consist of a certain number of elements, each element comprising one distributor and one moving wheel. A very interesting characteristic of the type of action turbines is the possibility which it allows of leaving very considerable play between the fixed parts and the moving parts, and this greatly facilitates construction and obviates the chances of dangerous friction if the bearings should become worn or the shaft somewhat bent. Besides this the wheels revolve in a chamber where the pressure is uniform. There is for that reason an absence of longitudinal thrust upon the moving parts and no necessity for the use of dash-pots for the purpose of overcoming the effect of this thrust, although such dash-pots are necessary in re-action or drum turbines. Finally, in the action type partial injection of steam is possible, that is to say, the steam may be directed upon a limited portion of the circumference.[45]

Rateau further claimed that his turbine suffered less steam loss through leakage than the Parsons design and that his turbine therefore had less need for very precise and expensive workmanship. Initially, Rateau turbines were used almost exclusively in central power stations.

With the publication of his *Treatise on Turbo-machines* (1898–1900), Rateau had established himself as a preeminent authority on steam and gas turbine design. He later pioneered in the development of ratary air compressors and turbosuperchargers. Rateau clearly was a highly sophisticated inventor-entrepreneur trying, successfully it turned out, to enter and exploit a rapidly expanding market already possessed of a well-developed, dynamic technological tradition.[46] Rateau seems to have been motivated in his turbine work both by a genuine desire for technical improvement and by a desire to establish his own unique design as a market strategy. Like any sensible businessman, other things being equal, he preferred to be a licensor rather than a licensee. In pursuit of those multiple goals, Rateau designed an original steam turbine.

The Rateau turbine was a significantly different mechanical device from either the De Laval or the Parsons turbine. Its purpose was to compete with the earlier designs in the rapidly maturing electrical generation market. Rateau was a man extraordinarily well-informed with regard to previous work; thus, his turbine, whatever its technical merits, hardly constitutes an independent, multiple invention.

Charles G. Curtis was born in 1860, the son of a well-known Boston lawyer. He earned a civil engineering degree from Columbia in 1881 and a law degree from New York Law School in 1883, after which he practiced patent law in New York City. In 1886, he and two partners established a firm to manufacture electric motors and fans; one of the partners had, in 1882, produced the first electric fan in the United States. The new firm quickly introduced the first standard-specification (as opposed to custom-built) electric motors in 1886. Curtis left the partnership in 1888 and the next year founded the Curtis Electric Manufacturing Company to make electric railway traction motors.[47]

Curtis patented his velocity-staged turbine, similar to the two-stage De Laval turbine, in 1896. Curtis conducted his own experiments on a moderate scale for four years, then sold the rights to his design to General Electric in 1901 for $1.5 million. G.E. spent something over $5 million more to fully develop the Curtis turbine. An engineer connected with Curtis turbine development at G.E. summarized its rationale:

> The general purpose of the Curtis design is to produce results with a reasonable number of simple parts and at moderate speeds, while the Parsons turbine requires a very large number of small parts, and the De Laval turbine employs excessively high speeds inapplicable to mechanical purposes without the use of speed-reducing gearing.
> The author's opinion given was that the Curtis invention afforded great possibilities, particularly in the matter of simplicity and economy of production; that the development of commercial machines was justified by the experiments and should be begun at once, and that the development of high degrees of steam economy was to be expected with further experience.[48]

The Curtis turbine as developed by General Electric was mechanically distinct from any of the other designs discussed here and was successfully marketed for central station use. The Curtis turbine was also adapted for marine use.

Curtis, another inventor-entrepreneur seeking a foothold in a rapidly developing market, a man thoroughly familiar with all the relevant, interrelated technologies, also produced a distinctive turbine design. But in the Curtis case, that design was only fully developed in the hands of a remarkably agressive major corporation which was trying to establish a vertically integrated hegemony in electrical power production, distribution, and utilization. Significantly, at the same time General Electric took up the Curtis turbine, the corporation was also founding its famous research laboratory at Schenectady.[49] Within two years, the firm would also begin support of Sanford Moss's internal combustion gas turbine experiments.[50] The development of the Curtis turbine is thus not separable from the market strategies and economic circumstances of the General Electric Company.

Many other steam turbine designs were patented and built after 1900. Two are of special interest here. First, the double-rotation (contrarotating) radial out-flow reaction turbine of Berger Ljungstrom established a recurrent design theme for later gas turbine proposals. Ljungstrom had begun his work before 1912, and his turbine, manufactured by his own firm in Liljeholmen, Sweden, enjoyed moderate success for producing medium-range powers. Ljungstrom's design embodied "several exceedingly clever ideas contrived to reduce tip leakage, gland leakage, conduction, and radiation losses."[51] The Ljungstrom turbine indicates the range of successful variation possible within a maturing normal technology.

Second, the unsuccessful Reidler-Stumpf turbine provides insight into developmental strategies within such a normal technological tradition. The Reidler-Stumpf was an impulse turbine which used a peculiar form of "reversing channel" to obtain a velocity-compounding effect. It was manufactured briefly by the Allgemeine Elektrizatasgesellschaft (A.E.G.) in Berlin, but was abandoned because of its high fabrication cost and relative inefficiency.

A.E.G.'s response to steam turbines is instructive. A.E.G. was a large, diversified company, but lagged in steam turbine development:

> The main steps in the development of the steam turbine types of the A.E.G. were according to the publications and the communications of Dr. P. Lasche, manager of the steam turbine works, the following. In 1902, when the A.E.G. decided to take up steam turbine manufacture, the turbine of *de Laval* was not built in sizes over 300 h.p., and the *Parsons* turbine in its then existing form did not satisfy the requirements as to simplicity, reliability, or steam consumption. Therefore, they began in 1904 with the manufacture of a tangential turbine according to the designs of *Reidler and Stumpf*, which as a "single wheel" turbine, offered the greatest simplicity and gave prospects of good economy. Experience, however, showed the guiding the steam jet through the pocket shaped buckets and through the long reversing channels of this tur-

bine was less economical than in the usual axial flow turbines. Besides, the manufacture proved to be very expensive. In the meantime, the *General Electric Co.* developed the Curtis turbine up to the largest sizes. Therefore, a community of interests agreement was made with this Company in the form of interchange of experiences, use of patents, with territorial sales limitations, and an ultimate type of turbine was decided upon having a Curtis turbine for the first stage and a multi-stage impulse turbine for the remaining stages.[52]

There would seem to be three stages in A.E.G.'s confrontation with the steam turbine. First, the firm realized that a new field existed and was potentially within the firm's technical and market domain. Second, in the face of technological uncertainty, A.E.G. attempted to establish proprietary rights in one, hopefully the best, design. Initially, the corporation tried to generate that design from internal sources. Third, A.E.G. attempted, in this case successfully, to stabilize and rationalize its position, with regard both to technology and to markets, by cross-licensing and cartelization. If the firm was to remain in the market at all, the third stage was a necessity whether or not the second was successful, as it was not in A.E.G.'s case. A.E.G. ultimately did make some significant improvements in the details of the basic G.E.-A.E.G. design.

Although many different steam turbine designs were tried, by 1920 the basic four types—De Laval, Parsons, Rateau, and Curtis (plus the Ljungstrom)—and developments based upon them, dominated steam turbine technology. Licensing agreements among major manufacturers spawned innumerable small variations on the central design themes as each manufacturer simultaneously sought incremental technical improvement and some degree of product differentiation for marketing purposes. That such cumulative improvements were usually patentable would imply that they were, by law at least, not obvious to a well-versed practitioner of the art in question.

By the 1920s, virtually every major heavy manufacturing firm in the United States and in Europe was producing some variety of steam turbine. The American Westinghouse Company worked out a licensing agreement with Parsons in 1895. Allis-Chalmers, after trying several domestic designs that proved unsuccessful, also took Parsons licenses. In Europe, Esher, Wyss and Company of Zurich undertook manufacture of the design of their chief engineer Zoelly, which was essentially similar to that of Rateau. The Zoelly turbine was, in turn, licensed to a number of firms, including Krupps and Maschinenfabrik Augsburg-Nürnberg (M.A.N.). Orlikon, another Swiss company, acquired Rateau licenses. The renowned Czech Skoda works evolved their own turbine from the Rateau design. Brown Boveri, a Swiss-German combine (Baden and Manheim) founded in 1892 by an Englishman and a German, began development of a Parsons-based turbine shortly before 1900. Thyssen and Company, a branch of the German steel cartel, produced a turbine which was a composite of the Parsons and Curtis.

In England, the Fraser and Chalmers Engineering Works of the General Electric Company, Ltd. (an affiliate of American G.E.) quite naturally produced a Curtis-type machine. The British Westinghouse Electric and Manufacturing Company at first built Parsons turbines, like its American parent company. After dissociation from American Westinghouse, when British Westinghouse became Metropolitan-Vickers Electrical Company, it turned to a Rateau-type turbine. The English Electric Company, an amalgamation of smaller firms formed in 1918, manufactured a Rateau-Curtis turbine. British Thompson-Houston (B.T.-H.) of Rugby produced Curtis turbines. Over time, the welter of combinations and "developments" were such that the original types virtually lost their identities. Furthermore, each company introduced its own distinctive features into its machines—different control systems, different means of lubrication, different bearings, and so forth.

The steam turbine is relevant to the turbojet revolution if for no other reason than that the steam turbine is in fact a gas turbine, just as a turbojet is. Moreover, the immense collection of data, of know-how, of technology, that constituted the steam turbine revolution was a basis upon which the turbojet could be built. All the work done on blade design, gas flow, shaft and bearing loads, temperature distribution, lubrication systems, governors, blade-cutting machines, test procedures and instruments, and countless other facets of design and production could be applied to gas turbine development. Indeed, the insights gained into gas flow through nozzles by steam nozzle experimenters from the time of De Laval is still fundamental to all reaction propulsion systems—even to those for space flight. Furthermore, virtually every company that later became interested in internal combustion gas turbines and many of the organizations that were to be most prominent in the actual development of the turbojet engine were precisely the ones that were leaders in the development and manufacture of steam turbines.

Yet not all the steam turbine's influences were positive. All the work done on gas turbines before 1930 assumed without question the steam turbine's fixation with shaft horsepower. The potentially thrust-producing exhaust of the gas turbine was regarded strictly as waste, with no inkling by anyone of its usefulness. Similarly, the assumptions about weight and temperature established in steam turbine practice persisted to haunt the early proponents of airborne gas turbines. Still, the steam turbine was the place to begin. With the exception of aerodynamics, the steam turbine heritage was perhaps the most important element in the unique historical environment in which the turbojet revolution was to occur.

TURBO-AIR COMPRESSORS

In the lineage of the turbojet, turbo-air compressors are linked backwards through centrifugal fans and turbine water pumps to water turbines, and forward through internal combustion gas turbines to the turbojet itself. Turbo-air compressors, furthermore, both co-evolved with steam turbines and further reinforced and more sharply defined higher-level traditions of technological testability.

The distinction between fans and rotary blowers, which were common by the turn of the twentieth century, and turbo-air compressors parallels the distinction between centrifugal water pumps and turbine pumps. The demarcation between fans and rotary blowers and turbocompressors may be operationally defined "as that point beyond which it is no longer convenient to measure the pressure developed in inches of water but to use instead pounds per square inch as the unit of measurement."[53] Such a rise in pressure indicates significant air compression, which in turn denotes conversion of kinetic into potential energy in the turbocompressor.

Rotary displacement-type air compressors were widely used and even more widely advocated during the last half of the nineteenth century. Most duplicated the configuration of various rotary displacement water pumps. Probably the most successful rotary displacement blower or air compressor was that of the Roots brothers. The origin of the Roots blower, still used for moderate pressures and volumes, is a model of serendipitous invention:

> About the middle of the nineteenth century, the brothers P. H. and F. M. Roots owned a textile mill in Connersville, Indiana. A water turbine was needed to drive the lineshafts by the fall of water from the canal which still runs by the Roots Plant into the Whitewater river. But a satisfactory water turbine wasn't built in those days, so F. M. Roots designed a two impeller contraption with a sheet metal case and wooden impellers. It was a failure because the wood swelled and jammed the turbine.
>
> After considerable scraping, the turbine was belted to the lineshaft and turned over to try it out. The local bewhiskered foundryman, curious about the new machine, looked in the top—and his hat blew off. "This will make a better blower than it will a turbine," he said. Thus was born the first Roots Blower in 1859.[54]

Even allowing for some romanticization, the Roots brothers appear to have been classic ingenious mechanics not exactly well versed in the latest European turbine practice. Nevertheless, the Roots blower proved to be the best of all displacement-type compressors. It is, however, incapable of high pressure or large-volume operation, and attains, at best, efficiencies of 35 to 40 percent. It is, furthermore, subject to thermal distortion and to lubrication and sealing problems.

The second half of the nineteenth century also saw widespread use of centrifugal blowers and axial fans, mostly for ventilation. Some units, however, were employed in industrial applications where relatively low pressures and large volumes were required. Before 1900, almost all such fans were single-stage units with little or no pressure rise.

The first true turbo-air compressor was developed by Auguste Rateau. Rateau not only successfully developed the compressor itself, but also made three other critical, related contributions: he evolved the first adequate theory of turbocompressor operation; he initiated turbocompressor manufacture in France and, through licensees, elsewhere, and spearheaded an aggressive marketing effort; and, pursuant to those activities, he developed novel testing techniques and hardware applicable not only to turbocompressors but to all compressor systems. Rateau had broad experience with axial fans and centrifugal blowers, as well as with steam turbines and turbine water pumps. He had published a definitive paper on centrifugal blowers in 1892,[55] and at the turn of the century was completing his multivolume masterwork on all forms of turbomachinery. By 1902, over 700 Rateau-designed axial fans were in use in Europe, mostly for ventilation of ships, theaters, and public buildings.[56] Over 200 Rateau-designed centrifugal blowers were in use, mostly for mine ventilation.

Rateau began design of his first high-pressure turbocompressor in 1898. That single-stage unit was fabricated by Sautter-Harle in Paris and put under test during 1901. Rateau published his test results in 1902.[57] In designing his turbocompressor, he was directly inspired by the success of turbine water pumps, which he had previously designed for commercial manufacture by Sautter-Harle. Rateau found the same fundamental theory applicable to both turbine pumps and turbocompressors, although he, like all other investigators, assumed an inviscid, irrotational fluid in his analysis.[58] Rateau's interest in the turbocompressor was further stimulated by his realization that by 1900 steam turbines and electric motors could for the first time provide power at the high rotational speeds necessary for the turbocompressor's effective operation without requiring speed-multiplying gears or belts.[59] Turbo-air compressors, like turbine water pumps, co-evolved with high-speed prime movers.

Rateau's test in 1902 of his first turbocompressor, powered by a specially constructed Pelton-type impulse steam turbine, showed a maximum efficiency of 56 percent at 20,000 rpm with a delivery pressure of one-half atmosphere (a compression ratio of 1.5 : 1). Rateau foresaw attaining pressures of 5 : 1 with four-stage turbocompressors.[60] Rateau's approach to advanced technology impressed the commentator in the *Engineer* (1902), who offered a remarkably perceptive interpretation of the emerging relationship between scientific theory, technological design, and technological experimentation founded upon rigorous application of scientific method:

> To the scientific mind there can be few things more interesting, in a way, than the study of inventions which are intended to improve the efficiencies of fans and pumps, especially the former; and it is refreshing to find an engineer who, in designing a machine, first forms an accurate theory of the subject, and perfects his work by systematic experiments which enable him to refine and confirm his theory, and also to decide upon certain points, such as the numbers of vanes that give the best manometric or mechanical efficiency, or to determine coefficients of contraction, velocity, and resistance which theory is unable to fix.[61]

Rarely has the ideology of scientific technology been better phrased.

Rateau quickly initiated commercial exploitation of his turbocompressor. Beginning in 1903, he designed single-stage units for use in steel works, sugar manufactories, and chemical works. Sautter-Harle installed the first Rateau-designed multistage turbocompressor in 1905. Later that same year, Sautter-Harle completed a large Rateau multistage turbocompressor for the Bethune mines. That unit comprised four sets of multicellular fans in series and attained an overall compression ratio of 7:1 with a total efficiency of slightly less than 50 percent. Intercooling (the provision for cooling the air, heated by compression, between the stages of the turbocompressor) was used for the first time in the Bethune mine compressor.[62] Although successful, the Bethune compressor did not meet Rateau's hopes for high-efficiency and high-compression ratios from a small number of stages.

Beginning in 1904, licenses for Rateau-designed turbocompressors were taken by Brown Boveri (Switzerland), the Charleroi Electric Company (Belgium), and Gutenhoffnungshutte (Germany). In 1906, in their Baden works, Brown Boveri constructed, to Rateau's and René Armengaud's designs, a very large compressor for the Armengaud-Lemale internal combustion gas turbine. Other firms in England, Germany, and the United States soon began manufacture of centrifugal turbocompressors.

For large volumes, turbocompressors had the usual numerous advantages over competing reciprocating compressors. Turbocompressors occupied one-twentieth the space of comparable reciprocating units, and an entire turbocompressor set might weigh less than the flywheel alone of a reciprocating compressor.[63] The turbocompressor offered virtually no vibration or metal-to-metal contact of parts, and therefore needed only simple lubrication provisions. Air from a turbocompressor was also uncontaminated by lubricants, since lubricating oil did not come in contact with the air passing through the compressor. There was no pulsation in air supply from a turbocompressor, so large receivers could be dispensed with. Electric motors or steam turbines could drive a turbocompressor directly, and in the latter case, relatively low-pressure exhaust steam from other engines could be utilized.[64]

As with adoption of other radical innovations, however, adoption of turbocompressors required radical redefinition of fundamental technological performance parameters. Indeed, Rateau's exegesis of the finer points of turbocompressor and turbine-pump theory and design set off a monumental row about basic definitions of efficiency and about the operational testing techniques by which those efficiencies were to be measured. Rateau's critique of conventional measures was cogent and acerbic:

> The figures for efficiency given above will no doubt appear very low beside those generally given for piston compressors, and one may be tempted to believe that the latter are much superior to the new machines. It should be remarked, however, that the efficiencies given for the reciprocating machines are far from showing the real value, such as we have defined it. Generally one finds diagrams for the steam and air cylinders, and the ratio of the work thus indicated is taken as representing the efficiency. The results obtained by this method are much above the exact value, for they take into account only the friction losses in the gearing (connecting rods, cranks, & c.) between the steam-engine and the compressor. The losses through wire-drawing of steam and air in the inlet and outlet ports, the exchange of heat between the walls and the fluid, the differences between the real diagrams and the theoretical diagrams in isothermal compression, do not enter into account in such an evaluation of the efficiency. It is necessary, therefore, to correct the figures indicated, and we shall see further on that this correction is always very important.[65]

Besides being inaccurate, the conventional efficiency definition based on indicator diagrams was simply physically inapplicable to turbocompressors: as Rateau said, "This definition has no significance whatever in the case of the centrifugal turbo-compressor."[66]

Rateau invented new efficiency definitions applicable to both centrifugal and reciprocating compressors and devised appropriate experimental apparatus for measuring those efficiencies. He preferred to define efficiency adiabatically, that is, with the work done in raising the temperature of the compressed gas included in the output of the compressor, since that definition was most favorable to his turbocompressor. In response to criticism from members of the reciprocating compressor community,[67] Rateau did acknowledge that figures for adiabatic efficiency could be adjusted to give isothermal efficiency (for compression of the gas at constant temperature, that is, with no heat added). For turbocompressors, isothermal efficiency was generally about 10 percent less than adiabatic efficiency (56 percent rather than 66 percent, for example). (Adiabatic efficiency is appropriate in some turbocompressor uses; in other uses, however, when the gas is needed at moderate temperature or when several compression stages are necessary, isothermal efficiency is a more appropriate measure.[68])

Significantly, Rateau himself developed and published a full description of a complex test apparatus for turbocompressors which would also yield commensurable results for reciprocating compressors. Prior technological testing practice provided easily utilized means of measuring three of the five critical parameters of turbocompressor operation: initial pressure and discharge pressure (measured by gauges) and power supplied to the compressor (obtained by dynamometer tests of the prime mover used, whether steam turbine or electric motor). Measurements of intake and discharge volumes or masses of air, however, proved more difficult. Rateau provided his test compressor with an

> intake casing of sufficient dimensions to obtain a low speed for the gas. At the entrance to this casing we fit a coverging nozzle, the final section of which is accurately known, and we measure the difference of pressure in the gas before and after its passage through the nozzle, as well as its temperature. The difference in pressure gives the speed of the gas on leaving the nozzle, and thus the volume through the final section. The same means are followed at the compressor discharge. The volume measured at discharge, brought back to the initial pressure and temperature, is generally a little lower than the suction volume. The difference is the leakage in the apparatus.[69]

Rateau's measured quantities corresponded to appropriate theoretical entities. His results could therefore be integrated into the corpus of prior thermodynamic theory. He also contrived suitable modifications for his apparatus so that it could be used to test intercooled turbocompressors. Rateau thus created a novel testing methodology which would prove to be, in much articulated form, essential to the testing and development of turbojet engines.

Furthermore, as so often seems the case, competition from the new system, turbocompressors, provoked considerable improvement in the prior normal technology, reciprocating compressors. Ironically, but in no way exceptionally, that progress in reciprocating compressor design depended upon a sharpened awareness of efficiency and of the sources of losses which arose from the refined theoretical and experimental innovations associated with the development of the turbocompressor.[70]

Contemporaneously with the development of the Rateau centrifugal turbocompressor and its derivatives, manufacture of axial turbocompressors was also begun. In 1884, Charles Parsons had patented an axial-flow compressor using reversed turbine blades. In 1887 he designed and successfully marketed a low-pressure three-stage centrifugal ventilating compressor for shipboard use. In 1897, he began extensive experiments on axial-flow compressors, and two years later built an 80-stage axial-flow unit which gave an adiabatic efficiency of about 70 percent (about 60 percent isothermal).[71] In the early Parsons units, "the guide-blades were flat on one side and curved on the other, the enlarging passage between each pair serving to transform the kinetic energy of the air into pressure." By mid-1907, Parsons and Company had "either built or on order" forty-one axial-flow compressor units.[72] Yet in 1908 Parsons ceased manufacture of axial-flow compressors: the crude blade-form then used rendered the complex, costly axial-flow compressor inefficient compared to centrifugal turbocompressors or even positive-displacement types. The then-unrecognized poor aerodynamic qualities of the Parsons blade design doomed his axial compressor. As A. I. Ponomareff, much later, noted:

> At that time the most fundamental difference between flow through the turbine and the compressor was not known, and use of turbine blading for a compressor produced very poor results. In a steam turbine the blading must be arranged for expanding or accelerating flow; in a compressor, for diffusing or retarding flow. In an expanding passage the pressure decreases in the direction of flow and the boundary layer is continuously supplied with pressure energy to accelerate the gas particles slowed down by friction with the confining walls, thus producing a stable flow. In a diffusing passage, on the other hand, the pressure forces are acting in a direction opposite to flow and tend to retard the boundary layer, producing eddying and back flow. The design of an expanding nozzle to produce 98 percent efficiency is not difficult, but to design a diffuser for 85 percent efficiency requires a strict adherence to certain rules as to rate of diffusion and change in direction of flow.[73]

Although Parsons did patent a somewhat improved blade shape in 1910, his firm did not again take up axial-compressor manufacture. In the United States, Westinghouse—at the time a Parsons-steam-turbine licensee—built an experimental axial-flow compressor during 1905–6. That unit yielded efficiencies of only 50 to 55 percent and was abandoned. After 1928, however, Westinghouse did manufacture single-stage axial-flow fans for cooling large steam-turbine-driven turbogenerators and multistage axial blowers for warship forced-draft ventilation. Those later axial blowers were designed in accordance with aerodynamic theory developed in the intervening years.[74] Clearly, early axial-flow compressors failed because the aerodynamic theory necessary to their efficient design was not then available.[75]

Successful centrifugal turbocompressors were founded upon the scientific theory and the technological practice of turbine pumps. The linkage of turbocompressors to that prior practice was both intellectual and personal. The turbocompressor in turn is the direct antecedent of one of the major components of the turbojet engine. Yet the turbocompressors of 1900 or 1910 were not the centrifugal compressors used in the early turbojets. The turbojet would require a highly developed, qualitatively improved compressor very different from those of early practice, however successful.

INTERNAL COMBUSTION GAS TURBINES

Intertwined with development of the first turbo-air compressors was the development of the first internal combustion gas turbines. If the steam turbine embodied the hope of the turbojet revolution, the internal combustion gas turbine symbolized its fear. Indeed, experiments with gas turbines between 1900 and 1910 had a schizoid impact on advocates of the turbojet. Early attempts to construct a useful internal combustion turbine failed: those experiments therefore constituted a tangible, potent, and enduring argument against any gas turbine system. Nevertheless, almost surreptitiously, early gas turbine work also engendered a subculture of turbine enthusiasts and a lineage of technology which at the critical moment was to provide technical sustenance to the embryonic turbojet revolution.

Like the water turbine or the steam turbine, the gas turbine has remote origins. Medieval Europe knew the gas turbine as a "smokejack" installed in a chimney: rising hot smoke turned a windmill-like turbine geared to a cooking spit. The first design for an internal combustion gas turbine proper is attributed to John Barber in England, who was granted a patent in 1791. The idea of the gas turbine was kicked around for most of the nineteenth century, but with little attention and no serious developmental effort devoted to it. The impetus for gas turbine development came, quite naturally, from the success of the steam turbine.

The first nonsteam gas turbine to be built did not incorporate internal combustion, however. The Stolze Hot Air Turbine, designed by Dr. F. Stolze of Berlin beginning in 1872 but not constructed and tested until 1900–1904, employed, as its name would imply, a hot air cycle. A multistage axial-flow rotary air compressor fed compressed air to a heat exchanger, the functional equivalent of a boiler, in which the compressed air was led, inside tubing, through a furnance. The heated, compressed air then turned a multistage axial-flow turbine which powered the compressor and was expected also to provide excess usable shaft output. The axial-flow compressor was among the first ever used, but, as noted, the lack of adequate aerodynamic theory at the time doomed all such systems to hopeless inefficiency.

In the early years of the twentieth century, any number of schemes for internal combustion turbines appeared. Innumerable patents were issued, chiefly in England, France, Germany, and, of course, the United States. The proposed modes of operation for the internal combustion turbine were as diverse as the number of inventors was large. Any who would hold eccentric creativity the province of painter and poet knows nothing of inventors. But all internal combustion turbine proposals were variations on one of two basic cyclic types, "constant volume" or "constant pressure." Constant-volume gas turbines employ combustion chambers that operate intermittently, each in turn exhausting upon some form of turbine. Constant-pressure gas turbines, since called "continuous combustion turbines," employ a turbine or turbines "actuated by the steady flow of the products of a continuous combustion under pressure in a combustion chamber." All successful gas turbines, and absolutely all turbojet engines, have been of the constant-pressure type.

But what is in retrospect fact was in 1900 only possibility. While all internal combustion turbines belonged to one or the other of the two basic types, the variety of proposals within the types was large. There were devices using rotating combustion chambers, contrivances employing contrarotating or reverse flow blading, self-compressing turbines, and so on. In deference to the difficulty of producing turbines with adequate heat-resisting qualities, there were even elaborate designs for hydraulic power transmission to the turbine by oscillating columns of water.[76]

Despite the hundreds of internal combustion turbine schemes, only three proposals were subjected to serious full-scale experimentation before the First World War: the constant-volume design of the German engineer Hans Holzwarth, and the very similar constant-pressure designs of René Armengaud and Charles Lemale in France and of Sanford Moss in the United States.

The first Holzwarth constant-volume, or explosion, turbine was constructed between 1906 and 1908 by Messrs. Korting of Hannover, Germany, and "operated on the explosion cycle without precompression."[77] Adolph Meyer reports, "On the basis of the results obtained with the experimental turbine, Messrs. Brown, Boveri built and tested, between 1909 and 1913, to the order of Dr. Holzwarth, a second gas turbine with a nominal rating of 1,000 h.p., which, however, gave a net output of only about 200 h.p."[78] This second turbine, known as the Mannheimer turbine, employed ten explosion combustion chambers which were charged by a small turbocompressor. The combustion chamber exhausted through a two-stage Curtis turbine wheel.

In the constant-volume, or explosion, cycle, air and fuel were admitted through intake valves into a combustion chamber and ignited. When the pressure within the chamber reached a maximum, a spring-loaded exhaust valve opened, and the products of combustion exhausted onto the turbine. The chambers fired successively, one exhausting while the others were in other phases of the cycle. The primary advantage of the explosion cycle was intermittent combustion, allowing the combustion chambers and nozzles to be adequately cooled by water, thereby mitigating high combustion temperatures (1,600°–2,000° C).[79] Furthermore, the constant-volume system required a much lower volume of air than would a constant-pressure system, making air compressor efficiency much less critical to the overall efficiency of the mechanism. Nevertheless, as indicated, the Mannheimer turbine was inefficient, a result Holzwarth ascribed "to the interference effects produced when more than five chambers were running together. The exhaust of the vessel just finishing its power 'stroke' stayed in the turbine blading sufficiently long to interfere with the scavenging stroke of the neighboring vessels."[80] In addition, the Holzwarth design suffered from forbidding mechanical complexity—at least three valves per combustion chamber were required, plus cooling pumps, oil pumps, and so forth. The machine was massive: the combustion chambers and the turbine alone weighed 25 ½ tons; the entire setup, including dynamo and steam-turbine-driven air compressors, weighed 53 ½ tons. Even if the Holzwarth turbine had attained its designed horsepower, its weight would not have improved "upon ordinary reciprocating steam-engine weights, including boilers."[81]

Despite the shortcomings of the Mannheimer machine, another Holzwarth design turbine was constructed by Maschinen-Fabrik Thyssen, Muhlheim-on-the-Muhr, in 1914 and was followed by several more units from the same company between 1914 and 1928.[82] In 1928, Brown Boveri constructed an advanced version of the basic Holzwarth design which operated on blast furnance gas and which was installed at the Thyssen steelworks in Hamborn. Even larger units were projected at the outbreak of the Second World War. While the Holzwarth design did enjoy limited success in specialized applications, its weight, complexity, and inefficiency precluded it from consideration for use in aircraft, and eventually doomed it to extinction as a prime mover.

Meanwhile, several years before 1905, two contemporary and remarkably parallel, although completely separate, constant-pressure internal combustion gas turbine projects were initiated: that of Armengaud and lemale, under the auspicies of the société Anonyme des Turbomoteurs in Paris, and that of Sanford Moss, which was conceived at the University of California, begun at Cornell, and extensively developed in the General Electric laboratories.

The efforts of Armengaud and Lemale were probably "the earliest elaborate gas-turbine experimental work." Charles Lemale had applied

Armengaud-Lemale gas turbine, with 25-stage centrifugal compressor (in three casings), designed by Rateau and built by Brown Boveri, 1907; turbine itself is at extreme left. (From René Armengaud, "The Gas Turbine: Practical Results with Actual Operational Machines in France," *Cassier's Magazine* 31 [1907].)

for a gas turbine patent in 1901, and he and Rene Armengaud began actual operation of a combustion gas turbine in France in 1903.[83] According to A. Barbezat, chief engineer for Société Anonyme des Turbomoteurs, "preliminary tests were first made with a 25 hp. de Laval turbine with compressed air of 71 lb. per sq. in. from the Paris compressed air mains. The turbine operated at constant combustion pressure, the fuel being petroleum which, atomized by a nozzle, was ignited by a glowing platinum wire."[84]

The promise shown by the initial Armengaud-Lemale experiments led to the construction of a larger, more ambitious turbine during 1905–6. That definitive Armengaud-Lemale turbine employed a "polycellular rotary compressor of the Rateau system" with twenty-five impellers in series arranged in three casings, all on the same shaft.[85] Rather than use separate intercoolers between the turbocompressor stages, Rateau and Armengaud jointly designed an internal water-cooling system for the entire compressor.[86] The turbocompressor delivered about 2,150 c.f.m. of air at a compression ratio of approximately 3:1; it ran at 4,000 rpm, required 328 h.p., and yielded "an efficiency of 65 to 70 percent depending upon load."[87] Armengaud and Lemale used a single pear-shaped carborundum-lined combustion chamber into which gasoline was injected and atomized. Combustion was continuous at a temperature of 1,800° C. A convergent-divergent nozzle was fabricated integrally with the combustion chamber. The nozzle was water-jacketed for cooling. The combustion chamber exhausted through a two-stage Curtis-type impulse turbine, the disc and blades of which were also provided with internal cooling-water passages. Water used to cool the turbocompressor, the combustion chamber nozzle, and the turbine was passed through coils downstream from the turbine itself, where the hot cooling-water was converted into steam by (otherwise) waste heat in the exhaust. The steam so generated was also exhausted, via separate nozzles, onto the same turbine as the products of combustion.[88] The definitive Armengaud-Lemale turbine, then, was a combination internal combustion gas and steam, or mixed, turbine. By full expansion of the products of combustion in the nozzle and by injection of the lower-temperature steam, turbine running temperatures were reduced to 400°C.

The definitive Armengaud-Lemale turbine did produce about 300 h.p. net of negative work of compression, versus 500 h.p. designed. It was, however, woefully inefficient, burning "3.9 lb. of petrol per brake-horsepower hour," compared to "0.5 lb. petrol per brake-horsepower hour for contemporary piston petrol engines."[89] The inefficiency of the Armengaud-Lemale turbine was attributed to the lack of appropriate high-temperature materials for nozzles and turbines, which necessitated the complex and efficiency-robbing cooling arrangements actually used. Yet no unambiguous efficiency figures for the turbine itself were ever published, and the 65 to 70 percent figures for turbocompressor efficiency appear to depict adiabatic efficiency, rather than the lower but more representative isothermal efficiency. It would seem that component (compressor and turbine) inefficiency, as well as materials difficulties, hindered development of the Armengaud-Lemale turbine. No weight or size figures for the Armengaud-Lemale turbine were published, but photographic evidence would suggest that both dimensions were substantial.[90] Certainly the necessity of using twenty-five impellers to attain a compression ratio of 3:1 does not indicate a system ever likely to be appropriate for airborne use.

The Armengaud-Lemale turbine was a rarity in that it worked at all. But René Armengaud died in 1909, and apparently, further experiments were abandoned. The Société des Turbomoteurs did use the experience gained with the Armengaud-Lemale turbine to develop a "paraffin turbine" naval torpedo which, rather than using only compressed air to drive the propeller turbine of the torpedo, as was then the custom, injected and ignited paraffin in a combustion chamber to raise the temperature and hence the pressure of the compressed air stream. Significantly, the Rateau-designed turbocompressor for the Armengaud-Lemale turbine was Brown Boveri's first commercial model. Its success induced Brown Boveri to remain in the turbocompressor business.[91]

Concurrent with the Armengaud-Lemale project, Sanford Moss in the United States was conducting almost identical, although totally independent, experiments on internal combustion gas turbines. Moss began his engineering career as a machinist's apprentice and then as a draftsman in the shops of Edward Rix, "an early master in compressed-air engineering," and proceeded to draftsman jobs in "various gas-engine shops."[92] Moss invented his gas turbine while "in the classes in thermodynamics and hydrodynamics of Prof. Frederick G. Hesse at the University of California in 1895." Hesse, then, like Burdin, though himself a relatively unsuccessful protagonist of technological change, was a mentor to more successful figures. Moss submitted a master's thesis on gas turbine design, including a proposal for a turbine-powered locomotive, to the University of California in 1900.

In 1901, Moss, now a doctoral student, began gas turbine research in the Sibley College Laboratory of Cornell University. It took a year of concentrated effort just "to get a continuous-combustion chamber in stable operation." Significantly, the combustion chamber "frequently went out and the oil on the red-hot firebrick lining filled the neighborhood with dense black smoke, so the Sibley College people well knew of the gas-turbine research." The initial turbine wheel that Moss used was from the first De Laval steam turbine in the United States, brought over in 1893 for the World's Columbian Exposition and for patent demonstrations. According to Moss, "This was the first turbine wheel actually operated by products of combustion in the United States, and possibly was the first such turbine wheel ever operated. The entire power was absorbed by a prony brake, and the air for compression was furnished by a steam compressor, with computation of the power required." The separate compressor Moss employed was a reciprocating type, and the complete internal combustion gas turbine visualized in his doctoral thesis (Cornell, 1903) also would have employed a turbine-driven reciprocating compressor. Moss's Cornell experiments were not successful: "As with many other experimental gas turbines, the power for compression was more than the turbine power. So except for the historical fact that the combustion chamber actually operated the turbine wheel, the experiment was a flat failure."

Nevertheless, in June of 1903, Moss, after receiving his doctorate, went back to General Electric, for whom he had previously worked as a steam turbine draftsman, and continued to pursue his gas turbine research. Under his direction, "a number of General Electric models of gas turbines successively were gotten into operation, first at Schenectady, and beginning in 1904 at Lynn, with consultation with Prof. Elihu Thomson and Richard H. Rice. There were combustion chamber, heat interchanger, nozzle, and single-stage impulse turbine wheel; but with power for compression from an independent air compressor and allowed for by computation."

As was rapidly becoming customary at G.E., their investigations into the various elements of internal combustion gas turbine design were the most thorough and comprehensive to that time. Extensive experiments were begun in the fall of 1903 on centrifugal compressor design, and a 1904 patent application in Moss's name demonstrated theoretically, "probably for the first time," the relation between the velocity of flow of a compressible fluid and diffuser shape. Essentially, Moss showed that for flow velocities below the local speed of sound in a compressible fluid (gas), compressor diffuser design could be treated "just as with an incompressible fluid" (water, for example), and that divergent diffuser passages were appropriate. Moss, then, seems to have independently duplicated Rateau's theoretical insights. It was on the basis of Moss's theory that G.E. began successful development of centrifugal compressors. Moss also performed fundamental investigations of energy conversion in nozzles, using compressed air, steam, or the products of internal combustion as his working medium.

General Electric experiments directed toward developing a complete internal combustion gas turbine continued until 1907, when "a fuel consumption was in sight of 4 lb. of kerosene per net hphr. At that time good oil engines were using 1 lb. of oil per net hphr." Moss ran into the same problem as Armengaud and Lemale: temperature limitation. "All early General Electric gas-turbine work was conducted with temperatures possible with materials then known. . . . So the temperatures were kept low by air excess or water injection, both resulting in inefficiencies. No way then seemed open to do better, and so the gas-turbine part of the research was postponed, but with vigorous commercial work on centrifugal compressors." In company with Brown Boveri and Rateau, General Electric found a good market for its centrifugal compressors in industrial applications, installing its first two blast furnace turbocompressors in 1910. Since Moss, like all other gas turbine experimenters before 1930, was seeking to maximize output of shaft horsepower from his turbine, the concept of exhaust thrust and any beneficial effects that excess air injection might have on it were quite beyond his consideration. The General Electric experiments therefore produced results very nearly identical to those of Armengaud and Lemale.

Moss remained at General Electric as design engineer for the turbo-air compressor business until it was sold in 1925 (never a large profit source, G.E.'s manufacture of air compressors conflicted with its more lucrative sales of turbines to power air compressors built by other firms). Moss thereafter became research engineer in the Lynn steam turbine division of G.E., although he did manage, as noted below, to keep a small piston engine turbosupercharger research and development program going.

Meanwhile, General Electric had hired another internal combustion gas turbine enthusiast, Glenn B. Warren. Warren had become interested in gas turbines even in high school and, after working summers first for Allis-Chalmers and then for G.E., wrote a B.S. thesis on gas turbines at the University of Wisconsin in 1919. Warren's proposal, for which he built and tested two combustion chambers while at Wisconsin, also visualized a combination internal combustion and steam, or mixed, cycle, in order to keep turbine inlet temperatures down to tolerable levels. After going to work for G.E. in 1920, Warren got permission to do a detailed design study for a 10,000 kW. gas turbine during 1921–22, in which he concluded that G.E. did not then have an adequate technological base, either in high-temperature materials or in compressor design, to build a successful gas turbine. Warren therefore concentrated on steam turbine problems, especially on materials resistant to high-temperature creep and vibration, turbine blade and disc design, and, significantly, nozzle design. For nozzle experiments, Warren designed a special testing device to determine nozzle efficiency, the lack of which was then plaguing large G.E. steam turbines.[93] In all of these areas, G.E. made major progress during the 1920s and 1930s, and this progress played a basic role in their ultimate capacity to adopt and develop the turbojet.

Several other gas turbine projects, less well developed, deserve mention. First, about 1908, the same A. Barbezat who had superintended the Armengaud-Lemale experiments supervised construction of a turbine of the Karavodine system. The Karavodine was an "explosion" turbine: its combustion chamber consisted of a long tube, a one-way shutter-valve, and a fuel nozzle. The mode of operation of the Karavodine combustion chamber, which exhausted onto a De Laval turbine to provide rotary power, was identical to that of the "pulse-jet" as used later in the German V-1 flying bomb. But what the English victims of the Nazi V bombing described as the V-1's "angry buzz," the gentler Edwardian Age termed the Karavodine's "continuous purr." As a system for producing shaft power, the Karavodine was hopelessly inefficient: a 1.6 net brake horsepower unit required 6.5 pounds of petrol per brake h.p./hr.[94] Second, an Austrian engineer, Adolf Vogt, in cooperation with S.A. des Ateliers Carel Freres, Ghent, Belgium, builders of diesel and steam engines, designed and built a series of experimental engines during 1904–5. He first tried a variable-compression-ratio diesel engine, and next a "water column" diesel, neither of which worked successfully. Then, in 1905, Vogt built a constant-pressure gas turbine. But he, like Armengaud and Lemale and Moss, could not evade temperature limitations, and his turbine too failed. Lastly, Hugo Junkers, professor of mechanical engineering at Aachen Technical University (and soon to be of aircraft fame), together with Otto Mader, worked before the First World War on a free-piston engine, a system in which crankless opposed pistons are used to produce exhaust gas to run a turbine. At the time, Junker's free-piston engine was no more successful than any of the other schemes.

Of the hundreds of gas turbine proposals current in 1900, only these few were developed, and only those of Holzwarth, Armengaud and Lemale, and Moss, seriously. Most of the projects were terminated outright. The Holzwarth design could in no sense be said to have been widely accepted. The first gas turbine revolution proved largely abortive. Ironically, its most successful progeny were centrifugal superchargers and turbosuperchargers for piston aircraft engines.[95]

The failure of the first gas turbine revolution serves to highlight two contradictory but equally central themes in technological practice: the rigor of community norms, especially testability, and the emotional commitment of the proponents of radical change. Virtually all of the early gas turbine investigators who published contemporary descriptions of their work omitted or garbled critical performance data—for obvious reasons. All, as a result, were criticized in the community journals. Armengaud got off lightest, apparently because he had the decency to die. Holzwarth, who published most, persisted longest, and enjoyed some modicum of success, was criticized most severely and tellingly. Commenting on Holzwarth's description of his nominal 1,000-h.p. turbine, the editors of *Engineering* wrote in 1911:

> The paper in question is interesting as a record of what has been done in this matter, but in other respects it leaves much to be desired. It consists in the main of comments on the advantages of turbines, of explanations of certain details of construction, and of a slight and imperfect sketch of the theory to which the machine has been constructed. We all know that a gas-turbine can be built to run, and also that a good gas-turbine would have enormous advantages, and hence at this date long-drawn generalities on such texts constitute very unattractive fare. Plain statements as to the gas consumed per kilo-watt-hour are entirely missing from Mr. Holzwarth's paper; an omission from which our readers will draw their own conclusions.
>
> As will be seen, the author does give some so-called efficiency curves, but in the case of all heat-engines there are so many ways of estimating efficiencies that in the absence of a definite statement as to which is intended, the term has no meaning.[96]

Holzwarth, then, stood accused of violating perhaps the cardinal norm of scientific engineering practice, that performance test results should be clearly and unambiguously reported. The invocation of such norms is clearly a means by which a community defends traditional practice against radical subversion. Such a defense would seem likely to be overcome only by a proponent who can marshall a sound theoretical explanation backed by replicable technical demonstration of a new system, as in the case of the Rateau turbocompressor above.

The deep emotional commitment of proponents of new systems to their ideas is perhaps best evidenced by Sanford Moss's almost bittersweet recollection in 1944 of his early gas turbine work. After describing the failure of his early machines to produce any usable output, Moss cautioned, "Before the reader is cruel enough to laugh at this result, let him put himself in the place of lots of us gas-turbine inventors who have sweated blood through years of research only to come out this way or worse." Moss then confessed "that the development of the gas-turbine has been the 'grand passion' of the author's engineering career. This career no doubt now is drawing to a close; but nevertheless the author expects to see many power-plant-gas turbines commercially developed by the younger generation."[97]

The prior technology discussed here formed an invaluable pool of technological capabilities for both intellectual conception and physical construction of the turbojet and therefore constituted primary elements of the historical milieu within which the turbojet revolution occurred. Much of this prior technology, however, had contradictory influences. Steam turbines were the paradigm to which protagonists of further revolutions could aspire, yet the parameters of its major practice were far from those of any airborne system. The abortive gas turbine revolution of the first two decades of the century was both argument against and inspiration for the proponents of the turbojet. The superchargers and turbosuperchargers which evolved from earlier turbo-air compressor and gas turbine work provided vital technical capability for the turbojet, but at the same time vastly improved the performance of the piston engine, thus making the proselytizing of the turbojet's advocates that much more difficult. Finally, however central these prior technologies might have been to the turbojet revolution as it occurred or might be to its historical reconstruction, nowhere in that prior practice was the turbojet implicit, inherent, or inevitable.

Aerodynamics as a science depended upon prior development of hydrodynamics, which in turn evolved from even earlier work in hydraulics and hydrostatics. As is true of most science, the ancients had positive ideas about fluid flow, and during the Renaissance and after, Galileo Galilei, Leonardo da Vinci, and later experimentalists such as Evangelista Torricelli made acute observations. Yet is was not until the work of Isaac Newton that a comprehensive tradition of fluid mechanics really began. In order to refute the Cartesian notion of a cosmic space filled with a material "ether" and thereby to validate the astronomical observations that showed no effects from the presence of such matter and thus supported his theory of gravitation, Newton, before 1687, undertook an extensive study of the properties and "resistance" of fluids such as water, air, and oil, as well as of hypothetical fluids which were considered for mathematical purposes to be "frictionless and rarefied."[1] Newton in his consideration of fluid motion made three basic assumptions: first, that motion of an object relative to a fluid and of a fluid relative to an object were identical (the foundation of all wind tunnel work); second, that the fluid was "composed of discrete particles without interaction among themselves"; and third, that given the first two assumptions, the motion of a fluid of discrete particles could be treated like any other problem in Newtonian mechanics.[2] Although Newton did consider viscous flow, the main emphasis of his treatment centered on this "frictionless" fluid. Early speculators on manned flight deduced from Newton, and wrongly attributed to him, the so-called sine square law, which gave an impossibly low lift-drag ratio for an inclined plane. There is no evidence that Newton himself thought his picture perfect, but with the mathematical tools avilable to him, it may have been the best possible. It was not until the next century that the development of differential equations capable of treating continuous change in space and time, equations in large part the work of the same individuals who were to use them so perceptively, permitted creation of more sophisticated hydrodynamics.[3]

Creation of that classical hydrodynamics rested principally on the work of four men: Daniel Bernoulli, Jean d'Alembert, Leonhard Euler, and Joseph Louis Lagrange. Daniel Bernoulli, scion of the Swiss family of mathematicians and a noted mathematician in his own right (the Bernoulli numbers), published his *Hydrodynamics* in 1738, the first use of that term meant to denote the combination of hydrostatics and hydraulics. As a relatively unemphasized part of that work, Bernoulli derived the theorem which bears his name and which is essential not only for hydrodynamics but for aerodynamics as well. Bernoulli's theorem holds that the pressure a fluid in motion exerts on the surface of an object lying parallel to the direction of flow (such as the interior of a pipe or the surface of a wing) is inversely proportional to the velocity of flow. In other words, other things being equal, the faster the fluid flows, the less pressure normal to the direction of flow it exerts.

In 1752 Jean d'Alembert introduced the idea of a continuous, incompressible fluid which could be analyzed in terms of "internal forces in equilibrium" and external forces acting upon the fluid. Following d'Alembert, Leonhard Euler, friend of Daniel Bernoulli and student of his father, produced in 1755 the most complete hydrodynamics to that date. Euler's equations, still a basic part of hydrodynamics, provide a "complete geometric and dynamic account of the field occupied by a fluid between which and the body there is a motion of translation." Euler perfected the artifice of the "ideal fluid"—an inviscid fluid, "a continuous medium with infinite mobility"[4]—which is essential to soluble hydrodynamic equations. Lagrange in 1781 produced a set of equations which treated a single particle of perfect fluid, in contrast to Euler's "field" of medium, and also correlated the work of Bernoulli, d'Alembert, and Euler. The fundamental hydrodynamics of an ideal fluid, irrotational, inviscid, and incompressible, was complete. In the first years of the nineteenth century, other experimenters developed hydrodynamics for other conditions: C.L.M.H. Navier formulated equations for motion of a fluid without equal pressure in all directions (1826), while Simeon Denis Poisson (1831) and George G. Stokes (1847) evolved expressions for viscous flow.

Hermann von Helmholtz made major contributions to hydrodynamics in two papers published in 1858 and 1868. The first paper dealt with "vortex motion" in a perfect fluid—that is, rotational or "swirl" motion. The second paper developed the idea of discontinuous flow, that with a "surface of discontinuity" two adjacent layers of a fluid might "slip" relative to one another with finite velocity. Helmholtz's ideas were further developed by William Thomson, Lord Kelvin, who published a paper on vortex motion in 1869. Lord Rayleigh, who would later become the first president of Great Britain's Advisory Committee for Aeronautics (1909), used Helmholtz's surface of discontinuity to explain the resistance of a plane submerged in a current (1876).

Meanwhile, other experimenters generated more empirical but no less vital hydrodynamic insights. In 1864 W. J. Macquorn Rankine published a rigorous mathematical two-dimensional analysis of streamline flow along a ship's hull. Between 1871 and 1881, R. E. Froude did extensive work on the resistance of hulls and the propulsive efficiency of propellers. In 1876 Osbourne Reynolds described, from his study of flow in pipes, the two modes of fluid flow—laminar and turbulent—and introduced the now universally used measure of flow similarity for different-sized pipes (or objects). For different fluids, the Reynolds'

Number $R = \rho v l/\mu$, where ρ is fluid density, v is flow velocity, l is a linear measure of pipe diameter, or alternatively, the length of an object subject to a flow, and μ is the viscosity of the fluid.[5] (For equal Reynolds' Numbers, flow about two similar objects will be identical; there are also two "critical" Reynolds' Numbers for various kinds of flow at which flow changes from laminar to turbulent or vice versa.) In addition to his work on fluid flow, Reynolds performed valuable experiments on cavitation of ship's propellers, model hull similarity, tidal and river models, the mechanical equivalent of heat, and, as mentioned earlier, centrifugal pumps. Thus, over approximately two hundred years, scientists had evolved, first, a mathematically profound, rigorous, and mature hydrodynamics based on the perfect fluid, and, second, an inclusive set of empirical insights into the behavior of real fluids.

Parallel to the development of a highly sophisticated hydrodynamics, other experimenters were taking the first steps toward the comprehension of aerodynamic phenomena. In 1743, Benjamin Robins, an English mathematician and ballistician, built the first "whirling arm" and with it discovered that differently shaped objects, although having the same cross-sectional area, offered different air resistance. In 1746, Robins employed the first "ballistic pendulum" to discover the large increase in resistance to the flight of a projectile at velocities of about 1,100 feet per second (now known to be the speed of sound in air). John Smeaton used a whirling arm in experiments on windmills, and discovered that a curved surface worked better than a flat one (in modern parlance, the curved surface created more lift).[6]

Throughout the late eighteenth and early nineteenth centuries men speculated on how flight might be achieved. Much attention was directed to balloon flight. One of the first to study heavier-than-air flight specifically was Sir George Cayley. In papers published in 1809–10, Cayley revealed, among other things, some remarkable studies of streamlining based on the shape of trout and, from whirling arm researches, revealed that on a plane the "center of pressure" was at the center of the surface only when the plane was at 90° to the airflow. Researchers following Cayley spent much time studying the shape of birds' wings and in making gliders similar to those shapes.

The first wind tunnel of which there is record was built by Herbert Wenham in 1871 for the Aeronautical Society of Great Britain. In 1884, Horatio Phillips built an improved tunnel and measured the lift of wings shaped like those of birds. Nikolai E. Joukowski built a tunnel at the University of Moscow in 1891. Between 1900 and 1910, wind tunnels were built by T. E. Stanton and Hiram Maxim in England, by Auguste Rateau and A. G. Eiffel in France, by Ludwig Prandtl in Germany, and by G. A. Crocco in Italy.[7] The Wright brothers built a small tunnel and tested many wing planforms and sectional shapes before building their gliders and finally their airplane. With a few exceptions, this early work on what is now called aerodynamics was purely empirical and mostly practical—trial and error tests to gather enough information to build a flyable airplane. The object was flight, not science.

During this period, one man did produce what were to prove especially valuable scientific insights. Ernst Mach, professor of physics (and later philosophy) in Vienna, made pioneer studies of supersonic motion. He was the first to use the Schlieren technique to photograph shock waves formed by objects moving at supersonic speed. He discovered the relationship between "Mach angle" (loosely, the angle of shock waves to the direction of motion) and velocity.[8] Mach's work, done mostly in the late 1880s, would provide the starting point for supersonic aerodynamics some forty years later.

The modern science of aerodynamics began with the theoretical innovations of two men, M. Wilhelm Kutta, a German mathematician who became interested in Otto Lilienthal's gliding experiments, and Nikolai E. Joukowski, the Russian aerodynamicist. Kutta (1902) and Joukowski (1902-1909) separately developed the mathematical theory of "circulation" about an aerofoil. Building upon Helmholz's theories of vorticity and discontinuity, the Kutta-Joukowski theorem (as it came to be known) demonstrated that for a two-dimensional wing (of infinite span), lift could be computed, or predicted, by computing the amount of theoretical circulation about the wing necessary to have equal velocities of flow off the upper and lower training edges. For the first time, generation of lift became subject to rigorous mathematical treatment, and students of aerodynamics realized for the first time that lift was mainly the result of lessened pressure above the wing due to accelerated flow (according to Bernoulli's theorem) rather than the result of air particles striking the lower surface of the inclined plane. The Kutta-Joukowski theorem was corroborated by wind tunnel experiments. Aerodynamics had reached the level of mathematical rigor and experimental testability essential to true science.[9]

The man probably most responsible for the full development of subsonic aerodynamics, however, was Ludwig Prandtl. Through his own work and the work of the Aerodynamische Versuchsanstalt (AVA) at Göttingen University, which he headed, and through the work of the men he trained there, aerodynamics became a mature, complete science. Prandtl was the son of a professor of surveying and was trained at the Technical University of Munich by the outstanding pioneer in applied mechanics, August Foppl. Prandtl's own doctoral dissertation concerned strength of materials. His first experiments with air currents resulted from his work on a suction system at the Maschinenfabrik Augsburg-

Nürnberg (M.A.N.) works in Nürnberg in 1900. After teaching briefly at Hannover, Prandtl moved to Göttingen in 1904.[10]

Prandtl himself made three major, revolutionary contributions to fluid dynamics. First, in 1904, he presented a paper to the Third International Congress of Mathematicians in Heidelberg in which he asserted "that for fluids of small viscosity, such as air or water, the viscosity will substantially affect the flow only in a thin layer adjacent to the surface. Outside this layer, viscosity can be neglected and the flow can be described to a high degree of accuracy by the mechanics of noviscous fluids." Further, Prandtl showed "that the small thickness of the boundary layer permits essential simplifications in the equations of motion of a viscous fluid, so that the problem of frictional drag becomes accessible to mathematical analysis."[11] Second, in 1914, Prandtl established that the transition from laminar (smooth) to turbulent flow specifically in the boundary layer, as well as flow breakaway from the surface, was a function of Reynolds' Number. Essentially, Prandtl had made it possible for the first time to harness hydrodynamic theory directly to aeronautical practice: "To put it briefly, the boundary layer concept allowed fluids on the whole to be treated as viscosity-free, but required that regard be paid to the effects of viscosity at the boundary layer. The boundary layer concept made it possible to specify general hydrodynamic theory into distinct theories, which covered particular types of technical operation such as lubrication, the airfoil, and the propeller."[12]

Finally, in 1918, Prandtl published the results of his general investigations of lift. By combining a set of brilliant, simplifying assumptions with the circulation theory of Kutta and Joukowski, Prandtl produced a complete theoretical treatment of any wing of high aspect ratio (span considerably greater than chord). With the Prandtl theory aeronautical engineers could with certainty for the first time solve two problems: "First, if the distribution of lift along the wing span is known, [they could] determine the flow pattern of induced velocities by a straightforward calculation, and also the energy necessary to obtain the lift distribution; second—and this is more interesting for the engineer—[they could] determine the lift distribution along the span when the geometry of the wings is given, i.e., when the distributions of chord, wing section, and angle of attack along the span are given."[13] For the first time, a wing could be scientifically rather than empirically designed and its aerodynamic lift and resulting induced drag precisely calculated. Prandtl's theory became the essential foundation for the whole accumulation of insight that was to be subsonic aerodynamics.

As important as Prandtl's work was, equally or perhaps even more important were the men he trained at the AVA. His students included Theodore von Karman, who was to head a similar aerodynamic institute at Aachen and, after 1930, the Guggenheim Aeronautical Laboratory at the California Institute of Technology; Max Munk, who joined the National Advisory Committee for Aeronautics (NACA) in 1921; Albert Betz and W. Encke, who were to remain at Göttingen; and Jacob Ackeret, who from 1927 headed an aerodynamic institute at Zurich.

Because of the dislocation caused by First World War and the sheer magnitude of his contributions, Prandtl's work was only slowly assimilated. After the end of the war, Hermann Glauert and R. McKinnon Wood of Great Britain's Royal Aircraft Establishment (RAE) were sent to Germany to study the work of the AVA. Glauert, who spoke fluent German, immediately grasped the general importance of Prandtl's theory and published a lucid English explanation of it in 1920. By combining Prandtl's techniques with his own highly ingenious approaches, Glauert produced a whole series of original contributions to aerofoil theory over the next eight years. Glauert's *Elements of Aerofoil and Airscrew Theory*, published in 1926, is still the classic English-language source on subsonic aerodynamics. Indeed, it was only Glauert's book that made Prandtl's theory readily accessible to the Anglo-American aeronautical community. Another Englishman, Frederick Lanchester, had anticipated many of Prandtl's ideas on circulation and vortices, but Lanchester's work was unclear and qualitative. It was Prandtl and his school, and those who followed them, who produced a quantitatively testable, truly scientific aerodynamics.[14]

One of the most important consequences of the mature development of subsonic aerofoil theory was the creation of a much more accurate and comprehensive theoretical understanding of propellers. Screw propellers originally had been conceived literally as screws, and their operation had been explained, like that of a mechanical screw, by reference to the simple principle of the inclined plane. W.J.M. Rankine, in 1865, developed a momentum theory of propeller operation based on the total change in momentum imparted by the propeller to the working medium (water or, later, air). William Froude, beginning in 1878, theoretically examined separate blade-elements (arbitrarily small propeller blade sections) and, by mathematically summing the behavior of each element, arrived at a theoretical description of the local fluid dynamics of each propeller blade. The two alternative theories were not completely compatible because the propeller induced rotation in the operating medium. Application of Prandtl's aerofoil theory permitted theoretical combination of the momentum and blade-element descriptions of the propeller. That synthesis was effectively accomplished between 1918 and 1924 by Albert Betz and Prandtl in Germany and by R. McKinnon Wood and Hermann Glauert in England. A comprehensive propeller theory was further developed by Sydney Goldstein (1929), first at Göttingen and later in England.[15] C.N.H. Lock at the National Physical Laboratory (NPL) in England rapidly developed Goldstein's theory "into a form suitable for routine engineering application."[16]

Two other major contributions to subsonic aerodynamics require mention here: those of B. M. Jones and of Theodore von Karman. Jones was professor of aeronautical engineering at Cambridge and did outstanding original work in several areas. One of his studies is of especial relevance. At the end of the First World War, there was considerable disagreement in England about scale effects, which in theory should have been predictable but in practice often were not, and therefore about the validity of wind tunnel data for full-size aircraft. As a result, R. V. Southwell of the National Physical Laboratory "was mainly responsible for initiating the so called international trials, consisting of tests of the same aerofoil and the same two airship models in the number of wind tunnels" in England and abroad. The results of the tests "gave smooth curves but did not coincide since the magnitudes of the turbulence varied greatly."[17] Jones analyzed the test results and "made the surprising observation that the drag of an airship model depended more upon the tunnel in which it was tested than upon its shape." The different turbulence level in each tunnel distorted the tests. Jones concluded that the "main ingredient in the drag of well streamlined bodies—not excluding aerofoils—" was skin friction, much higher for turbulent than for laminar flow.

This insight led Jones, in an Aeronautical Research Council Reports and Memorandum published in September 1927 to the concept of an ideal streamlined airplane.[18] Jones first computed the theoretical, irreducible induced drag necessary to generate lift (Prandtl's theory) and the irreducible skin friction drag for laminar (smooth) flow for a number of aircraft. Using known theoretical and empirical information on propeller efficiency, he could therefore compute the engine power necessary to overcome unavoidable drag. The residual—the difference between that power theoretically essential for flight and the power actually expended by a given aircraft—had to be absorbed in overcoming unnecessary turbulent drag which resulted from improper streamlining. By this reasoning, Jones discovered that then-current airplanes wasted from two-thirds to three-quarters of their power in overcoming such unnecessary turbulent drag. Aeronautical engineers had always paid lip service to streamlining, but had usually ignored it when expedient because there was no positive, quantitative measure of the performance penalty for a given breach of streamlining. After Jones's report, given wider circulation by a lecture to the Royal Aeronautical Society in January 1929,[19] the aeronautical community began to realize how much power was actually being wasted by contemporary aircraft or, alternatively, what increases in speed were theoretically possible with existing power plants.

Theodore von Karman developed two major theoretical innovations in this earlier period. In 1913 he published his analysis of what was to become known as the von Karman vortex street. He not only constructed a mathematical explanation for the stable, asymmetrical arrangement of vortices shed by an object subject to viscous flow, but also "connected the momentum carried by the vortex system with the drag and showed how the creation of such a vortex system can represent the mechanism of wake drag."[20] In 1930, he published a comprehensive theoretical treatment of skin friction drag, laminar and turbulent, of transition from one type of flow to the other, and of flow separation. Von Karman's flow theory permitted direct theoretical analysis of the generation of turbulent drag with which Jones had dealt by the method of residuals.

Jones's and von Karman's work on turbulent drag stimulated intensive theoretical and empirical investigations of streamlined and turbulent flow. At Cambridge in the early 1930s, Jones himself developed the novel pitot traverse method of ascertaining flow and lift patterns across wing sections. Other investigators began extensive use of "tufting" (fastening small tufts of cloth to the surface of an aircraft to indicate flow direction and smoothness) both on wind tunnel models and on full-scale prototype aircraft in order to improve actual streamlining. Such investigations led directly to the Townend Ring and NACA full cowlings for radial engines and to Cal Tech's discovery of the value of wing fillets (concave sections added at the juncture of wing and fuselage to provide smooth flow transition). In general, Jones's and von Karman's theoretical insights and the work deriving from them led to radical redesign of aircraft and to the formulation of the essential design desiderata underlying the revolution in aircraft structures during the first half of the 1930s.[21]

By the end of 1930, through the work of Prandtl, von Karman, and a cadre of other brilliant and devoted practitioners, aerodynamics had reached a mature understanding of subsonic phenomena. The early 1930s, the gestation period for the turbojet revolution, saw rapid application of that full understanding to aeronautical practice.

In contrast, supersonic aerodynamics was just attaining rigorous theoretical formulation and secure empirical foundation during the emergent years of the turbojet revolution. Mach and his work, which was fundamentally ballistics, not aerodynamics, has already been noted. Work in ballistics had gone forward between Mach's time and the early 1920s, but the velocities with which ballisticians commonly dealt (Mach 3 or better) were so high and the shapes with which they experimented so different that their results were but marginally useful to aerodynamicists. Furthermore, ballisticians, usually associated with ordnance works or army proving grounds, constituted a different community from aerodynamicists, and aerodynamicists were as a result generally unfamiliar with ballistics.[22]

The first direct aerodynamic investigations of near-sonic velocities occurred during the years immediately after the First World War. Propeller tips had reached speeds at which they encountered compressibility "burble," the sudden large increase in drag and corresponding loss of lift characteristic of shock wave formation. During the 1920s and early 1930s, extensive experimental investigations specifically directed toward attaining an empirical understanding of propeller-tip compressibility "burble" were undertaken in the United States by Hugh L. Dryden, L. J. Briggs, C. F. Hull, John Stack, and others at the NACA, and in Great Britain by Hermann Glauert, T. E. Stanton, G. P. Douglas, and others at the RAE.[23]

The same period saw the initial theoretical development of supersonic aerodynamics. Glauert (1928) and Prandtl (1931) showed that high subsonic speeds in a compressible fluid only marginally affected the behavior of flow across a thin aerofoil so long as shock waves did not form. G. I. Taylor and C. F. Sherman (1928) developed an electrolytic analogy (using an electrolytic-fluid-filled tank) for analyzing such flow.[24] Aurel Stodola (1927) theoretically examined shock wave formation in De Laval convergent-divergent nozzles.[25] Prandtl and Adolf Busemann developed a rather complex graphical technique for describing shock wave behavior in such nozzles. Prandtl presented a more complete treatment of supersonic flow in nozzles in 1931.[26]

Jacob Ackeret published the first rigorous theoretical treatment of supersonic lift and drag for a two-dimensional (infinite-span) aerofoil in 1925.[27] Ackeret's theory was the supersonic analogue of the subsonic Kutta-Joukowski theorem. Herman Schlichting gave "an exact method of calculating induced wing-drag" at supersonic speeds in 1937.[28] In the intervening years, a small group of outstanding aerodynamicists made significant contributions to solving important parts of the supersonic riddle. With the exception of Taylor and Maccoll in England, those contributions were made exclusively by German or German educated scientists, Prandtl, Busemann, Meyer, and Ackeret, especially.[29] Still, even in 1940, the lift and drag of a given aerofoil at supersonic speed could not be confidently predicted. Much work was done during the Second World War, but it was not until 1947 that Theodore von Karman produced a comprehensive supersonic theory equivalent in breadth and precision to Prandtl's 1918 subsonic theory.

During the critical period for the turbojet revolution, 1925-35, supersonics, though undeveloped, still held several important implications. First, experiments with compressibility burble indicated that there was a definite practical limit to the speeds at which propellers could be used. Second, existent insights into supersonic flow behavior emphasized the importance of overall aircraft streamlining and smoothness. Third, both theoretical and empirical evidence suggested that even with exceptionally fine streamlining and thin-section aerofoils, flight into the transonic and supersonic regimes would require much greater power or thrust to overcome the intrinsically much greater drag in those regimes. That realization led some members of the larger aeronautical community to seek alternatives to the piston engine and propeller.

Although Prandtl published his comprehensive subsonic theory of lift and drag in 1918, it was, as noted, some time before the community of aerodynamicists outside Germany fully assimilated Prandtl's work. That process of assimilation, like the assimilation of any new scientific theory, was characterized not only by propagation of the theory through the community but also by its refinement, development, extension, and corroboration. Prandtl's aerofoil theory, and other insights derived from it, by 1926 led four men—A. A. Griffith in England, Jacob Ackeret in Switzerland, and Albert Betz and W. Encke in Germany—to investigations of axial compressors and turbines which would prove of tremendous moment for the turbojet revolution.

A. A. Griffith was the son of the English journalist and science fiction writer George C. Griffith and had received an honors bachelors degree in mechanical engineering from the University of Liverpool in 1914, after which he went to work on the shop floor of the Royal Aircraft Factory (later the Royal Aircraft Establishment, or RAE). As a result of his obvious talent, Griffith advanced rapidly, becoming senior scientific officer in the Physics and Instrument Department of the RAE in 1920; he meanwhile took his master's (1917) and doctorate (1921) from Liverpool.[30] Among his early achievements was a method, developed with Geoffrey I. Taylor and based on earlier work by Prandtl (1903), for using soap film structural analogues to analyze torsional stress (1917). Griffith also wrote a still-classic paper on strength of materials, "Theory of Rupture," which analyzed the mechanics of crack formation (1920). Griffith probably would have continued his research on strengths of materials—he had discovered, for example, the extraordinary strength of unmarred fine-drawn glass fibers—but an accidental fire caused by a glass-melting torch led to inquiries, and Griffith was instructed to undertake other work, since aircraft faced "no fatigue problems."[31] Griffith had done early investigations on the shape of piston engine cooling fins

(1916–17), and his work on torsional rigidity led to studies of propeller shape and stress (1917–18), which in turn led to his work on axial-flow compressors.

In 1926, Griffith proposed a new aerofoil theory of axial compressors. Previously, flow in axial compressors and turbines had been treated as hydrodynamic flow through passages, which assumed no blade to medium or medium to blade transfer of energy. Griffith argued that since the primary purpose of a compressor was to transfer energy to the working fluid (to apply mechanical energy to compress it), the design of the blading of the compressor should be approached through aerofoil theory in order to get the maximum transfer of energy with minimal losses (the analogues of lift and drag, respectively). (The converse of the same argument would be valid for turbines.) As Griffith wrote in his equally classic (but unpublished) report for the RAE:

> These considerations at once suggest the basis of a new method in designing turbo-machinery. The blades, instead of being regarded as the walls of channels, whose shape determines the velocity and pressure changes taking place in the fluid, are to be regarded as aerofoils and the changes in velocity and pressure are to be calculated from the blade reactions. These, in turn, can be found with the help of the known aerodynamic characteristics of the blade sections.[32]

An appendix to the paper applied the new theory to the design of a hypothetical turboprop engine.

The Aeronautical Research Council (ARC) considered Griffith's paper, and authorized construction of a small single-stage test rig, which was built under Griffith's supervision during 1927 and tested and reported on, again in an unpublished paper, by W. C. Clothier in 1928. The test turbine and compressor stage, which employed "free-vortex" blading (blading "twisted" to compensate for differential flow along its length), yielded stage efficiencies of better than 90 percent. The ARC also authorized wind tunnel cascade experiments (a "cascade" is simply a row of aerofoil sections or blades used to simulate a compressor or turbine stage); and in a report published in September 1928 R. G. Harris and R. A. Fairthorne of the RAE fully confirmed Griffith's theory.[33]

Griffith, meanwhile, was promoted to principal scientific officer at the Air Ministry Laboratory, South Kensington. He continued to develop his axial-flow compressor ideas, however, and in November 1929 submitted a memorandum (unpublished and secret) to the ARC containing a design study for a very complex contrarotating, contraflow, 500 h.p. turboprop engine. In his proposal, Griffith sought to evolve an engine fully competitive with contemporary piston engines in power, weight, and, most importantly, fuel consumption. Such desiderata implied that both

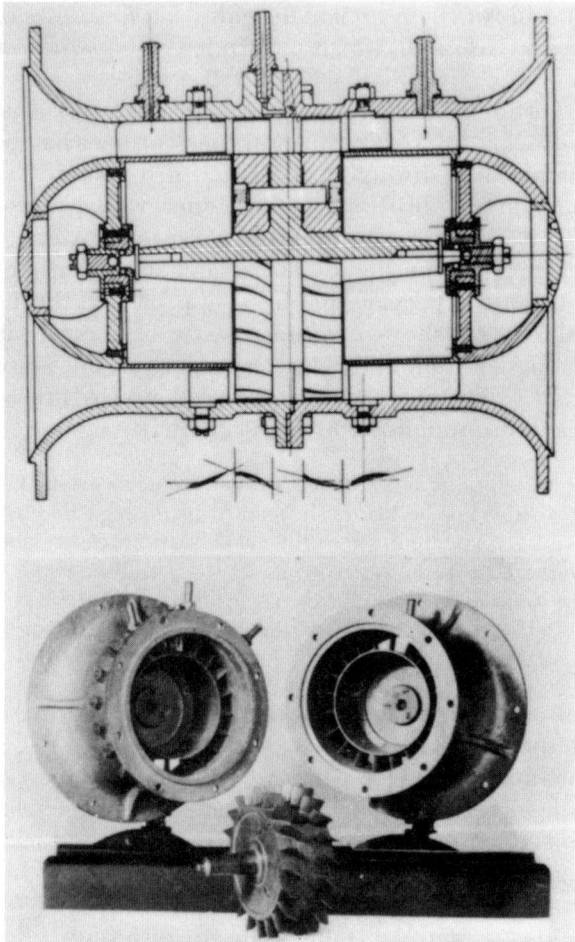

Griffith's RAE turbo-compressor experimental unit, 1928-29. (From F. W. Armstrong, "The Aero-Engine and Its Progress—Fifty Years after Griffith," *Aeronautical Journal* [December, 1976]. Reprinted by permission of the author.)

compressor and turbine had to achieve very high efficiencies (on the order of 90 percent) and that those components therefore would be extraordinarily sensitive to operating "off-design" (not at precisely intended rpm and pressure, such as during start-up or at high altitude). In light of these anticipated difficulties, Griffith reasoned:

> If now, the mechanical sub-division be carried to the limit so that each mechanically independent element consists of a single-stage turbine driving a single-stage compressor, the effect of compressibility on the flow pattern (apart from the effect on the aerodynamic properties of individual blade rows) may be for practical purposes entirely eliminated. The range of effi-

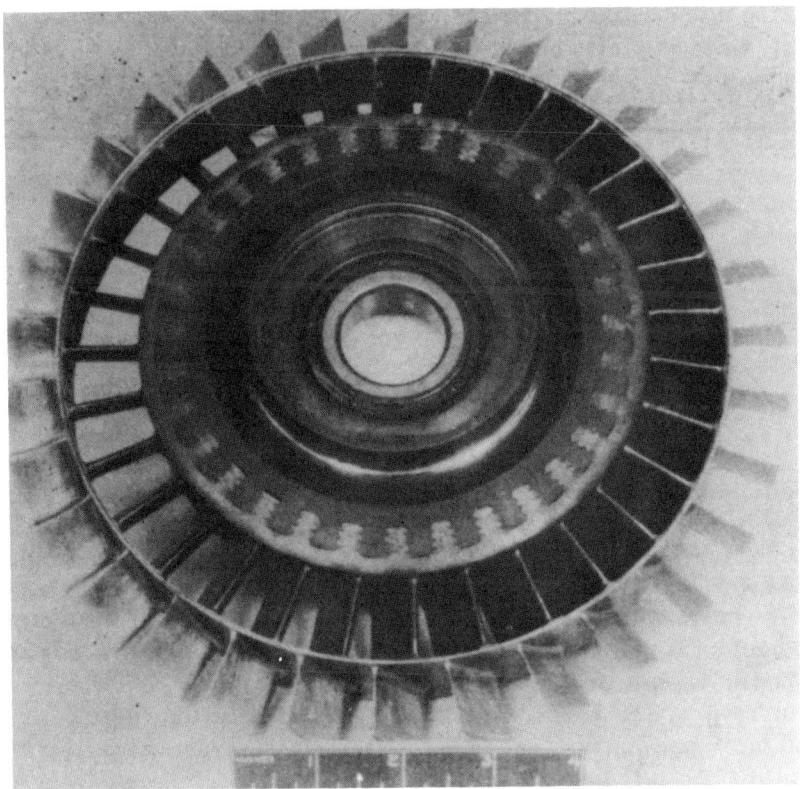

Rotor from Griffith-designed nine-stage contra-rotating, contra-flow experimental turbine-compressor test unit, built 1939 according to Griffith's 1929 concept. (From F. W. Armstrong, "The Aero-Engine and Its Progress—Fifty Years after Griffith," *Aeronautical Journal* [December, 1976].Reprinted by permission of the author.)

cient running conditions will then be limited at the lower end only by the adverse scale effect and at the upper end only by the onset of the compressibility stall.[34]

The result was a design in which compressor and turbine were concentric, with each successive turbine stage independently driving the compressor stage it surrounded; no stator blades were to be used in either compressor or turbine, each stage revolving in opposite directions (hence obviating any need for gearing between stages or between compressor and turbine). As a first step, Griffith proposed building an experimental fourteen-stage contrarotating, contraflow compressor and turbine unit.

A special engine subcommittee of the ARC, chaired by Henry Tizard, did approve construction of the fourteen-stage test unit early in 1930, but was apparently less than enthusiastic about the engine's prospects.

By the time Griffith returned to the RAE, Farnborough, in 1931, however, the economy measures of the MacDonald government, taken in response to the depression, seem to have precluded construction even of the test rig.[35] In any event, the RAE did not take up the turboprop again until 1936. Griffith was one of the first to apply aerofoil theory to compressors, and although his particular design for a turboprop engine would prove impracticable, he was the first, in marked contrast to the experimenters in Switzerland and Germany, to realize the implications of the aerofoil theory for construction of a gas turbine aero-engine.

Griffith's approach represents a nearly ideal pure science to technological practice sequence: he first developed a sound scientific theory, derived from prior science, then subjected it to careful experimental test; once the theory was corroborated, he proposed building a prototype test unit before attempting to actually develop a practical engine. At the specified hierarchical level, that of gas turbine internal component efficiencies, Griffith's program was flawless. All he missed were higher level considerations affecting aircraft speed which would have made a much simpler turbojet practical. Ironically, and in singular distinction to early turbojet protagonists, in the intervening years Griffith made major contributions to the further development of aircraft piston engines.[36]

Jacob Ackeret, the second person to develop the aerofoil theory for compressors, was, as noted, one of Prandtl's students at Göttingen and had produced the first two-dimensional theory for supersonic aerofoils. In 1927, Ackeret left Göttingen to return to Zurich to organize a new aerodynamic institute at the Swiss Federal Institute of Technology. Included in the equipment for the new institute was to be a closed supersonic wind tunnel, the world's first. For the supersonic tunnel Ackeret designed and asked Brown Boveri to construct a thirteen-stage axial blower that gave a pressure ratio of just over 2:1 at an efficiency of 80 percent. Ironically, the supersonic tunnel was built "only secondarily" for aeronautical purposes; its major use was to be the investigation of aerofoil compressibility phenomena in relation to the construction of large steam turbines and turbocompressors.[37]

Brown Boveri had independently become interested in axial blowers. Between 1925 and 1931, its French concessionaire, Electro-Mecanique, had built, under the direction of M. Darrieus, a series of windmills for the generation of electric power. Darrieus had designed the windmill vanes according to aerofoil theory, and had recommended application of the theory to axial blowers. He had independently discovered free vortex blading, and had in mind the eventual construction of a multistage compressor. Brown Boveri therefore built a small, experimental four-stage compressor in 1927. Meanwhile, Brown Boveri had been using single- and two-stage axial blowers as cooling fans in its generator and electric motor sets, and had done thorough wind tunnel work on single-stage units.[38]

Nevertheless, the unit that Ackeret designed and ordered for the Zurich tunnel was the first large, multistage axial blower built by Brown Boveri. That Ackeret designed the Zurich blower would seem to indicate that his understanding of the axial compressor was somewhat more advanced than Brown Boveri's. (So far as precedence is concerned, the application of aerofoil theory to turbomachinery was probably first discussed by Bauersfeld in 1922,[39] but Griffith seems to have been the first to work out the idea fully, followed by Ackeret, Brown Boveri, and the German experimenters.) At any rate, by 1931, Griffith and the RAE and Ackeret and Brown Boveri were about equally advanced.[40] Ackeret's design for the Zurich tunnel served as a prototype for succeeding closed-circuit supersonic tunnels. Brown Boveri constructed a supersonic wind tunnel at Guidonia for the Italians in the late 1930s and several exceptionally good tunnels in Germany during the war, including the phenomenal Bavarian Motor Works B.M.W. Oberweisenfeld/Munich high-altitude engine test facility.[41] Ackeret himself would become involved in development of a closed-cycle gas turbine in the late 1930s.

Still, Albert Betz and W. Encke of the AVA carried out the most complete investigation of axial compressors done prior to the turbojet revolution. Betz was one of Prandtl's closest collaborators, working with him on the general theory of 1918. From 1918 to 1926, Betz worked (but not exclusively) on the application of aerofoil theory to propellers, including windmills. In 1927, Betz and Encke began experiments on the behavior of axial-flow compressors. Their investigations were based upon aerofoil theory and were conducted using a specially constructed test rig at the AVA. Their explicit objective was to gain the information necessary to design an axial-flow piston-engine supercharger. By 1929, Betz had in operation a six-stage test compressor built by Junkers that gave a 2:1 pressure ratio with 85 percent efficiency at 37,000 rpm. In fact, Griffith heard of Betz's favorable results and cited them in his 1929 proposal. Betz meanwhile had by 1931 fully investigated the use of airfoil lattices (cascades) as an experimental aid in compressor design.

By the end of 1930, Betz and Encke could get the same 2:1 ratio out of five wheels instead of six, and by about 1936 needed only four wheels, giving a pressure rise per stage of just over 1.2:1. Betz and Encke also experimented with various configurations for the axial compressors. They considered not only straight-through units, but also counter-rotating units (as separately proposed by Griffith). Betz and Encke encountered mechanical difficulties in counter-rotating sets with more than two stages and therefore emphasized mechanical simplicity in their designs, which complemented their quite valid penchant for straight-through airflow.[42] Betz and Encke were probably more advanced than Griffith or Ackeret even in 1930, and their experiments went on uninterrupted right through the Second World War. By 1936-37, the critical period for the turbojet revolution in Germany, Betz and Encke at the AVA had a uniquely comprehensive knowledge of axial compressors. Although the 1927 test rig was driven by a compressed-air turbine (fed from a tank), and although Betz must surely have been aware of his work's value to possible gas turbine design, there is no evidence that he sought to exploit his knowledge in that direction. Unlike Griffith, who pushed for a turboprop from 1926 on, Betz and Encke would wait for others to make use of the insight they had generated.

As indicated, the aerodynamic research portrayed here represents a small subset of all the aerodynamic research undertaken between 1915 and 1940. The theoretical developments included here meet at least one of two criteria: they are of great importance to the overall theoretical development of aerodynamic science, or they have a specific retrospective relevance to the turbojet revolution. It is directly from this selective matrix of progress in theoretical aerodynamics that some individuals deduced the presumptive anomaly which led to creation of the first turbojets.

Yet, also as noted, the linkage from aerodynamic progress to presumptive anomaly to turbojet was neither determinant, teleological, nor singular. The theoretical progress that led to the turbojet also had profound impacts on related and even competitive technologies. Furthermore, distinctive national patterns in the pursuit and utilization of the science surveyed here internally clearly biased its impact on the turbojet revolution.

5

Regnant Normal Technology: Aircraft Piston Engines, Aircraft Structures, and Alternative Propulsion Systems

Measured on any plausible dimension—gross power output, power per pound or per cubic inch displacement, specific fuel consumption, altitude or speed attained, or total numbers in service—development of the aircraft piston engine achieved stunning and rarely paralleled technological success between 1910 and 1945. That development program comprised not only the vast improvement of basic piston engine components (cylinder blocks, pistons, valves), but also the adoption of revolutionary subsystems (internally cooled valves, variable-pitch propellers, superchargers) and the cooperative promotion of revolutionary progress in other fields, notably fuels and lubricants. Progress in aircraft piston engine development, moreover, co-evolved with and circularly and cumulatively reinforced radical new departures in the design and construction of aircraft structures. These interrelated spearheads of technological progress produced, within less than two generations of the Wright brothers' flight, a modern aviation system no pre-twentieth-century child of Icarus would have dreamed possible.

Yet that immensely successful aviation system was not a gift of the gods or a bounty of nature, but was wrested from a recalcitrant aerosphere by years of painstaking, frequently brilliant, and always persistent developmental effort. Solutions to the problems of aeronautical normal technology were neither easy nor obvious. Each increment of normal system performance was dearly bought in terms of time, effort, and money. The turbojet revolution occurred, then, not in the face of a failed or discredited normal technology, but within a deeply committed community which by its own efforts had created one of the most spectacular and successful of modern technologies.

The very difficulty of the problems encountered in the development of aircraft piston engines and the great personal and professional ingenuity shown in resolving those problems did induce within the aeronautical community at large recurrent consideration of basic alternatives to the conventional gasoline piston engine and propeller combination. Those alternatives ranged from different forms of piston engines, such as diesels, to gas-turbine-driven turboprops and reaction propulsion systems. Each alternative, however, was in turn vitiated by the ultimate, unquestionable success of gasoline piston engine development.

The functional antecedent of the turbojet engine, the aircraft piston engine, together with the new streamlined airframes and the failed alternative propulsion systems, constitute the full, rich, and hostile milieu in which the turbojet revolution was to occur.

AIRCRAFT PISTON ENGINES

The overall development of aircraft piston engines is adequately covered in the literature, notably by Roy Fedden's long 1944 article[1] and by Robert Schlaifer's definitive *Development of Aircraft Engines* (1950) and Sidney Heron's companion *Development of Aviation Fuels* (1950). All that is required here is to indicate the gross outlines of the success of piston engine development and to give some indication of the power and complexity of the developmental efforts which produced that success.

Between 1912 and 1945, the maximum power output of aircraft piston engines increased by a factor of twenty-five, from 120 h.p. to 3,000 h.p. Specific weight declined by a factor of six, from 4.8 to 0.82 pounds of engine weight per horsepower. Output per cubic inch of cylinder displacement increased between 1915 and 1944 by a factor of four. Between 1928 and 1944 alone, specific fuel consumption (pounds of fuel used per horsepower per hour) dropped by 25 percent as a result both of better engines and of higher octane fuel.[2] Table 1 shows power, specific weight, and related data for representative aircraft piston engines.

Piston engine progress was founded upon a whole array of fundamental innovations. During the 1920s, stronger, lighter, more precisely shaped forged or extruded aluminum pistons replaced cast steel pistons. Individual cylinders with separate water jackets screwed into cast crankcases were replaced by cast aluminum monoblocks in which the cylin-

Table 1. Piston Aeroengines

Mfr.	Desig.	Country	Date	Cyl.	Config.	Power	Displ. (cu.in.)	Wt. (lbs.)	lb./h.p.
Wright		U.S.	1909	4	in-line	30	410	176	6.9
Le Rhone		Fr.	1915	9	rotary	80		253	3.16
Benz		Ger.	1917	6	in-line	108		425	3.94
Consortium	Liberty	U.S.	1918	12	V	400	1,650	856	2.14
Napier	Lion	G.B.	1918	12	W	400	1,461	890	2.23
Rolls-Royce	Kestrel	G.B.	1928	12	V	520	1,295	900	1.73
Curtiss	D-12	U.S.	1929	12	V	435	1,145	687	1.47
Wright	Cyclone	U.S.	1930	9	R	550	1,823	947	1.9
Rolls-Royce	R	G.B.	1931	12	V	2,530	2,239	1630	0.73
Pratt & Whitney	Wasp	U.S.	1933	9	R	550	1,344	883	1.6
General Motors	V-1710	U.S.	1940	12	V	1,090	1,710	1325	1.2
Rolls-Royce	Merlin	G.B.	1942	12	V	1,750	1,649	1665	1.01
Pratt & Whitney	Double Wasp	U.S.	1942	18	R (2-row)	2,000	2,800	2300	1.15
Daimler Benz	603L	Ger.	1943	12	V	2,100	2,720	2120	1.0
Napier	Sabre	G.B.	1944	24	H	2,400	2,240	2360	0.98
Wright	R-3350	U.S.	1944	18	R (2-row)	2,200	3,350	2670	1.2
Bristol	Centaurus	G.B.	1945	18	R (2-row)	2,700	3,270	2920	1.08
Pratt & Whitney	Wasp Major	U.S.	1945	28	R (4-row)	3,000	4,363	3482	1.15
Wright	R-3350 Turbo Compound	U.S.	1951	18	R (2-row)	3,400	3,350	3443	1.01

ders and water passages were cast integrally with the crankcase. The monoblocks provided much stronger, more rigid, and lighter engine structures. Cooling and bearing alignment were both simplified. With the use of separate cast iron cylinder liners, maintenance was simplified. Similar progress in cast and machined finned cylinders for radial engines kept the radial type competitive. Development of supercharging and turbosupercharging, both founded upon earlier experience with turbo-air compressors and internal combustion gas turbines, played an especially critical role in piston engine progress. Indeed, the most highly developed post–World War II piston engines, the turbocompounds, used exhaust gas turbines both to power a turbocharger and, through direct gearing to the crankshaft, to furnish additional power to the propeller.

Of all the improvements in piston engine technology, probably the single most important was that in fuel chemistry. Fuel progress is commonly credited with at least 50 percent of the power increase of gasoline piston engines between the wars. The First World War brought the first awareness of problems of detonation (knocking) and preignition in aero-engines. As a result, beginning in 1916, the Royal Dutch Shell Oil Company began support of researches by H. R. Ricardo into the performance of fuel in internal combustion engines. Further work was done in France during 1918–19. From these experiments and those of subsequent researchers, the effect of various compounds in fuel was gradually understood. In 1921, Thomas Midgeley and T. A. Boyd, sponsored by Charles F. Kettering of General Motors, discovered the detonation-preventing properties of tetraethyllead, which had originally been studied, during hostilities, for possible use in chemical warfare. In 1925, Dr. Graham Edgar formulated the system of octane ratings which allowed various fuels to be accurately compared.[3] Progress in fuels naturally demanded progress in other areas of piston engine design, since the improved fuels generated higher temperatures and greater pressures and stresses. Engines and fuels, too, co-evolved.

Development of better engines also entailed development of better accessories. Some were minor but vital: fuel pumps, magnettos and electrical systems, fuel tanks and sealants, lubricants, and controls. Progress in propeller design, however, played a singularly major role in preserving the piston engine system. The uncertain qualities of wood propellers resulted in experiments with metal propellers during the First World War, but because of weight and hub attachment difficulties, little of use was accomplished. As airplanes became heavier and attained higher altitudes, however, demand grew for some means of varying propeller pitch. Both the Royal Aircraft Establishment (RAE) and the U.S. Army Air Service experimented with variable-pitch (v.p.) propellers during the closing months of the First World War. Mechanical complexity, unreliability, weight, and the relatively small benefits accruing at the low speeds then attained prevented v.p. prop adoption at the time. H. S. Hele-Shaw and T. E. Beacham in England built a very good v.p. prop during the mid-1920s, but it was again not adopted. The U.S. Air Corps likewise tested a number of v.p. props during the 1920s, but for the usual reasons rejected them for service use. It was left to Hamilton-Standard, owned by Pratt and Whitney, to develop a truly practicable hydraulic v.p. prop, used commercially for the first time in 1933 on Douglas's DC-1. After 1933, v.p. props, both hydraulic and electric, were increasingly widely used. V.p. props made it possible for monoplanes with high wing loadings to take off within reasonable distances and to land slowly but still to maintain high efficiencies at high speeds and altitudes.

Manufacturing techniques for piston engines also underwent revolutionary development. Roy Fedden contrasted the situation in 1944 with that in the early years of aviation development:

> It must be remembered that in the first few years of the present century, engineering materials were in an entirely different state of development from what they are to-day: aluminum alloys were crude and quite unknown for highly stressed parts working at high temperature, such as pistons; high quality alloy steels, free from impurities, had not yet come into general use, and nitriding and other means of high precision hardening were unknown. Modern precision machine tool operations ensuring accurate limits and the elimination of vibration, so important to the light-weight, highly stressed aero engine, were also in their infancy, and accurate gear cutting and subsequent grinding of the teeth, honing, lapping, and fine boring were all unknown, while gauging, inspection and testing procedures, all of which have now been standardized, were then much more haphazard.[4]

Even in 1920, for example, little was known of crankshaft flexural vibration. As late as 1928, Frank M. Owner was using his own rather eccentric testing technique:

> Preliminary tests of frequency were then made in the Experimental Fitting Shop by striking a suspended crankshaft and listening to its musical note (I am blessed with a fairly accurate sense of absolute pitch). This was too much for the shop foreman, who reported to his astonished superior that I had finally gone "round the bend"—"He's down there knocking hell out of a Mercury crankshaft and moaning to himself!"[5]

Owner soon devised an electric strain gauge. Similar advances were made in other areas of engine testing.

Two major triumphs in piston engine subsystems development illustrate the diversity, complexity, and uncertainty associated with all major developmental efforts, as well as the labyrinthine interconnections that hierarchical systems promote among various traditions of technological practice. Supercharging and turbosupercharging were vital to all aspects of improved piston engine performance and were absolutely essential for high-altitude operation, while valve design proved to be one of the most persistent and intractable barriers to piston engine progress.

Supercharging refers to the precompression of air ingested by a reciprocating piston engine. By supercharging, the power developed by a given piston engine at a specified rpm can be increased. Since more air is ingested, more fuel can be burned within the same size cylinder. Superchargers, like other air compressors, may be of the reciprocating piston, positive displacement (Roots), or centrifugal types. Superchargers may be geared directly to the engine they supercharge, be chain driven, be driven by an auxiliary electric motor, or be powered by an entirely separate engine. Turbosuperchargers or turbochargers are simply superchargers driven by a gas turbine installed in the exhaust system of a piston engine, thereby making use of otherwise wasted energy. Supercharging and turbosupercharging not only represent crucial and revolutionary subsystems within piston engine development but also are closely related, theoretically, technically, and through individual researchers, to internal combustion gas turbine investigations.

Supercharging by means of piston-type air compressors was used before 1900 to insure proper mixture distribution to and adequate exhaust scavenging of the cylinders of large gas or diesel engines.[6] Dugald Clerk, the English inventor of the two-stroke piston engine, conducted perhaps the first comprehensive investigations of supercharging. His initial experiments, using a reciprocating compressor, were done during 1901–2.[7] From 1909 to 1912, Hugo Junkers in Germany and Sulzer Brothers in Switzerland used forms of supercharging on diesel engines. Some experiments also were conducted on supercharged automobile engines before the First World War.[8]

Alfred Buchi, who had previously worked on early diesel engine development at Sulzer's and on the Voigt gas turbines in Belgium, conducted large-scale experiments on diesel turbosupercharging at the Sulzer works during 1911. Although his turbosupercharging work was promising, it was abandoned because the Sulzer firm regarded it "as only a novelty."[9] After the First World War, Buchi successfully developed the Buchi-Duplex turbocharging system for diesel engines. The Buchi-Duplex system employed tuned-length exhausts and inlet and exhaust valve timing modifications to assure complete cylinder scavenging and smooth turbocompressor operation. But while the Buchi system was technically successful, it required complex, heavy exhaust piping and was efficient only over a very narrow range of engine operating conditions. It was therefore not widely adopted, and was not suitable for use in aircraft.[10]

Supercharging aircraft piston engines was first seriously considered during the First World War. The airplanes with which the powers began the war in 1914 were scarcely more highly developed than those the Wright Brothers had flown a decade before. Under the pressure of wartime exigencies, however, especially during 1916 and after, more and more performance was demanded—greater speed, heavier loads, longer ranges, greater dependability, and higher altitudes. With few exceptions, the form and construction of the airplane itself was fixed. The only quick way to greater performance was more powerful engines, which usually meant larger and heavier ones, and which also meant trading range and load for speed and altitude. Yet with larger and better engines airplanes soon encountered a new problem: the less dense air at high altitudes crippled the performance of naturally aspirated engines. Supercharging or turbosupercharging could provide a solution: by precompressing the less dense and colder air at altitude, piston engine power could be maintained.

The first person to apply supercharging to aero-engines was Auguste Rateau. The Lorraine-Dietrich Company approached Rateau in 1915 to design a gear-driven supercharger. Upon consideration, however, Rateau decided that the necessary multistage fan and gearing would be too large and heavy for aircraft application. Additionally, unless some change-speed gear were incorporated, variation in altitude and hence pressure would not be compensated for. Rateau therefore set the project aside. But performance deficiencies in piston engines at higher altitudes continued to plague the French Air Force, and in 1916, a pilot, Bastiou, again approached Rateau. This time Rateau turned to the turbosupercharger. Freed from the necessity of gearing, Rateau could employ very high rotational speeds for the compressor impeller which meant that sufficient pressure could be gotten out of a single-stage compressor. This factor obviated the weight and size problem. Also, as the airplane climbed into less dense air, atmospheric back-pressure on the turbine would be lessened, permitting it to speed up and automatically compensate for increasing altitude. Rateau began tests on his design in the early part of 1917, and some of his turbochargers were used by the French Air Force during the First World War.[11]

W. F. Durand, chairman of the U.S. National Advisory Committee for Aeronautics (NACA), was in Europe during 1917 and learned of Rateau's work through the Interallied Commission.[12] Durand had been a professor at Cornell and "remembered the black smoke from the [Moss] gas-turbine research and knew of the extensive business that the

General Electric Company then had in steam turbines and centrifugal compressors." When he returned to the United States, Durand asked Sanford Moss to undertake development of a turbocharger for the U.S. Army Air Corps. Drawings of Rateau's turbocharger were provided, but G.E. apparently did its own detail design.[13] The first G.E. turbocharger ran in 1918, using an impulse turbine provided by the De Laval Steam Turbine Company. The development of the turbocharger by G.E. (under Moss's direction) and the Army Air Corps was continuous, but very modestly funded, after 1917. The G.E.-Air Corps program culminated in the turbocharger used in American B-17, P-47, and P-38 aircraft during the Second World War. Ironically, Sanford Moss, who thus did so much to advance aviation, disliked airplanes and never flew in one.[14]

Beginning in 1917, G.E. also built geared centrifugal superchargers for Curtiss-Wright and, later, for Pratt and Whitney. The army also tested a turbocharger designed by E. H. Sherbondy in 1917 and built by the De Laval Company for Fergus Motors. The Sherbondy machine was essentially similar to that of Rateau. Although it showed "a number of ingenious details," it was dropped after construction of three models in favor of the G.E. unit.[15]

In England, the Royal Aircraft Establishment (RAE, then called the Royal Aircraft Factory) began work on supercharging in 1915, considering first piston and displacement (Roots-type) blowers. The work at the RAE was under the direction of James E. Ellor in cooperation with Metropolitan-Vickers. The RAE had turned to the turbocharger before it heard of Rateau's work. Since Rateau was ahead, however, the RAE began actual experiments with Rateau machinery. Experiments continued for some four years after the war, but, because of the fire hazard caused by the intake and exhaust ducting of the turbocharger in the fabric covered aircraft of the time, it was abandoned. The RAE developed instead a geared centrifugal blower, the first of which (and the first of its kind in the world) went into service in the Armstrong-Siddeley Jaguar engine in 1926. British Thomson-Houston, a G.E. affiliate, also worked on superchargers. Geared, multistage superchargers were produced by Siemens in Germany and Brown Boveri in Switzerland during the First World War. Geared, integral centrifugal superchargers were normally used after the First World War in radial aero-engines (Curtiss-Wright, Pratt and Whitney, Armstrong-Siddeley, Bristol) because the supercharger provided better mixture distribution to the cylinders than did any other means. Development of superchargers for altitude performance also continued, both for radial engines and for in-line engines (G.M.'s Allison with the G. E. turbocharger and with a supercharger, and Rolls Royce's engines with geared superchargers).

The development of superchargers and especially turbosuperchargers obviously generated a great deal of technological experience useful to the turbojet. In the turbosupercharger, mechanical engineers had to deal with exceptionally high rotational speeds (to 40,000 rpm) at high temperatures (600° C). All the work on stress and resistance to creep in high-temperature materials, compressor and turbine design, gas flow, bearings, and ducting done in connection with piston engine supercharging could be applied directly to the turbojet. But even the turbocharger was not a turbojet. By 1935, single-stage centrifugal compressors for aircraft piston engines could attain maximum compression ratios of 1.85:1 with efficiencies of 65 to 70 percent.[16] Contemporary turbosuperchargers normally had compressor efficiencies of at most 70 percent and turbine efficiencies of roughly 60 percent (neither of which was all that critical because of the large amount of energy available in the exhaust of aero-engines), for a net efficiency on the order of 40 percent, a long way from a practicable turbojet. Yet the supercharger and turbosupercharger did stand as a unique repository of technical proficiency for the turbojet revolution, even as they permitted the piston aero-engine to attain otherwise impossible levels of performance.

The evolution of valve design, in contrast to supercharging, illustrates the range of solutions possible within a well-defined tradition of practice. As piston engine cylinder sizes (and therefore valve sizes) increased, and as cylinder pressures and temperatures also increased, adequate cooling of poppet valves became an increasingly intransigent problem for engine designers. Even by the end of the First World War, valve difficulties probably constituted the principal limitation to greater piston engine power. The development of sodium-cooled valves by Sidney D. Heron and the development of a whole new line of sleeve-valve engines by the engine department of Bristol Aircraft represent two contrasting solutions to that fundamental problem.

Heron's was the less radical of the two innovations. The principle of an internally cooled poppet valve is essentially simple: "An internally cooled valve is one with a closed, hollow stem containing a certain amount of some substance which is liquid at the operating temperature of the valve. This liquid moves back and forth with the motion of the valve, absorbing heat when it is at the head end and transferring this heat to the stem when it is at the other end."[17] To be effective, however, the internal coolant must have properties such that it "wets," or coats, the metal interior of the valve sufficiently well to insure high rates of heat transfer.

During 1916–17, Heron had worked as a designer on air-cooled cylinder development under the direction of A. H. Gibson at the old British Royal Aircraft Factory (later RAE). After 1921, Heron was employed as a civilian designer at the U.S. Army's aeronautical laboratories at McCook Field, Dayton, Ohio. British engineers had tried valves internally cooled by mercury as early as 1913–14, and at the Royal Aircraft Factory Gibson himself had experimented, unsuccessfully, with an internally water-cooled valve with small external radiators for an otherwise air-cooled cylinder. Kettering and Midgeley at the Dayton Electric Company (Delco, owned by G.M.) had learned the importance of "wetting" for internally mercury-cooled valve operation and had used the expedient of internal tinplating to promote it. Although used in some quantity, their valves were not highly developed.

Heron encountered severe valve troubles with a new air-cooled cylinder he was developing for the army in 1921. He attributed his valve problems to inadequate cooling. Ironically, he later discovered that the valves he had been supplied were simply defective. In the meantime, he happened to see a heat-treating bath using a mixture of nitrates of sodium and potassium "which wetted the pot so well that it crept out over the sides." After trying a nitrate mixture, Heron "thought of using metallic sodium or potassium or a mixture of the two when he read that such a mixture was used in high-temperature thermometers."[18] Heron applied for a patent on the two modes of internal cooling in 1922. Graham Edgar of the General Motors Chemical Corporation (later Ethyl Corporation) furnished developmental advice to Heron before commercial development of the sodium-cooled valve was put in the hands of the Rich Tool Company. Heron himself went to work for the Wright Engine Company in 1926, and the type-K Wright "Whirlwind" radial was the first production engine to use sodium-cooled valves. After 1930, virtually every high-powered poppet-valve aircraft engine used some form of internally cooled valve.

Heron's major innovation, then, permitted the continued progress of aircraft piston engine technology. Viewed from the perspective of valve design, the sodium-cooled valve is revolutionary; viewed from the hierarchically higher level of piston engine development it is the essence of successful normal technology.

The development of sleeve-valve radial air-cooled engines by the Bristol engine department represented a more radical solution to the problem of piston engine valve design. Indeed, sleeve-valve engines constitute a distinctive type of piston engine. H. Roy Fedden, while working for Brazil Straker before the First World War, first encountered a sleeve-valve engine when he visited the 1914 Naval and Military aero-engine trials, which were won by an Argyll Burt-McCollum sleeve-valve design.[19] After the war, the aero-engine business of Brazil Straker was taken over by the Bristol Aeroplane Company. Since Brazil Straker had entered into an agreement with Rolls Royce during the war forbidding the former's development of large liquid-cooled engines, the new Bristol engine department began work on large air-cooled radials. Fedden initiated a sleeve-valve aero-engine development program at Bristol in 1927 for the purposes of attaining higher cylinder pressures and better internal gas dynamics and of obviating the problem of valve cooling.

The development program involved not only engine design and testing but also the creation of novel production and materials-fabrication techniques. After trying many different porting arrangements and after "many thousands of hours" of running single-cylinder test units, Bristol had an operable sleeve-valve engine. Acceptable oil consumption could be achieved, however, only by squeezing "every sleeve round after hardening" and then grinding "each piston and cylinder to fit the particular sleeve." Bristol could hand-build an engine that would run, but could not build one using interchangeable parts suitable for series production. Firth Vickers (a specialty steel firm) undertook development of a centrifugal casting process to produce sleeves of consistent quality. That subsidiary development program necessitated an extensive materials investigation: "In all some fifty to sixty varying compositions associated with approximately 1,100 heat treatment procedures were investigated and tested both metallographically and physically before the final production sleeve was produced."[20] In addition, production of sleeves required development of special machine tools. In total, Bristol and the British Air Ministry together spent over £2 million solely on sleeve development. Even new subsystems do not come cheaply.

The Bristol sleeve-valve program was ultimately successful. First series production of sleeve-valve engines began in 1936, with the introduction of the first two-row model in 1937. The big two-row eighteen-cylinder Centaurus was running by mid-1939. The fully developed Centaurus would produce 3,000 h.p. by 1945. Ironically, the sleeve-valve program was not without opportunity cost: Frank Whittle approached Bristol in 1930 with his early turbojet design, but after Bristol technical director Frank Owner's pessimistic assessment of likely component efficiencies, Bristol decided that they simply did not have the manpower or capital for the requisite "design and research effort." Bristol's commitment to the radial sleeve-valve engine precluded their simultaneous investigation of other, even more radical alternatives.

Supercharging and turbocharging, together with Heron's sodium-cooled valves and Fedden's sleeve-valve engines, illustrate the great if costly range of fundamental variation possible within the piston aero-engine tradition. But superchargers and valves are only two among a multitude of piston engine components, each susceptible to redesign and further development to give increased performance to the whole aircraft-engine system. Indeed, development strategies undertaken to obtain higher performance commonly involve a holistic, multilevel assault on every aspect of system design. Such strategies are well exemplified by the modifications made to the Rolls Royce model R racing engine between the Schneider Trophy contests of 1929 and 1931.[21] Rolls Royce racing engine development is admittedly a somewhat special case, but the company's program does represent, in condensed format, the more general practice of normal technological development.

The V-12 R engine of 1929 produced 1,900 h.p. at 2,900 rpm and was itself a highly developed version of the 825 h.p. Rolls Royce Buzzard H commercial aero-engine. The 1931 R engine, in contrast, produced 2,330 h.p. at 3,200 rpm.[22] To extract that additional 430 h.p. (a more than 20 percent increase) from the same basic engine displacement and weight required great and ingenious effort. The additional power came from increased engine speed (300 rpm) and a higher supercharger compression ratio. But those expedients necessitated a whole array of ancillary improvements. A larger supercharger compression ratio and greater air throughout would have ordinarily required a larger supercharger diameter, which in the finely streamlined racing application envisioned was unacceptable. Instead, Rolls Royce developed a novel two-sided impeller which gave greater pressure and volume flow in a smaller overall supercharger diameter. The greater supercharger flowrate required, in turn, a redesigned "ram-effect" air intake.

Higher engine speeds, cylinder pressures, and power necessitated numerous internal changes in the engine. New, stronger, and heavier connecting rods were designed. A new crankshaft with larger bearings and new balance weights was installed. Larger and stronger exhaust valve springs and sodium-cooled valves were used. A larger oil pump was fitted, as well as additional piston oil control rings. A completely new carburetion system was used, and a completely redesigned propeller employed. Carefully tested special fuel mixtures, concocted by F. R. Banks of the Anglo-American Oil Company (an ESSO) affiliate), gave the desired compromise between specific heat content, antiknock properties, and volatility.[23]

The development program for the R engine also required evolution of advanced testing procedures. Spark plug tests were run on a special single-cylinder test "engine." A special supercharger test rig powered by a separate Rolls Royce Kestrel engine was used. Complete R engine tests were done by the "run to fail" technique—running the engine until something broke, at first about 15 to 20 minutes. But the R engine was so powerful that a redesigned hydraulic dynamometer had to be specially constructed.

The R engine program was successful. England won the Schneider Trophy in 1931 and set a new world's speed record. Furthermore, the outstanding success of the R engine development program induced Rolls Royce to continue development of large, high-powered V-12 liquid-cooled engines. That derivative development program culminated in the Rolls Royce Merlin engines used in British Spitfires and U.S. P-51 Mustangs during the Second World War, as well as the larger RR Griffon used late in the war.

The remarkable achievements in aircraft piston engine development between 1915 and 1945 signify the evolution of a well-defined, highly specialized, and well-integrated research, development, and implementation system. Government research establishments, aero-engine manufacturers, accessory manufacturers, airframe manufacturers, fuel producers, and user communities cooperated to make the piston engine–propeller system an unparalleled success. It was within that technological and institutional context that other alternative systems failed and the turbojet alone would succeed. The turbojet not only would overthrow the piston engine, but also would capture and exploit the vast pool of technological capabilities, the array of production and quality control techniques, that were equally essential to high-powered piston engines or to turbojets. The turbojet would usurp and make its own the very institutional structure that had produced aircraft piston engines.

THE REVOLUTION IN AEROSTRUCTURES

During the 1930s, piston engine and propeller design co-evolved with revolutionary developments in aircraft structures. That revolution in aerostructures, founded upon the aeronautical design desiderata implicit in the work of Jones, von Karman, and their peers, was essential to the coming of the turbojet. It is indeed ironic that the successful co-evolution of piston engines and aerostructures should have, in effect, defined or created a new environment in which the turbojet could thrive.

The magnitude of co-evolutionary progress in aircraft structures and its importance to the turbojet revolution are obvious. If a fully developed turbojet had been delivered to the Wright brothers or to any one of the belligerent air forces during the First World War, and if they had had the electrical system and the fuel system to get the engine into operation, which they did not, the engine would have promptly jerked apart any aircraft in which it might have been installed. Even in 1919, aircraft, with

Sewing fabric on Boeing MB-3A fighter wing, Seattle, 1921. (From Boeing Co., *Pedigree of Champions* [Seattle, Washington, 1963].)

very few exceptions, were still wood and canvas and piano-wire concatenations, with fixed landing gear, wood propellers, at least two wings, and a vast propensity for self-immolation. There is currently no adequate history of the development of modern aircraft structures, although Ronald Miller and David Sawers do offer an informative and suggestive

Ford Tri-motor, 1926. (Courtesy of the Smithsonian Institution.)

chapter in their *The Technical Development of Modern Aviation*.[24] The gross outlines of the revolution in aerostructures, however, are clear from the contemporary literature and from Miller and Sawers.

The most obvious and significant change between the wars was from wood and cloth to metal construction. Wood had always been an awkward material with which to work: no two pieces of wood have identical densities or stress properties, wood absorbs moisture, and, of course, wood is combustible and subject to rot. But initially, the only alternative to wood was steel, which, in the gauges acceptably light, offered insufficient structural rigidity. Although numerous steel-framed (and even steel-skinned) airplanes were built, they were an unhappy compromise at best.

At first aluminum was rejected as a material for aircraft because, in its common form, it was deficient in strength. Although a German, Alfred Wilm, who was seeking a way of producing an extruded aluminum cartridge case, in 1909 accidentally discovered the large increase in aluminum strength obtainable by quenching, "duralumin" (or "dural"), as it was called, was not widely used until after 1934. Part of the resistance to use of duralumin seems to have come from the innate conservatism of the aeronautical community, a conservatism common to all practicing communities, and part from the problem of aluminum corrosion. Not until 1927, when ALCOA introduced the cladding of alloys with pure aluminum ("Alclad"), was the corrosion problem licked. Ironically, ALCOA had developed Alclad in connection with the U.S. Navy's airship program.[25] Airplanes with tubular aluminum frames were common in the late 1920s, and Junkers and Rohrbach, followed by Fokker, Ford, and others, produced aircraft with corrugated aluminum skins on tubular frames. As aeronautical engineers began to apply knowledge of cantilevered structures, monoplanes also became more common during the same period.

But the really critical revolution in airframes was the development of monocoque, stressed-skin structures in which the aluminum skin itself was made to carry a significant portion of the structural load. Rather than a separate, complete frame, a system of longitudinal stiffeners and bulkhead rings, both formed from sheet dural, when covered with stressed skin proved sufficiently strong for fuselages and offered both a cleaner design and lighter weight. Similarly, stressed-skin single or double spar wings, also with lateral stringers and longitudinal frames, could be designed to exact streamlined aerodynamic shapes (sections) and be expected to hold their original, manufactured shape without warping or deteriorating. Herbert Wagner in Germany in 1925 first theoretically analyzed the diagonal tension field of a thin metal sheet rigidly sup-

Rohrbach Ro II all-metal seaplane with emergency sail kit deployed, 1924. The aircraft reportedly handled splendidly under sail. (Adolf Rhorbach, "Large All-Metal Seaplanes," *Jl. RAeS.* 28 [1924]: 655–75.)

Wing structure of Rohrbach Ro II all-aluminum seaplane; entire structural load is carried by stressed-skin center box-spar. (Adolf Rohrbach, "Large All-Metal Seaplanes," *Jl RAeS.* 28 [1924].)

ported along its edges. Although his work was not published until 1928 and not widely recognized until the early 1930s, it did provide the theoretical foundations for future progress in stressed-skin structures. In the United States in the late 1920s, John K. Northrop independently and empirically developed stressed-skin structural methods similar to those implied by Wagner's theory. Northrop's designs were heavier and less elegant than Wagner's but somewhat stronger and, initially, cheaper to fabricate. Modified Northrop structures were used in the DC-1, 2, and 3 series aircraft.[26] A tremendous amount of subsequent theoretical and experimental investigation went into the development of the new type of airplanes.[27] Monocoque, stressed-skin structures provided the airframe—light, strong, and clean—without which the turbojet would have been, at best, a laboratory curiosity.

Numerous other innovations in airframes, most requiring years of painstaking development, were also necessary to jet-worthy aircraft. Without retractable landing gear, leading-edge slots, and Fowler flaps, and without highly developed techniques of control-surface balancing and careful structural prevention of control-surface reversal, turbojet-powered aircraft could never have flown in any reasonable or useful sense. For example, aircraft wing loadings had increased only from 8 to 10 pounds of aircraft weight per square foot of wing area between 1918 and 1925. By 1933, the combination of higher engine power, v.p. props, stronger and better designed wings, and flaps and slots permitted ordinary commercial transport aircraft to use wing loadings of 15 pounds per square foot. Common wing loadings doubled again, to 30 pounds per square foot, by 1940.[28] In contrast, one of the fastest piston-engined fighters of the Second World War, the Hawker Tempest V, had a wing loading of 45 pounds per square foot, while the best of the first-generation turbojet fighters, the Me 262, had a loading of 63 pounds per square foot, albeit at the price of a very high landing speed.

Furthermore, without the fuel pumps, fuel lines and sealants, lubricants, electrical and control systems, and insulating materials developed for piston-engined aircraft, turbojets could not have run. The same comprehensive and complicated production technology—the systems of jigging; the explosive rivets; the punches, drills, and lathes; the quality control tests; the managerial techniques—necessary to high-speed piston-engined aircraft was equally necessary to turbojet-powered aircraft. Yet, prior to the turbojet revolution, the holistic airframe-piston engine-propeller system produced such astounding increases in total system performance that the aeronautical community, correctly, rejected all alternative systems and was ultimately very nearly blinded to the promise, and to the necessity, of the turbojet itself.

Regnant Normal Technology

ALTERNATIVES: DIESELS

During the interwar years, diesel, or compression-ignition piston engines, represented perhaps the most thoroughly explored and highly developed alternative to the gasoline aircraft piston engine. All significant diesel aircraft engine schemes would have used a conventional propeller, and hence were less radical than some other proposals. For conventional aircraft speeds, diesels offered an array of hypothetical advantages. Diesels were safer: diesel heavy oil was much less volatile than gasoline and thus posed less of a fire hazard. Diesels had fewer things to go wrong: they required no ignition system once they were running, and they used direct fuel injection rather than carburetors, which, on gasoline engines, required careful adjustment and were subject to icing. Some diesels (two-stoke designs especially) used intake and exhaust ports rather than valves, and thus could dispense with complicated valves and timing gear as well as escape all valve-cooling difficulties. Since diesels ordinarily employed very high compression ratios (17:1 versus 6:1 for ordinary contemporary gasoline aircraft engines) and high internal cylinder pressures (50 percent greater than gasoline engines) and used direct fuel injection, they suffered none of the preignition, or "knocking," associated with high-performance gasoline engines. For the same reasons, diesels were readily adaptable to high degrees of supercharge, which made them potentially well-suited to high-altitude operation. Finally, diesel engines offered significantly better specific fuel consumption than contemporary gasoline aircraft engines: 0.35 pounds of fuel per horsepower-hour for diesels compared to 0.43 to 0.45 pounds per horsepower-hour for gasoline engines, representing 38 percent brake thermal efficiency for diesels versus 28 to 30 percent for gasoline engines.[29] Diesel fuel was considerably cheaper to boot. Better fuel consumption could hypothetically be translated into greater aircraft range and larger payloads.

In 1940 Paul H. Wilkinson presented an adequate if overly enthusiastic summary of diesel aircraft engine development through the end of 1939.[30] Large-scale development of diesel aircraft engines was initiated before 1926 by Hugo Junkers, with the cooperation of Otto Mader, among others, at the Junkers Engine Company. The Junkers two-cycle opposed-piston diesels remained the most highly developed and successful of all diesel aircraft engines. Junkers diesels were widely used during the 1930s on Junkers aircraft belonging to Deutsche Lufthansa and to the still-surreptitious Luftwaffe. Junkers diesel engines were also installed in aircraft manufactured by Focke-Wulf, Dornier, and Blohm and Voss. During the Second World War, Junkers 207-series diesels powered especially modified Junkers Ju-86P and Ju-86R high-altitutde photoreconaissance aircraft. Junkers, however, remained the only widely adopted diesel aircraft engines.[31]

Nevertheless, most other aero-engine firms at least experimented with diesels during the interwar period, among them, in England, Rolls Royce, Bristol, and Napier; in France, Clerget, Coatalen, and Salmson; in the United States, Guiberson and Packard; and, in Germany, besides Junkers, the Bavarian Motor Works (B.M.W.) and Mercedes-Benz (which built the huge V-16 1,320 h.p. diesels that powered the *Graf Zeppelin* and the *Hindenberg*). Both two- and four-cycle engines were built, using both liquid and air cooling. A wide variety of configurations were tried: both upright and inverted V, vertical-opposed piston, H, single- and double-row radials, and "Barrel" engines. Some aircraft diesels had precombustion chambers. The French Salmson diesel, built to Szydlowski designs, was a two-row water-cooled radial in which paired front- and back-row cylinders shared the same combustion chamber.[32] In short, diesel aircraft engines enjoyed great ingenuity in their design and great effort in their development.

Yet Diesel aircraft engines failed. Diesels had one great intrinsic disadvantage: because of their inherently higher cylinder pressures, they had to be heavy. The best power-to-weight ratio of any developed diesel aircraft engine was 1.43 pounds per horsepower (for the 1,000 h.p. Junkers Jumo 207), although 1.72 (Clerget 16H) or 1.82 (Guiberson A1020) is perhaps more representative. In contrast, contemporary gasoline piston aircraft engines offered power-to-weight ratios of from 1.2 pounds per horsepower (Rolls Royce Merlin V-12 liquid-cooled, 1940) to 1.3 pounds per horsepower (Pratt and Whitnew R-2800, 2,000 h.p., 1940).[33] For a 2,000 h.p. engine, such a difference in specific weight might mean a total diesel weight disadvantage of over 800 pounds, or 10 percent of the all-up weight of a high-performance pursuit airplane.

The diesel engines' intrinsic disadvantage became a fatal flaw for one extrinsic reason: the immense success of gasoline aero-engine development. As Fedden put it, "The introduction of new fuels has always given the petrol engine the advantage."[34] H. R. Ricardo likewise observed: "During the late 1930s ... the rate of increase, actual or potential, of the octane number of petrol steepened, with the result that the performance of the petrol engine gained a lead such as the compression-ignition engine could not reasonably hope to catch up with, for it must be remembered that, unlike the petrol engine, the compression-ignition engine, with its immunity from detonation, had little to gain from any improvement in its fuel."[35] The ability of the gasoline engine to handle the greater power developed with better fuels depended, in turn, upon the whole array of major innovations in gasoline engine design noted

Regnant Normal Technology

Boeing Model 200, "Monomail," 1930. (Courtesy of the Smithsonian Institution.)

Boeing 247 D, 1934. (Courtesy of the Smithsonian Institution.)

Heinkel He 111 C, 1935. (Courtesy of the Smithsonian Institution.)

Boeing Model 307 Stratoliner, 1938. The 307 was the world's first pressurized commercial aircraft. It was a parallel development of the Model 299 (B-17) and shared that aircraft's wing, engines, landing gear, and tail surfaces. (Boeing Co., *Pedigree of Champions* [Seattle, Wash., 1963].)

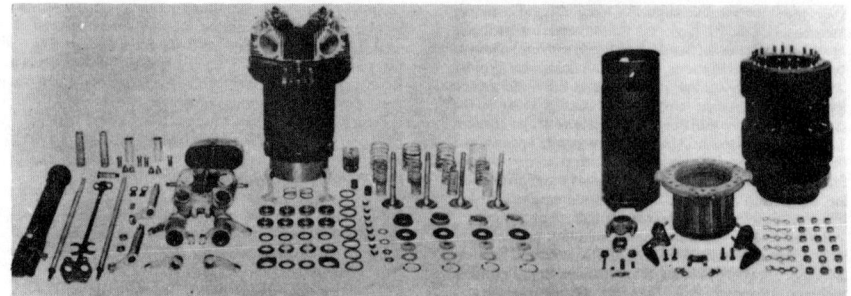

Cylinder components for poppet-valve (left) and sleeve-valve piston aero-engines of comparable power. (M. Roy Fedden, "The First Twenty-five Years of the Bristol Engine Department," *Jl. RAeS.* 65 [1961].)

above. Diesel aero-engines did not fail outright, but failed in competition with the dominant gasoline aero-engine technology.

RADICAL ALTERNATIVES: REACTION PROPULSION AND TURBOPROPS

As discussed above, conventionally powered aircraft encountered severe altitude performance limitations during the First World War. In addition to deficient power output from unsupercharged piston engines, propellers also presented difficulties. Fixed-pitch propellers appropriate for takeoff and low-altitude performance were wrong for high altitudes. Furthermore, larger-diameter propellers, fitted in order to absorb greater engine output, led to greater propeller-tip velocities. Since the speed of sound decreases with altitude, large propellers with high tip speeds soon encountered "compressibility burble," or the formation of sonic shock waves on the propeller tips. These problems elicited a dual response. First, as detailed above, there were intensive and ultimately successful attempts to further articulate and develop the conventional piston engine and propeller. Superchargers and turbosuperchargers, together with variable-pitch propellers, ultimately resolved the altitude performance deficiencies of piston-engined, propeller-driven aircraft. But while the outcomes of those conventionally oriented projects were still problematical, attempts were also made to develop radical alternative systems. Serious investigation of radical systems embraced two primary categories of devices: direct reaction propulsion and internal combustion turboprops.

Reaction propulsion proper, like so much else, is of antiquarian origin. "Firebolts," arrows boosted by powderlike propellants, appear "in the old literary sources of almost all peoples,"[36] although such references may be merely to burning fire arrows, not to rockets. The great pre-Roman armies—Chinese, Persian, Arabian, and Greek—used rockets for signaling. Weapons rockets were apparently first highly developed by the Chinese in the thirteenth century. Probably the first reaction propulsion device for carrying a man was designed by Isaac Newton about 1680: "On a four-wheeled vehicle he mounted a spherical boiler over a fire, the top of the boiler having a rearwardly directed nozzle. By the reaction of the jet of steam on the atmosphere [sic] the carriage was expected to be propelled in a forward direction. Speed was to be regulated by the driver's control of a steam cock in the issuing pipe."[37] Needless to say, Newton's design was never constructed. Rockets, and with them reaction propulsion, were consigned to an unexceptional career as weapons of war during the eighteenth and early nineteenth centuries. By the middle of the nineteenth century, however, artillery had supplanted rockets for almost all uses. Artillery was much more accurate, and also more likely to injure the enemy than its own crews.

Yet, despite the rocket's eclipse, the empirical work of De Laval and the theoretical analysis of Osborne Reynolds did provide engineers by the turn of the twentieth century with the essential design data necessary for an efficient reaction propulsion nozzle. The convergent-divergent nozzle which De Laval had constructed and which Reynolds had hydrodynamically described provided the means for maximum conversion of heat-pressure energy into velocity or kinetic energy, that is, into thrust. The pioneer work of De Laval and Reynolds, together with the vast body of design data on steam turbine nozzles and the more generalized knowledge of gas flow and temperature change which had been formulated during the subsequent steam turbine revolution, provided the essential "state of the art" from which an accurate comprehension of the requirements of rocket nozzle design could be drawn.

Still, no significant rocket project reached fruition before the turbojet revolution. There was, however, considerable discussion of the possibilities and problems of interplanetary travel by rocket and some serious theoretical and even experimental work. Moreover, several rocket societies were founded, a number of publicity stunts were staged, and one serious program was initiated. Rockets, and the ideology of speed associated with them, thus constitute important but not directly causal elements in the milieu within which the turbojet revolution occurred. Early rocket experiments and the formation of the first rocket societies is egocentrically but exhaustively covered in Willy Ley's *Rockets, Missiles, and Men in Space*, and patchily but more objectively portrayed in the proceedings of the International Academy of Astronautics historical symposia published by NASA. The initiation and execution of the German Army program which culminated in the V-2 rocket is adequately covered by General Walter Dornberger's *V-2*.[38] Thus, all that is required here is a brief outline of rocketry activities.

Modern concepts in rocketry were the creation of a number of enthusiasts, three of whom—Konstantin Ziolkovsky (Russia), Robert Goddard (United States), and Hermann Oberth (Austria and Germany)—made notable early theoretical contributions. Ziolkovsky, however, did no experimental work and seems to have had little influence on the later development of the art. Goddard, after being embarrassed in 1919 by hideously sensationalized press reports of his hypothetical proposal to fire a signal rocket to the moon, worked in almost total secrecy and thus similarly had little influence on later developments. In contrast, the 1923 publication of Oberth's *Rocket into Interplanetary Space* led to the formation, in 1927, of the Verein für Raumshiffahrt (VfR), the Society for Space Travel, commonly called the German Rocket Society, among whose members was Wernher von Braun. A Soviet society for "reaction motion" was formed in 1929, in part through the efforts of Professor Nikolai A. Rynin. The American Rocket Society was founded in 1930, and the British Interplanetary Society in 1933.

Only the activities of the VfR produced notable results. Some of those results were more trivial, although widely publicized, than substantive. For example, Fritz von Opel, the German car manufacturer, funded construction of several black-powder rocket-propelled automobiles in 1928 and of a rocket-propelled airplane in 1929. Von Opel's projects were mere publicity stunts and of no technical significance, although they did focus some public attention on rocket propulsion. In 1929, Oberth himself participated as technical advisor in making director Fritz Lang's film *By Rocket to the Moon*. While only entertainment, the film did contain some remarkable special effects suggested by Oberth. The result again was to popularize the idea of high-speed, extraterrestrial rocket flight.[39]

The activities of the VfR did have one genuinely serious consequence. The Treaty of Versailles forbade the German Army to manufacture or possess long-range artillery. As a result, in 1929, the army began investigating the possibility of artillery rockets with a three- to five-mile range. Dornberger joined the program in 1930 and recruited Wernher von Braun in 1932. The ultimate outcome was the German V-2 ballistic missile of World War II. The only other significant rocket propulsion experiments before 1940 were also instigated by von Braun and were conducted by the Heinkel aircraft firm. That work, which resulted in flying both a conventional He 112 fighter and a specially built rocket plane, the He 176, on rocket power alone, was done in parallel with Heinkel's work on the turbojet, and will be discussed later.

In general, rocket propulsion activity prior to the turbojet revolution did focus attention on the possibilities in high-speed, high-altitude flight. But there were no significant technical achievements in rocketry until after the turbojet revolution had begun, and at no point did rockets compete with conventional aircraft piston engines.

Not all nonturbojet reaction propulsion systems, however, are rockets (which are distinguished by carrying their oxidizing agents internally). A number of reation propulsion systems using atmospheric air were proposed between 1900 and 1940. Probably the first sophisticated jet engine design was the Lorin scheme of 1908, in which

> a conventional engine exhausts directly through a discharge funnel to produce a propulsive reaction. Air is compressed in the cylinder, there is combustion of fuel and the effluent expands in the discharge funnel. Apart from the energy used in driving the usual auxiliaries, the engine functions solely to produce the reactive-jet. No power is taken off the engine crankshaft. Lorin visualized multi-cylinder units of this type installed in the wings of aircraft.[40]

In 1917, another Frenchman, O. Morize, proposed a system comprising a conventional reciprocating engine driving a reciprocating air compressor which delivered compressed air to a combustion chamber, where fuel was to be added and continuously ignited. The combustion chamber was to exhaust into a Morize "ejector," which consisted of a convergent-divergent nozzle structure with air inlets for introducing additional air at the forward end of the nozzle tube.

In the same year, 1917, an Englishman, H. S. Harris, devised a not entirely dissimilar propulsion plant. The Harris scheme employed a reciprocating engine to drive a low-pressure single-stage centrifugal blower, which was to deliver into an extended discharge nozzle tube designed rather as an inverted Morize ejector, but with the same object, to induce a flow of additional air.[41] The Melot "thrust augmentor," with which the French Air Corps experimented during the First World War and which was patented by its inventor in 1920, was also similar to the Morize ejector. In the Melot proposal, a combustion chamber charged with air by ram effect exhausted into a convergent-divergent nozzle through a series of concentric nozzle structures, or venturi, thus inducing a flow of excess air into the propulsion nozzle. Excess air was necessary in all these proposals both to reduce the temperature of combustion products in deference to materials limitations and to increase mass flow through the propulsion system to enhance propulsive efficiency.

By 1920, the principle of reaction propulsion, both in its pure form of the rocket and in its potentially more useful form of the reaction jet engine, was the object of acute if circumscribed interest. Efficient reaction propulsion for aircraft was dependent, however, upon the creation of some mechanism for the production of large quantities of high-pressure, high-temperature gas. The Lorin scheme and especially the Morize proposal were attempts to meet this fundamental requirement. The internal combustion piston engine was then the preeminent form of aircraft power, and the natural first step toward replacing propeller drive with reaction propulsion seemed to be the adaptation of the piston engine to the new system.

The definitive study of reaction propulsion devices in the post–World War I period was published by the National Advisory Committee for Aeronautics (NACA) in 1923.[42] The study was done by Edgar Buckingham of the U.S. Bureau of Standards at the request of the Army Air Service. Buckingham considered reaction propulsion, but did not consider any form of gas turbine. He dealt only with systems using atmospheric air as an oxidant, some form of compressor driven by a reciprocating piston engine, and a simple De Laval nozzle. His analysis indicated that at a compression ratio of "below 7 to 1 fuel increases so fast that it is quite useless to consider such compression ratios as can be obtained with a turbo-booster, and a reciprocating compressor must be used." He thus was left with both a reciprocating engine and a reciprocating compressor, each highly complex. The limited size of the reciprocating compressor limited air flow and therefore propulsive efficiency. Buckingham concluded:

> The relative fuel consumption and weight of machinery for the jet decrease as the flying speed increases; but at 250 miles per hour the jet would still take about four times as much fuel per thrust horsepower as the air screw, and the power plant would be heavier and much more complicated.
>
> Propulsion by the reaction of a simple jet can not compete, in any respect, with air screw propulsion at such flying speeds as are now in prospect.

At 200 mph, the highest speed then attained, the jet would use five times more fuel than the conventional propeller. Even at 350 mph, the highest speed for which computations were done, the jet would have required three times as much fuel.

In 1927, Eastman N. Jacobs and James M. Shoemaker of the National Advisory Committee for Aeronautics did a follow-up series of experiments on "thrust augmentation," basically the system proposed by Melot.[43] They concluded that "although it is possible to increase the thrust of a jet, the increase is not large enough to affect greatly the status of the problem of the application of jet propulsion to airplanes." In 1930, G. B. Schubauer of the Bureau of Standards, at NACA's request, undertook a somewhat more thorough study.[44] He acknowledged that at some future time when extremely high altitudes and speeds could be attained, "jet motors or rockets will find an important applicaion." He also concluded, however, that at then-current performance levels jet propulsion schemes were hopelessly inefficient.

Still, interest in such systems, as well as in pulse-jets and ram-jets, did persist, although very little serious experimental work was done. For the most part, investigations consisted of relatively competent, scholarly aerodynamic and thermodynamic analyses of various reaction propulsion systems using external air. For example, at the Volta High Speed Conference in 1935, both Professor Maurice Roy of the Ecole Nationale Superieure de l'Aeronautique and Professor Renin (Rynin) of Leningrad presented papers on atmospheric reaction propulsion. In both cases, they limited themselves to theoretical analyses of thrust augmentation and of ram-jets at supersonic speeds. Each could cite a small but orthodox literature. Neither considered a turbojet.

Analyses such as Roy's and Renin's represent the most respectable of the speculations on high-speed, high-altitude flight. At the other extreme are proposals like Robert Goddard's 1932 "Turbine Rocket Plane for the Upper Atmosphere."[45] Goddard did have his fundamentals right. He realized that while a rocket could be efficient at extremely high speeds and altitudes, some other arrangement, one providing high mass-flow at relatively low velocity, was needed for efficient takeoff and climb. Ordinary propellers provide such mass-flow. Therefore, Goddard proposed equipping a rocket plane with large propellers fitted within large hoops which would carry turbine blades. A movable frame would place the turbine blades in the rocket exhaust, thus turning the propellers, for takeoff and climb. Goddard's proposal was and is absurb.

What is important is that Goddard, as well as Roy, Renin, and others, did grasp at least part of the presumptive anomaly upon which the turbojet would be based. All realized that flight at exceptionally high speeds and altitudes was not possible with a conventional piston engine and propeller. Yet none combined that realization with new performance assumptions about gas turbine components to imagine a turbojet.

Turboprops, using an internal combustion gas turbine to drive a conventional propeller, were also considered and rejected during the interwar years. The definitive study of the turboprop was done in 1920 in England at the behest of the Aeronautical Research Committee (ARC) by W. J. Stern of the Air Ministry Laboratory, South Kensington.[46] Stern considered various forms of turbines, but concluded that only a constant-pressure machine employing a rotary centrifugal air compressor with intercooling and a single-stage or multistage impulse turbine could be even remotely capable of use in aircraft. His computations suggested that a large Rateau compressor or the similar Zoelly compressor of at least seven stages turning at 4,000 rpm to produce compression ratios of 10:1 would be necessary. He assumed maximum compressor and turbine efficiencies of 70 percent. The best turboprop unit Stern thought possible with then-current technology would give 1,000 h.p. for a weight of 6,000 pounds and dimensions of 12 feet in length by 3 feet in diameter, or about three times as heavy and twice as big as comparable piston engines. The turbine's fuel consumption would also be about three times higher than that of the best piston engines. Stern concluded that "in its present state of development, the turbine has thus no chance of adoption in aircraft."

Stern did point out, correctly, that the key to a usable turbine was a more efficient and lighter-weight compressor, that the combustion system would require a great deal of development, and that in fabricating turbine blades of appropriate quality the "experience of firms like the General Electric Company, Parsons, and Brown Boveri, will be most useful." Stern, like all who studied the internal combustion gas turbine before the turbojet revolution, thought only of the turboprop, and made much of the "negative work" done and lost by ingesting excess air to keep temperatures within tolerable limits. Later critics took Stern to task for considering only then current industrial practice, especially in regard to weight of machinery. Nevertheless, for the time in which it was written, Stern's report seems intelligent and open-minded, though not prophetic.

By the middle of the 1920s, then, both jet propulsion and the internal combustion gas turbine as means for driving a propeller had been well explored and had been decisively rejected. The attitudes of the conventional aeronautical community, official and industrial, were for the most part fixed, negatively, for at least the next ten years. Those attitudes were uniformly based on assumptions of limited aircraft speeds (at most, 250 mph) and altitudes (15,000 feet), assumptions that were completely plausible and completely reasonable given the scientific and empirical data base circa 1925. Similarly, those investigations which considered the use of turbo-air compressors or gas turbines assumed component efficiencies based upon established industrial experience—again, perfectly reasonably. Given that milieu, which in 1925 no rational person could have questioned with logical coherence, no radical propulsion system—neither any form of reaction propulsion nor the turboprop—could have been expected to compete effectively with the piston engine and propeller.

THE SECOND INTERNAL COMBUSTION GAS TURBINE REVOLUTION

Ironically, no sooner had the attitudes of the conventional aero-engine community hardened against reaction propulsion and the turboprop than aerodynamic insight literally breathed new life into the internal combustion gas turbine. A number of people had continued working on gas turbines after the First World War, but their efforts were largely abortive. Yet by 1940 a very few projects had brought the gas turbine to the verge of general acceptance, and had in fact secured its wide adoption for special purposes. The main burden of this incipient revolution fell to Brown Boveri, although other experimenters also made significant contributions.

Brown Boveri was ideally situated to create a revolution in internal combustion turbines. The company had an excellent background in steam turbine manufacture; it had constructed the multistage centrifugal compressor for the Armengaud-Lemale turbine and had subsequently marketed centrifugal compressors; it had constructed a number of Holzwarth intermittent combustion (explosion) turbines, and had also manufactured centrifugal turbosuperchargers for diesel engines and geared centrifugal superchargers for aero-engines during the First World War. And, as noted above, it was in the forefront of scientific investigation of axial compressors.

Still, Brown Boveri came to market an internal combustion gas turbine in a rather roundabout fashion. In 1927, Aurel Stodola, a professor at the Zurich Polytechnic Institute and one of the earliest gas turbine enthusiasts, tested a Holzwarth explosion turbine at the Thyssen steelworks in Mulheim. Among other results, he noticed very high (and dysfunctional) rates of heat transfer into the water jacket surrounding the turbine. Noack of Brown Boveri "drew unexpected conclusions from Stodola's reports: 'If heat transmission is so disturbingly high at high velocity,' he said, 'let this be turned to advantage in a steam boiler of extremely reduced heating surface.'" Brown Boveri therfore developed the "Velox" boiler, the first versions of which, appearing in 1932, operated on the "explosion" cycle also. The explosion cycle entailed "delicate problems of scavenging, of forming an explosive mixture, and of ignition"; so, on the basis of its experience with axial compressors and turbines, Brown Boveri changed to a constant-pressure cycle, with an axial compressor, a high-intensity combustion system, and an axial turbine operating in the furnace exhaust to run the compressor—in effect, a turbocharged boiler. The first constant-pressure Velox units, introduced in 1933, required electrical supplement to the power developed by the turbine. But by 1935-36, the turbines and compressors were so highly developed that the turbine not only delivered enough power to run the compressor but also could run a small auxiliary generator.[47]

In 1936, the Houndry catalytic cracking process in petroleum refining was just coming into use. The Houndry process required large quantities of compressed air for burning off carbon residues left on the catalyst beds. Henry Thomas, chief engineer of Sun Oil Company in Philadelphia, which was the first American adopter of the Houndry process, realized that a combination of a compressor and an expansion turbine working in the cracker outlet would provide not only the necessary air but electrical power to boot. Because of its Velox boiler experience, Thomas turned to Brown Boveri. After overcoming considerable opposition within Sun Oil, Thomas ordered the first catalytic cracking turbocompressor unit in 1936. The units were an unqualified success, and several score of them, some manufactured by Brown Boveri, some by Allis-Chalmers under license, were eventually installed. Generally the units were quite large: twenty-stage compressors and five-stage turbines, for an airflow of 40,000 cubic feet per minute. But the large compressors gave efficiencies of 86 percent.

The cracker turbocompressors had another use, even more important. "For shop tests, at Brown Boveri, the charging sets for the Sun Oil Co. were fitted with an oil-burning combustion chamber, and not until the entire equipment was actually running was it realized that a regular combustion turbine had been born, and was in fact standing there showing what it could do."[48] Under the leadership of Adolf Meyer, Brown Boveri immediately constructed a true internal combustion gas turbine. Meyer secured an order for a stand-by generator set to be installed in a bomb-proof bunker for the town of Neûchatal, Switzerland. That unit was displayed at the Swiss National Exhibition in Zurich in 1939. A similar gas turbine unit was built for a locomotive of the Swiss National Railways in 1942. The Brown Boveri gas turbines were so good that in the same year the British Air Ministry ordered one just "to play with." The gas turbine's time had come.[49] The Brown Boveri Velox boiler, the cracking turbocompressors, and finally the internal combustion gas turbine all exploited compressor and turbine-blade shapes designed in accordance with aerofoil theory. Thus, Brown Boveri designers had grasped one of the critical, radical assumptions essential to a practicable turbojet.

There were other gas turbines besides those built by Brown Boveri, but none were so highly or effectively developed. Probably the best of these other turbines resulted from a joint effort by Jacob Ackeret and Curt Keller, director of research of the Escher-Wyss concern. Escher-Wyss completed construction of the first AK turbine in the summer of 1939. It was a very large, complicated, closed-cycle gas turbine for the generation of electricity which used air as its working medium. It comprised four axial compressors with three intercoolers, two heat exchangers (to provide heat in place of internal combustion), two turbines, and a water-cooled regenerator to reduce the gas to lower pressure at the end of the cycle. Although about a dozen AK units manufactured by

Escher-Wyss were in central-station use by 1966, the AK turbines proved to have real advantages over comparable steam turbine generating plants only where a supply of adequately pure water was unobtainable or prohibitively expensive.[50]

Another Swiss firm, the diesel engine manufacturer Sulzer Brothers, also undertook development of gas turbine systems. Sulzer, beginning in 1936, investigated three distinct types of engines employing a gas turbine. The first was a highly supercharged two-cycle diesel engine with a multistage axial compressor driven by a multistage exhaust turbine. Second, Sulzer constructed and tested a very large three-cylinder free-piston gas producer for use with a multistage axial power turbine. Finally, Sulzer turned to a continuous-combustion semi-closed-cycle internal combustion gas turbine with multistage axial compressors and turbines. Few of the Sulzer turbines were built, and they do not seem to have been a commercial success.[51]

A. J. R. Lysholm, chief engineer of the Ljungstrom Steam Turbine Company, Stockholm, who had done a great deal of work on the contrarotating, radial, outward-flow Ljungstrom steam turbine, began investigations of internal combustion gas turbines in 1928. Lysholm designed a large number of different types of gas turbines between 1928 and 1932, but it was not until the latter year that actual construction was undertaken by the Bofors concern.[52] The Bofors unit, originally intended as an airplane turboprop, was under test between 1933 and 1935. Severe surging problems (irregular and sometimes reversed flow) were experienced in the multistage centrifugal compressor then used, so in 1935 Lysholm began experiments on positive displacement compressors, initially of the Roots type. The Roots compressor was incapable of handling the volume necessary for gas turbine work, and Lysholm designed a rather extraordinary helical lobe compressor. Construction of the Lysholm compressor was in hand before the outbreak of the war, and his turbine was under test during the war years. The Elliot Corporation in the United States built a similar experimental gas turbine for the U.S. Navy. Unfortunately, the helical lobe compressor proved subject to thermal distortion and was abandoned after the war.[53]

Christian Lorenzen of Berlin began experiments on axial turbines with internally air-cooled blades shortly before 1930. His hollow-blade design apparently followed a Brown Boveri patent used under license.[54] Lorenzen was a pioneer in aeronautics and in automobiles. He had participated in early variable-pitch airscrew and turbocharger development in Germany at the end of the First World War. In 1926, a turbocharger of his design was tested at the German Aeronautics Test Laboratory (Deutsche Versuchsanstalt für Luftfortforschung, or DVL), Berlin-Adlershof. In 1928, he installed a turbocharger on a Mercedes-Benz

Regnant Normal Technology

automobile engine in place of the Roots blower normally used. In his experiments, Lorenzen encountered blade warping as a result of overheating. He therefore began work on air-cooled blades. By 1930, he had not only done considerable work on the air-cooled turbine, but had also designed a complete internal combustion gas turbine around it. The cooling air was introduced into the hub of the turbine and whirled outward through the blades, cooling the blades and compressing the air at the same time. At its exit from the blades, the air was fed to a combustion chamber, which exhausted through the turbine. The Lorenzen gas turbine proved impracticable, probably because of internal flow losses, but his work on air-cooled blades proved invaluable to the Germans during the Second World War.[55]

In 1939, George Jendrassik of Budapest tested an internal combustion gas turbine of his own design built with the assistance of the Royal Hungarian Ministry for Industrial Affairs. The Jendrassik compressor used ten axial stages to get a 2:1 pressure ratio and employed a seven-stage axial turbine. Although the Jendrassik turbine seems to have given acceptable efficiency when tested, its design apparently was not up to the standards set in Switzerland. In any event, the war caused an end to its development.[56] Another Hungarian, Coloman Cukor, had designed a sulfur dioxide closed-cycle turbine for an automobile in 1931, but there is no evidence of its ever having been constructed. G. Bertin reportedly designed and built a small turboprop in Paris in the early thirties; Giuseppe Belluzzo, of the Royal School for Engineers in Rome, allegedly constructed and tested an internal combustion turbine in 1933. There were others, some just designs, some actually tested, none successful.

In 1940, the internal combustion gas turbine as incarnate in the Brown Boveri machines and, to a lesser degree, in the turbines of Ackeret-Keller and of Sulzer, was just on the verge of becoming a really practicable proposition. The turbojet revolution would overtake, overshadow, and by its theoretical and materials progress, remake the incipient revolution in gas turbines for general use. But in the critical period of the turbojet's birth, the work on gas turbines provided essential encouragement and information. What that work did not, and could not, do for the turbojet was to successfully impugn conventional aero-engine practice.

THE RESPONSE OF NORMAL TECHNOLOGY

Practitioners of normal technology could not have been oblivious to the new opportunities provided by the second internal combustion gas turbine revolution or to the new perspectives offered by supersonic aerodynamics. They did respond to the challenge of high-speed, high-altitude flight. That response, nevertheless, was traditional, unenthusiastic, and ill-fated. The response of the conventional community is perhaps well represented by Harry Ricardo's paper at the last session of the Volta High Speed Conference. Ricardo, the man who had done so much to make high-performance aero-engines possible, turned his attention to the problem of high-altitude flight.[57]

Ricardo proposed a highly refined, very complex development of the traditional piston engine and propeller to meet the demands of flying at 40,000 feet. Naturally, he would use a v.p. prop to meet the varying conditions of takeoff, climb, and high-altitude cruise. He would adopt a bi-fuel system for the engine. An alcohol-base fuel could be mixed to obtain a higher octane rating, greater resistance to detonation, and cooler maximum combustion temperature, and would be used for takeoff and climb. Even though the alcohol-base fuel would have a lower specific heat content (a lower caloric value per pound) than gasoline, its short-time use for climb and descent would be justified by its other advantages. Petrol, with its higher specific heat and higher combustion temperature, would be used for high-altitude cruise, where colder ambient air temperature would simplify engine-cooling problems. Ricardo recommended liquid-cooled engines, since radiators would be more efficient at altitude than air-cooled cylinders.

To derive maximum benefit from the bi-fuel arrangement, Ricardo proposed using direct fuel injection, with its precise fuel metering and complete atomization. To obtain maximum thermal efficiency, he proposed using the stratified charge principle, which would encourage complete fuel combustion and would permit precise control of cylinder pressure and temperature. Stratified charging could most easily be obtained using sleeve valves, which would permit essentially any desired porting configuration and would also permit higher supercharging, since the cylinder would have no "hot spots" around conventional poppet exhaust valve-seats. To obtain optimal supercharging, Ricardo proposed a "stratified supercharging" system comprising a turbosupercharger, a large intercooler, and a smaller-volume, higher-pressure Roots-type gear-driven supercharger in series with the turbocharger. At altitude, most of the ingested air would be introduced into the cylinders directly from the turbocharger. To obtain higher supercharge, however, and to insure a properly stratified mixture, a separate set of induction ports near the top of the pistons' travel would admit a smaller volume of high-pressure air from the Roots blower. By such an arrangement, Ricardo expected to obtain high supercharge together with low cylinder exhaust backpressure from the turbocharger turbine and high overall efficiency for the supercharging system. The same supercharging system would supply air for passenger cabin pressurization.

Actual in-flight operation of the system Ricardo proposed would be complicated. Pilots or flight engineers would have to carefully manipulate the two interlocking supercharger circuits as well as the bi-fuel systems to insure proper engine operation. The size, cost, weight, and complexity of the Ricardo system are self-evident. But what is critical here is that in 1935 the piston engine–propeller system was in no technical sense moribund. It still could generate an ambitious research and development program which could plausibly promise high altitude (40,000 feet) and high speed (350–400 mph) flight. It could promise pressurized comfort and high thermal efficiency, albeit at considerable cost in mechanical and operational complexity.

Yet about Ricardo's proposal there is the echo of late Ptolemaic astronomy, the weary addition of epicycle to epicycle to produce an increasingly complex and clumsy, if still functional, system. Perhaps Ricardo's concluding paragraphs adequately capture the mood of the conventional aero-engine community on the eve of the turbojet revolution:

> I realize that there must be a great many ways of treating this problem of flight at very high altitude; rightly or wrongly I have considered it from the aspect of long distance high-speed passenger travel.
>
> Whether flight at such altitudes will ever appeal to the travelling public is another question altogether. I fear that boredom, and the extremely restricted quarters, will be a serious obstacle.[58]

6

National Patterns in the Pursuit and Utilization of Scientific Knowledge

The myth of an undifferentiated, homogeneous, international invisible college of practitioners does not adequately portray either the pursuit of aerodynamic science or the utilization of that science in aeronautics during the period 1920 to 1940. Rather, the pursuit of aerodynamic science and its utilization exhibit marked national differences that would seem to reflect diverse national scientific traditions and fundamentally different political, economic, social, and geographical exigencies. In stark terms, Germany and England led in the exploration of high-speed aerodynamic phenomena; the United States lagged. Germany, England, and Switzerland theoretically and empirically investigated axial turbo-compressor phenomena; the United States did not. The United States shared with Britain, Germany, and other countries rigorous investigation of subsonic flight problems. The United States built the world's best commercial airliners and developed the world's largest and best commercial airline system. Great Britain and Germany used commercial aviation as "chosen instruments" of national politics and built mediocre airliners. Germany prepared for war after 1933, as did Britain after 1936. The United States dawdled. Britain and Italy competed for the Schneider Trophy after 1926; the United States did not. Britain, Italy, and finally Germany held the world absolute speed record for aircraft during the 1930s; the United States did not. The United States, the Soviet Union, and other countries indulged in other forms of spectacular and well-publicized aeronautical competition. England and Germany produced indigenous turbojet engines before 1940; the United States did not.

Prior technology (both structural and functional antecedents) and fundamental science (aerodynamics, operating through the mechanism of presumptive anomaly) were the intellectual substance of the turbojet revolution. Yet the impact of these essential factors on the turbojet revolution was mediated by the specific cultural milieux in which the protagonists and would-be protagonists of that revolution worked. The sharp divergencies in those cultural milieux are expressed in distinctively different national patterns in the pursuit of aerodynamic science, in the development of commercial aviation, and in participation in international aeronautical competition.

AERODYNAMICS, 1920-1940

National differences in the focus of basic aerodynamic research are evinced by three categories of convergent evidence: qualitative assessment of actual contributions to scientific knowledge, quantitative indices of those contributions, and an analysis of experimental facilities deployed, in this case, wind tunnels. An examination of the central theoretical innovations in high-speed aerodynamics described in Chapter 5 will reveal that American scientists, with the exception of the special case of Theodore von Karman, made no such contributions. Before 1940, none of the major theoretical developments in transonic or supersonic aerodynamics were the product of American science. Instead, such advances were the work of Germans or of scientists with German training (Ackeret, von Karman). Some important work on supersonics was done in Britain, but it did not approach the quality of the German investigations. Ronald Smelt of the Royal Aircraft Establishment (RAE) summarized the prewar distribution of supersonic effort in 1946:

> Even before 1939 the problems of high-speed flow were receiving a great deal of attention in the German aeronautical world. Their leading aerodynamicists, Prandtl, Busemann, and Schlichting among others, had contributed largely to the theory of compressible fluid flow, and basic experimental research in the high-speed field had begun both at Aachen and at Göttingen. The activity in Germany in this particular field before the war was, in fact, much greater than in Britain, where Mr. Lock and his co-workers at the National Physical Laboratory carried practically the entire responsibility for such work.[1]

Some supersonic research was done in Italy, France, and the United States, but in each case results were meager.

In the United States, supersonic research was tied exclusively to development-oriented investigations of propeller-blade-tip compressibility burble. According to Eastman N. Jacobs, "The first experiments applying modern wind-tunnel methods to airplane parts at high speeds" were done at McCook Field in 1920 on "airfoil sections suitable for use as propeller blade sections." Maximum velocities in those tests were under 450 mph.[2]

In 1927, the National Advisory Committee for Aeronautics (NACA) began development of a small-diameter (11-inch) induction jet high-speed wind tunnel for investigating subsonic compressibility phenomena. The design of that tunnel, and of a larger 24-inch-diameter tunnel completed in 1934, derived, ironically, from Jacobs' and Shoemaker's 1927 experiments on thrust augmentors for jet propulsion. Thus, the investigation of an alternative technological system was turned to use as a development tool for the current normal technology, an inspiration attributed to George W. Lewis, director of aeronautical research at NACA since 1924 (and staff director since 1919). Jacobs himself described the uses to which the new experimental technology was put:

> The first and most important problem investigated in the 11-inch high-speed tunnel was concerned with the study of the effects of compressibility for airfoil sections simulating propeller-blade sections, particularly with regard to the first appearance of the compressibility effects and their subsequent development with increasing speeds. With this object in view several commonly used propeller-blade sections and later other airfoil sections were tested to determine experimentally the nature of the air force variation with speed. Later other objects including fundmental shapes and some streamline wire section were investigated.[3]

Initial experiments in the 24-inch tunnel continued investigation of propeller-blade-tip compressibility burble.[4] The tenor of the American high-speed research effort is perhaps best summarized by Jacobs' conclusion to his paper on American experimental methods for the 1935 Volta High Speed Conference: "It is desired to avoid drawing further conclusions until more experimental data are obtained. In the meantime engineers in dealing with problems relating to the onset of compressibility phenomena may utilize the qualitative concepts and the approximate formulae herein presented, together with available high-speed force-test data."[5] The purpose of American high-speed research was empirical engineering development data for normal technology, not theoretical progress in aerodynamic science.

Similarly, the United States lagged in investigations of cascades of aerofoils and axial compressors and turbines. The work of Griffith in England, Ackeret at Zurich, and Betz and Encke at Göttingen has already been described. Americans did nothing comparable. Eastman N. Jacobs and Eugene W. Wasielewski did begin investigations in 1938 at Langley Field to determine "the performance of an axial-flow compressor based on the current information gained from extensive research on

Langley 30 × 60 foot full-scale tunnel, 1929. (George W. Gray, *Frontiers of Flight: The Story of NACA Research* [New York: Alfred A. Knopf, 1948]. Reprinted by permission of Carl Mydans, photographer.)

airfoils." That investigation was begun in response to "greatly increased" interest in such compressors "since 1935," the year of the Volta Conference. The first NACA eight-stage experimental compressor was not complete and under test until 1941-42.[6] NACA did not get around to translating Betz's and Encke's fundamental papers until 1942-43.[7] Thus, in the two areas of aerodynamic science critical to the turbojet revolution—the investigation of high speeds and of turbomachinery components—the United States was deficient before 1940.

In addition to this qualitative evidence on national patterns in the pursuit of aeronautical science, some limited but convergent quantitative indices may be derived. The Volta High Speed Conference, held in Rome from 30 September to 6 October, 1935, comprises an apparently complete and representative survey of high-speed aerodynamics on the eve of the turbojet revolution.[8] Sponsored by the Italian Academy of Sciences, the Volta Conference drew the world's preeminent aerodynamicists (Ludwig Prandtl, Theodore von Karman, Jacob Ackeret, Adolf Busemann, G. I. Taylor, and Eastman Jacobs), as well as outstanding aeronautical practitioners (H. E. Wimperis, D. R. Pye, and Harry Ricardo).[9] The papers were divided into three major topics— achievement, aerodynamics, and thermodynamics, the last group deal-

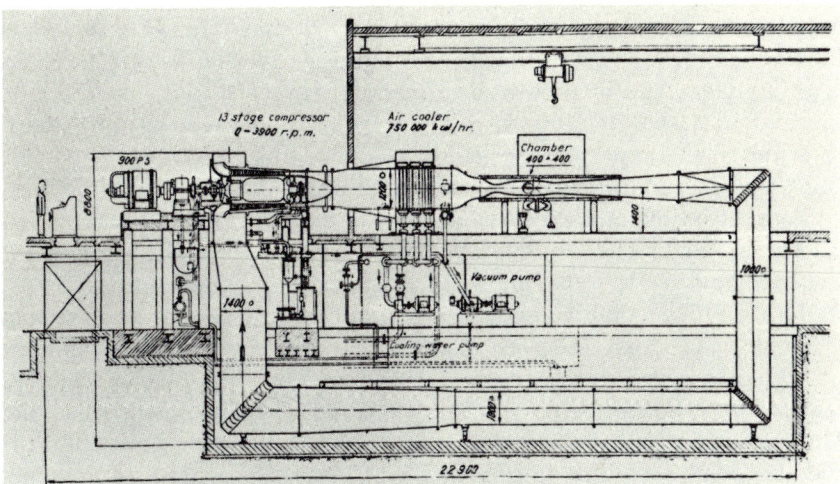

Ackeret's supersonic wind tunnel at Zurich, 1935. (Jacob Ackeret, "High-Speed Wind Tunnels," NACA TM No. 808.)

ing with engine design. The nine papers on aerodynamics, which constitute almost half (nine of nineteen) of the conference papers, seem to represent the informed judgment of an international peer group as to what was at the time scientifically significant.

One of the nine papers specifically concerns airscrews; of its 21 citations, only 3 are to theoretical papers. Of the remaining eight papers, five contain footnote references, 115 in all. Of those, 14 are to pre-1915 papers, and 11 are to empirical papers (9 of which are American). Of the 90 remaining citations to essentially theoretical works, only 3 are to papers published in countries other than Great Britain, Germany, France, the United States, or Italy. Of the 90 core theoretical citations from those five countries in all six papers, then, 29 are British, 26 German, 13 Italian, 12 American, and 10 French. Assuming those citations to be a representative random sample of the total population of scientific output in high-speed aerodynamics, the chance of observing that distribution if all five countries were in fact equal in their output is less than 5 in 1,000. Furthermore, application of a difference in true means test (at a 95 percent level of confidence) indicates that Germany was producing on the order of twice the significant theoretical high-speed aerodynamics that the United States was.[10]

In contrast, while not a large enough sample to be statistically significant, the citation of 9 American works among 11 citations to empirical investigations does serve to indicate American preeminence in those areas. That impression is corroborated by qualitative evidence and would probably be corroborated by a more broadly based statistical inquiry. For example, G. V. Lachmann, chief designer for Handley-Page in England, in a 1937 Royal Aeronautical Society lecture entitled "Aerodynamic and Structural Features of Tapered Wings," cited a total of 27 articles and memoranda.[11] Of those 27, 11 were British, 8 American, and 6 German. American investigators in general seem to have done first-rate empirical or experimental work on problems with direct contemporary technological relevance, but to have done minor work, or no work at all, on fundamental theory. NACA was widely recognized for the excellence of its experimental data and for little else.

Thus, both qualitative and limited quantiative evidence suggest that German or German-educated scientists led in theoretical investigations of high-speed and turbocompressor phenomena, that British scientists lagged only slightly behind as late as 1935, and that scientists in the United States, Italy, and France lagged far behind. American scientists did, however, produce unmatched empirical design data for normal subsonic aircraft and for piston engine–propeller propulsion systems.

A third indicator of distinctive national patterns in the pursuit of aerodynamics is the choice of types of wind tunnels built in different countries. Unfortunately, there is no comprehensive history of wind tunnel development. Nevertheless, at least a representative picture of facilities built in Germany, England, and the United States can be assembled from available sources.[12] Although T. E. Stanton in England and Alexandre Gustave Eiffel in France built relatively sophisticated wind tunnels before the First World War, the first large, modern low-speed tunnel, with an open-jet working section 15 by 15 feet, was designed by Prandtl and built at Göttingen during 1916–17. A more advanced open-throat tunnel, built at Göttingen in 1931, became the prototype for most modern low-speed tunnels.

Busemann and Ackeret built a small 2.4 inch by 2.4 inch compressed air supersonic tunnel at Göttingen under Prandtl's direction in 1928. In the early 1930s Busemann designed a 3.5-inch closed-circuit high-speed tunnel powered by a 300 h.p. centrifugal blower for construction at Dresden. A larger high-speed tunnel (with a working section 4.3 inches by 5.1 inches) was built at Göttingen before the Second World War. The Deutsche Versuchsanstalt für Luftfortforschung (DVL) completed what was then the world's largest and best high-speed (1.0 Mach) closed-circuit wind tunnel at Berlin in 1939. That tunnel had a working section 9 feet in diameter and required 17,500 h.p. to run. Mention has already been made of Ackeret's superb Zurich supersonic (2.0 Mach) tunnel, begun in 1930. The Zurich tunnel, prototype for succeeding tunnels, had a working section 15.7 by 15.7 inches and was powered by a thirteen-stage Brown Boveri axial-flow compressor. Ackeret, whose work was closely associated with that of the German aerodynamicists, also designed the Brown Boveri–built Italian tunnel at Guidonia, as noted. During the Second World War, the Germans built the largest and most sophisticated high-speed and supersonic wind tunnels up to that time. For example, in 1942, they began construction of a 26-foot-diameter 100,000 h.p. Mach 1.0 tunnel at Otztal. German aerodynamic research establishments generally were the most completely and lavishly equipped of those of any nation.[13]

The principal prewar British wind tunnels were located either at the National Physical Laboratory (NPL) or at the Royal Aircraft Establishment, although B. M. Jones did do remarkable experimental work in a 28-by-20 inch NPL open-return tunnel at Cambridge. In addition to a series of large-section (7 to 9 feet) low-speed open-return tunnels, the NPL also built, in 1921, a very large full-scale 14-by-7-foot-section closed-circuit Duplex tunnel. (The Duplex design employed what were essentially two parallel 7-foot tunnels joined at the working section.) In 1932 the NPL built a compressed air tunnel (CAT) patterned on the American NACA variable density tunnel. A 12-inch-diameter high-speed tunnel (HST) was completed at NPL in 1935, Britain's first. The high-speed tunnel was intended primarily for airscrew investigations, but early on did give rise to one of aeronautic's more persistent myths. Asked by a newsman in 1935 what he did with the HST, W. F. Hilton, pointing to an aerofoil drag plot, replied, "See how the resistance of a wing shoots up like a barrier against higher speed as we approach the speed of sound." The next day, Hilton "was crucified by all the leading dailies for having coined the phrase 'sound barrier.'"[14] A new, larger 20-by-8-inch HST was designed in 1939, using a 5-by-2-inch scale model. The large tunnel was completed in 1940 and proved capable of Mach 1.3. The NPL had begun supersonic work in 1922 with a very small 0.8-inch-diameter tunnel designed by Stanton. By 1940, the NPL had also constructed for the Ordnance Board an 11-inch-diameter continuous-flow variable-density supersonic tunnel patterned after Ackeret's Zurich design.

The RAE had built several NPL-type low-speed tunnels during and after the First World War. In 1931, it built a 12-foot-diameter open-jet vertical tunnel specifically to investigate aircraft spinning. The RAE began construction of a 24-foot-diameter full-scale low-speed tunnel in 1932, completing it in 1935. In 1939, it began work on a large 10-by-7-foot 4,000 h.p. compressed air high-speed tunnel.

In the United States the NACA completed its first wind tunnel, a low-speed 5-foot-diameter design, at Langley Field in 1920. By 1928, it had completed a full-scale 20-foot-diameter closed-channel propeller research tunnel with a working cross-section of over 300 square feet. An even larger but similar tunnel with a working section of about 1,300 square feet was completed in 1932. Both full-scale tunnels could take complete engines and fuselages for test. The NACA also built a vertical tunnel in 1930 to investigate spinning. Max Munk designed a novel variable-density wind tunnel for NACA in 1925. The variable-density tunnel employed a closed-return channel with a 4.5-foot-diameter working section. Its novelty derived from the fact that the entire tunnel was pressurized to 20 atmospheres, which gave Reynolds' numbers (indicating flow similarity) closer to those encountered by actual aircraft. The NACA variable-density tunnel was the model for the later English CATs.

The first American high-speed compressibility studies of propeller-blade-tip burbling were conducted by Dryden and Briggs in 1923, as noted. Those experiments used nozzles of 2 inches to 12 inches in diameter and were run in the Lynn works of General Electric during regular acceptance tests of a large three-stage centrifugal compressor. Later, similar studies were done using 2-inch De Laval nozzles fed by a compressor which had been used during World War I to fill gas shells. The NACA built its first injection-type 12-inch-diameter high-speed tunnel in 1928 and a larger 24-inch-diameter tunnel of the same design in 1934. Virtually all of the NACA equipment, including that for supposedly high speeds, offered test velocities of less than Mach 1.0. Even as late as the first years of American participation in the Second World War, the entire U.S. test establishment for true supersonic speeds comprised one 1-inch-square tunnel.[15] Indeed, the NACA did not have a truly useful supersonic facility until completion of John Stack's slotted-throat tunnel at Langley in late 1947.[16]

Put bluntly, even American high-speed research was tied to empirical investigation of development-relevant phenomena, which in turn were tied to the capabilities and underlying assumptions of the conventional aircraft propulsion system. NACA's in-house historian, George W. Gray, reported somewhat defensively in 1948:

> In 1934, the twenty-four-inch tunnel, with a speed of 760 m.p.h., was completed. At the time these tunnels were going into operation, most of the European studies of compressibility effects were directed at experiments in the wholly supersonic speed range, though there were two notable exceptions in Britain and Italy. But in the United States—and that meant the Langley Laboratory—the effort in high-speed aerodynamics was concentrated on problems of transonic flow. It was recognized that before an airplane could fly at supersonic velocity it would have to pass through the transonic region. Solving supersonic problems would be of little avail if the airplane were unable to negotiate successfully the abrupt and erratic changes in airflow which usually accompany the transition from subsonic to supersonic flight.[17]

The object was flight, not science, practice, not theory.

If nothing else, that policy was certainly, in the short run, cost-effective. In his 1940 contribution to *Research: A National Resource,* Jerome C. Hunsaker noted the cost benefits of the NACA-developed cowling for radial piston engines: "It is estimated that the fuel bill in 1939 for United States domestic air lines was about $5,000,000 and for the Army and Navy at least $6,250,000. Removing the N.A.C.A. cowls from a typical transport plane or bomber would increase the drag approximately 30 percent or reduce the speed 10 percent. To maintain the same speed, the national fuel bill would be increased $3,375,000. This sum represents an annual recovery of many times the cost of the research."[18] Such was aerodynmics' service to man.

Von Karman did attempt to infect NACA with a different ideology. After attending the Volta Conference and having seen what European aerodynamic practitioners were accomplishing, he warned both the Army and NACA of the progress being made abroad and asked NACA to sponsor construction of a large supersonic wind tunnel. The Army, citing budgetary constraints, took no interest in von Karman's intelligence. George Lewis, speaking for NACA, responded "that he did not see why anybody would need a major wind tunnel that developed a speed greater than the speed of the propeller blade-tips—about 500 to 600 mile per hour, the point of maximum propeller efficiency."[19] (The same unimaginative George Lewis reportedly had called a meeting of all NACA executives after the DC-3 had first flown the year before to ask "what else was left to do.")[20] The United States had no adequate high-speed research facilities until the closing years of the Second World War, which, of course, was entirely appropriate to a research establishment which had no interest in fundamental aerodynamic science.

In physical construction, tunnel design in the Western countries did not vastly differ. Up until the late 1930s, Germany seems to have emphasized more, smaller, theory-relevant wind tunnels, although she later built superior, huge transonic and supersonic tunnels when scientific inquiry and technological development demanded them. The United States, in marked contrast, possessed unequalled full-scale, low-speed experimental equipment and also developed adequate high-speed subsonic tunnels. The orientation of American effort throughout, however, was toward solution of normal technological problems. Great Britain seems to have pursued a median policy, with adequate but not extraordinary programs both in full-scale low-speed research and in transonic and supersonic investigation.

The distinctive choices of experimental equipment and of research programs among the major Western powers bespeak fundamental differentiations in aerodynamic science. Although they were members of an international community of aerodynamic scientists, researchers in different countries had different research tools, different research technology development capabilities, and different objectives. In some disembodied sense what they knew may have been the same; but what they did was different.

COMMERCIAL AVIATION

National diversity in the pursuit of aerodynamic science reflects and is reflected in highly differentiated national patterns in the utilization of aerodynamic knowledge. The 1930s saw the creation of modern commercial aviation. In 1929, all airlines in the world flew a total of 96.3 million passenger-miles. In 1939, all airlines flew a total of 1,395.5 million passenger-miles, a better than thirteen-fold increase in ten years. The growth of commercial aviation was geopolitically biased. In 1929, all of Europe accounted for 48.9 million passenger-miles, the United States for 38.1 million passenger-miles. In 1939, all European airlines flew 434.4 million passenger-miles, while all American airlines flew 754.5 million passenger-miles. The single most highly developed European national airline system, that of Germany, flew 79.5 million passenger-miles in 1938, the last complete prewar year. Germany was followed in Europe by Great Britain, with 53.4 million passenger-miles (1938); France, with 45.6 million passenger-miles (1938); and the Netherlands, with 37.7 million passenger-miles (1939). In contrast, American airlines flew 554.2 million passenger-miles in 1938, almost seven times the German total and more than ten times the total of any other nation. Between 1938 and 1940, American passenger-miles doubled to 1,150.3 million.[21] In 1939, American flag airlines flew an additional 103 million passenger-miles internationally.[22] The American lead in commercial aviation passenger-miles reflected American leadership along every other dimension by which commerical aviation could be measured: total plane-miles flown, total number or capacity of commerical aircraft, and quality and frequency of service. The United States unquestionably developed the world's largest and best commercial aviation system.

The American triumph in commercial aviation resulted from the interaction of a variety of historical factors, including, at minimum, private enthusiasm and entrepreneurial commitment to aviation, governmental support (both in terms of direct subsidy and in terms of research and development support), general economic affluence, geography, and effective but not invulnerable competing transporation systems, notably railroads.

In addition, vigorous American manufacturers developed the world's best commerical aircraft. Yet those aircraft were the product of conscious developmental choices and a quickly established developmental lead, not the result of any intrinsic or unique American technological capability. American practitioners, at the least, had full access to complete subsonic theory, and American empirical investigation of subsonic phenomena was second to none. As noted, the development of the essential components of the modern airliner (stressed-skin all-metal wing and fuselage structures, variable-pitch propellers, retractable landing gear, flaps and slots) was international, with German, British, and American practitioners making approximately equal contributions. The basic technology drawn upon for the great American transports of the 1930s constituted an international pool of technological capabilities.

The new transport aircraft introduced during the 1930s provided revolutionary gains in performance compared to predecessor aircraft (see Table 2). If the Ford Tri-motor (steel frame, corrugated aluminum skin) with its 105 mph cruising speed is taken as representative of transport aircraft performance in 1926 (actually it was one of the best available), then the performance gain provided by such types as the DC-3, with its 194 mph cruise, is striking. What are not equally obvious from the figures but were equally important to the growth of commercial aviation were corresponding improvements in comfort, reliability, and all-weather flight capability.

The revolution in transport aircraft performance depended upon the simultaneous introduction of the whole set of modern innovations (v.p. props, retractable gear, flaps, and so forth) in one holistic system. Each of the major changes had been tried independently in the 1920s without success: in mathematical metaphor, their benefits were multiplicative rather than additive. For example, as noted, variable-pitch propellers had been proposed and tried as early as the end of the First World War. The value of a v.p. prop is that the pitch can be changed in flight to adjust the propeller precisely for differing speeds, altitudes, and loads. The same propeller pitch necessary for takeoff at full load is not suitable for high-speed cruise at 15,000 feet. Early experimental v.p. props installed in biplane aircraft with fixed landing gear were technically successful. But the very moderate overall gain in performance for the aircraft in which the props were installed did not justify the added weight, complexity, and expense of the v.p. provision.[23] It was not until v.p. prop could be applied to an all-metal, well-streamlined monoplane with retractable landing gear that it was worth the trouble. For example, the Boeing 247D transport enjoyed a higher cruising speed than the top speed of the original 247 (189 versus 182 mph). The only differences between the two versions were the installation of Hamilton-Standard v.p. props and NACA full cowlings on the 247D—and most of the boost in performance was due to the new props. The same argument applies to the other novel features of the new transports: retractable landing gear without all the other innovations was not worth the added weight, and so

Table 2. Inter-war Commercial Aircraft

Nation	Date	Mfr.	Model	Name	Type	Power*	V (mph) max.	V (mph) cruise	Passengers	Construction Material
U.S.	1926	Ford	4-AT	Tri-motor	pass.	3 @ 220	115	105	10 + 2	corrugated aluminum
U.S.	1927	Lockheed	Vega		pass.	1 @ 220	135	110	6 + 1	wood
G.B.	1931	Handley-Page	H.P. 42		pass.	4 @ 550		100	40 + 2	corrugated aluminum
Ger.	1931	Junkers	Ju52		pass.	3 @ 770		125	15 + 2	corrugated aluminum
U.S.	1932	Lockheed		Orion	pass./mail	1 @ 550		200	6 + 1	wood
U.S.	1930	Boeing	200	Monomail	mail	1 @ 575	158	135		stressed aluminum
U.S.	1931	Boeing	215	YB-9	bomber	2 @ 600	173	147		stressed aluminum

Country	Year	Manufacturer	Model	Name	Type	Engines*				Construction
U.S.	1933, 1934	Boeing	247, 247D		pass.	2 @ 550	182, 200	155, 189	10 + 2	stressed aluminum
Ger.	1932	Heinkel	He70	Blitz	mail	1 @ 630	221	193	4 + 1	stressed aluminum
U.S.	1935	Douglas	DC-3		pass.	2 @ 1000	220	194	21 + 2	stressed aluminum
Ger.	1935	Junkers	Ju86		pass./bomber	2 @ 600	202	177	10 + 2	stressed aluminum
Ger.	1935	Dornier	Do17		pass./bomber	2 @ 750	220	196	6 + 2	stressed aluminum
Ger.	1935	Heinkel	He111		pass./bomber	2 @ 950	230	214	10 + 2	stressed aluminum
Ger.	1937	Focke-Wulf	Fw200	Condor	pass./bomber	4 @ 820	224	170	26 + 2	stressed aluminum
U.S.	1938	Boeing	307	Stratoliner	pass.	4 @ 900	246	220	33 + 5	stressed aluminum
U.S.	1938	Boeing	314	Clipper	pass. (flying boat)	4 @ 1200	193	183	74 + 10	stressed aluminum
G.B.	1936	Short	S-23	Empire	pass. (flying boat)	4 @ 910		140	24	stressed aluminum

*number of engines and power of each

on. The modern, streamlined airliner was a total system and had to be designed and built as such.

The first modern, all-metal commercial monoplane, with all the features characteristic of the modern airliner except the v.p. prop, was the Boeing Monomail of 1930, a single-engine, high-speed mailplane. Boeing used the experience gained in monocoque and stressed-skin construction in the Monomail to design a prototype Army bomber, the YB-9, in 1931. On the basis of that experience, Boeing built the Model 247 transport, the world's first truly modern airliner, in 1933. Donald Douglas, who also had military experience, countered with his DC-1 in 1934, produced a slightly enlarged version, the DC-2 later the same year, and came out with a developed and enlarged form, the DC-3, in 1935. Boeing produced the world's first four-engined airliner, which was also the first pressurized commercial aircraft, the Model 307 Stratoliner. The 307, built in 1938, was a commercial development of the Model 299 bomber, the famous B-17 "Flying Fortress" first flown in 1935. The Stratoliner represented the zenith of prewar commercial aviation.

Although the United States thus clearly had a lead in development of commercial transport aircraft, it had no technological monopoly. In Germany, Heinkel's He 70 of 1932 was technically superior to the Monomail. Similarly, the Ju 86, the He 111, the Do 17, the Ju 90, and the Fw 200 were in no technical sense inferior to American contemporaries. The German machines generally carried fewer passengers at higher speeds less economically—but that was a matter of choice, not of technical deficiency. British commercial aircraft during the 1930s were markedly inferior to German and American types, but the British certainly had access to, indeed helped create, the technological base for the more modern machines. Likewise, the other European nations, while they did not develop competitive aircraft, probably could have had they chosen to do so.

In general, neither America's lag in aeronautical science, which did exist, nor any peculiarly American technical advantage, which did not exist, accounts for the supremacy of American commercial aviation. America had adequate science and, at most, only slightly more than equal access to technology.

American commercial aviation, then, would seem to have developed in response to older and broader factors in the American experience: a vast contiguous land mass, and a large, mobile, wealthy, and more or less egalitarian population. America was the classic continental market. More specifically, commerical aviation was able to develop in the United States because great overland distances separated large and wealthy population centers. Table 3 shows airline distances between selected American

Table 3. Airline Distances (miles)

New York–Chicago	713	London–Paris	210
Chicago–San Francisco	1,858	Paris–Berlin	540
New York–San Francisco	2,571	London–Berlin	577
New York–Miami	1,092	Berlin–Moscow	1,000
London–Cairo		2,175	
London–Bombay		4,468	
London–Melbourne		10,476	

cities, between major European capitals, and between London and Cairo, Bombay, and Melbourne. Clearly, speed is at considerably less of a premium on the 210-mile-run between London and Paris than it would be on the Chicago-to-San Francisco flight (for 210 miles, the difference in flight time for a 200-mph versus a 140-mph average block speed is less than half an hour). Oddly, moderate speed differentials (200 versus 140 mph) also matter less over very great distances requiring many stops, especially when competing modes of transportation are very slow. For example, Imperial Airways introduced seven-and-one-half-day service to India (via Cairo, Basra, and Karachi) using relatively slow Short Calcutta flying boats and D.H. 66 biplanes in 1929. Seven and one-half days was a lot better than two to three weeks on a packet steamer.

Moreover, Table 4 presents times and distances for crack train service in the United States, circa 1951; if anything, the times here are slightly better than the train schedules of the late 1930s. Table 5, by contrast, shows airline schedules of the decade from 1930 to 1940 over comparable routes. Under special circumstances, mid-1930s commercial aircraft could perform even better than the times indicated in the table: in February 1934 a new Transcontinental and Western (later TWA) DC-2 set a Los Angeles-to-Newark airmail record of 13 hours 4 minutes—compared to 90 hours by rail—and the flight was made at night.[24] Ironically, the very excellence of American passenger train service, combined with the great distances within the continental United States, seems to have technologically defined a niche ideally suited to the potential of the first modern transport aircraft. The railroads provided just enough competitive pressure to demand the best that aircraft manufacturers could then design.

Three other circumstances enhanced the hospitality of the American aeronautic environment. First, no questions of international boundaries arose: the entire flight took place within the United States. In the 1930s, principles of free overflight similar to freedom of the seas had not been established—as they still have not been in many parts of the world. Second, the entire flight was over land. It was possible to construct (and

Table 4. Crack Train Service in the United States about 1951

Train	Railroad	Route	Miles	Time (hrs.)	V (mph) max.	V (mph) av.
Daylight	Southern Pacific	San Francisco–Los Angeles	470	10	75	47
Broadway Ltd.	Penn.	N.Y.–Chicago	908	16	80	57
20th-Century Ltd.	N.Y. Central	N.Y.–Chicago	961	16	80	60
Sunset Ltd.	Southern Pacific	New Orleans–Los Angeles	2,070	42	75	50
Empire Builder	Great Northern	Chicago–Seattle	2,211	45	...	49
City of San Francisco	Ch. & N.W., U.P., S.P.	Chicago–San Francisco	2,256	40	79	51

SOURCE: Henry Sampson, ed., *World Railways, 1952–1953* (London: S. Low, Marston, 1952).

the U.S. government had constructed) navigational aids, both high-intensity visual beacons and radio beacons, plus a series of emergency landing fields and weather reporting stations. Such facilities were not possible in over-water situations or in relatively backward areas over land. Furthermore, such navigation systems require high utilization rates to be economically feasible. Even though it is farther from San Francisco to New York than from London to Cairo, the political and geographic conditions in the United States were incomparably more favorable. Third, the American population was large enough, and the economy rich and diverse enough, to provide sufficient demand for frequent airmail and passenger service.

The development of commercial aviation in Great Britain and in Germany highlights these uniquely American conditions. Britain, with her compact geography, poor weather, and excellent railway system had little need for domestic air service. Except for a few major routes, Britain ignored continental service, partly by choice, partly because of the rampant nationalism in Europe. Britain's principal interest in commercial aviation was as an instrument of imperial policy—a means of binding the empire together. British airlines were seminationalized, generously subsidized "chosen instruments" of foreign policy. This circumstance does much to explain the character of British commerical aircraft during the 1930s. The imperial routes were long distances over water and through underdeveloped colonial areas. Traffic was, to say the least, sparse. Emphasis was therefore upon large, reliable, sturdy, easily maintained aircraft. Because of a lack of NAV-COM aids, flying could be done only in

Table 5. *Typical Airline Service in the United States, 1930-40*

Aircraft	Year	Route	Miles	Time (hrs.)	Stops
Ford Tri-motor	1930	N.Y.–Chicago	713	8:00W 6:30E	1
Boeing 247D	1935	N.Y.–Chicago	713	4:30W 4:00E	0
DC-3	1940	N.Y.–Chicago	713	4:20W 3:30E	0
DC-3	1940	Chicago–Seattle	2,055	14:20W 13:15E	7W, 6E
Ford Tri-motor	1930	N.Y.–L.A.	2,797	22:00*	11
DC-3	1936	N.Y.–L.A.	3,797	17:40W 16:00E	3
Boeing 307 Strato-liner†	1940	N.Y.–L.A.	2,797	15:30W 13:45E	3

SOURCE: Donald Miller and David Sawers, *The Technical Development of Modern Aviation* (London: Routledge & Kegan Paul, 1968), pp. 214–16.
*Thirty-four hrs. E; 36 hrs. W, with overnight stop at Kansas City.
†Pressurized.

daylight; distances were so great that speed lost some of its urgency. Lack of traffic precluded the purchase of large numbers of aircraft, so that economies of scale, especially with regard to development costs, were not possible. Thus, the character of British commercial transports was determined by considerations not scientific and technical but political and economic.[25]

In contrast, Germany in 1939 had the most highly developed commercial aviation system in Europe. Lufthansa, however, was purely an instrument of political and military policy and was totally under government control. The structure of the German economy, and her excellent rail system, limited the domestic air passenger market. Emphasis was therefore on fast mail planes carrying a few passengers. Later, with the Ju 90 and Fw 200, Germany used her airlines to facilitate political penetration of the Balkans and South America. By 1939, however, German objectives in Europe had altered from political and economic hegemony to outright military conquest. Every German transport aircraft from the He 70 on was either designed or later developed as a bomber. Certainly Germany had the technical and scientific capability to equal if not surpass American accomplishments in commercial aviation. Her government was otherwise inclined.

The impact of political factors on British and German aviation calls into question the completeness of the "people of plenty" argument for the development of American commercial aviation. Geographic and economic factors made the American system possible; commercial aviation, however, was not something that merely happened, but something that had to be made. And it was made through direct, if occasionally circumspect, government intervention.

Airmail, the principal instrument of American governmental support, began as a fashionable Long Island stunt in 1911. By 1915 a small group of enthusiasts had gotten an appropriation through Congress for a Post Office experiment with airmail. The First World War delayed the experiment until 1918, when the first true airmail flight was made by Army pilots. Apparently, Warren Harding was impressed by airmail, and his administration continued its funding. More importantly, the Post Office began creation of a network of NAV-COM stations and emergency landing fields. By 1923, the first stretch of "lighted airway" was completed between Chicago and Cheyenne, Wyoming. Regular overnight airmail delivery was instituted between New York and Chicago in 1924.[26]

Although the government support for commerical aviation became enmeshed, in turn, in the quarrels centering around Billy Mitchell's proposal in the mid-1920s for a National Air Service (essentially a quasi-government-owned air system closely linked to military aviation), in disputes concerning Air Corps procurement in the late twenties, and in early New Deal political brawls about alleged airline monopolies and Hoover administration malfeasance,[27] it never entirely lost its efficacy. Except for a brief hiatus in 1934, during which an ill-equipped Army Air Corps was required to try to fly the mails, government policy consistently used airmail contracts to support development of commercial aviation. After 1930, the postmaster general could contract with airlines on a space-available rather than piece-rate basis, which permitted the government to subsidize purchase of larger, more advanced aircraft.

The magnitude of direct governmental support for commercial aviation in the United States was enormous. Between 1929 and 1939, the United States Post Office Department disbursed $148,895,000 in airmail subsidies. During that period, the government recouped $75,877,000 in airmail revenue, for a net subsidy of $73,017,000. In contrast, the British government's subsidies to all British commercial airlines between 1921 and 1939—an eighteen-year period—totaled £7,810,000, or approximately $37,488,000. Between 1929 and 1939 alone, the British government paid out £6,522,000 ($31,306,000), or about half the United States' net expenditure. For the period 1931 to 1939, airmail constituted a major portion of American airline revenue: $134,276,000 in airmail receipts versus $144,467,000 in passenger revenue. Such immense government expenditures clearly indicate American priorities: for the whole of its existence, from 1915 to 1940, NACA's total budget came to

$24,915,000; and most of that, as noted, went for development-related investigations. Between 1931 and 1939, the NACA budget totaled $15,009,000, or less than one-fourth the $63,672,000 spent in net airmail subsidies.[28]

The American policy of supporting commercial aviation was awesomely successful. The growth of American commercial aviation has already been described. Even though airmail subsidy payments fell from a high of $19,938,000 in 1931-32 to a low of $8,813,000 in 1934-35, before climbing again to $16,625,000 in 1938-39, the postal subsidy policy did create a viable, self-supporting commercial aviation system. In 1938-39, the airmail subsidy was only $300,000 more than airmail revenue, and passenger revenues had climbed to $34,174,000—more than twice the postal revenue. Given the fundamentally hospitable geopolitical and economic environment in America, government subsidy policy could create the world's best commercial aviation system.

The development of commercial aviation in the United States, in the British Empire, and in Germany represents the differential utilization of the same fundamental scientific and technological foundation. Subsonic technological capabilities in all advanced countries were or were potentially about the same. But a variety of geographical, political, social, and economic factors intervened to produce sharply divergent choices in the development of commerical aviation.[29]

INTERNATIONAL COMPETITION

Differential participation in international aviation competition reveals further distinctively national patterns in the utilization of scientific knowledge. The 1920s and 1930s were the great age of aeronautical competition and aviation spectaculars. Airplanes were new; not much had been done, and aviation enthusiasts, frequently with governmental support, quickly began to explore the continually widening limits of the new technology with all the elan of children with new toys. Despite an atmosphere that often bordered on that of a carnival sideshow, aviation feats during the interwar years did make serious contributions to technological development. Yet enthusiasts could compete along a number of different dimensions; speed, altitude, endurance, distance, and flights across novel geographical areas. The consequences for technological progress of accomplishments along these independent dimensions were quite often divergent if not contradictory.

Between 1919 and 1939, the only aircraft for which maximum speed was virtually the only design objective were those built for the Schneider Trophy competition. The Schneider Trophy races are especially critical here because they provided the only overt, nontheoretical evidence suggesting that the aircraft speeds necessary to efficient reaction propulsion were in fact possible. The Schneider Cup was given to the Aero Club of France by Jacques Schneider, of the Schneider-Creusot family, with the proviso that whatever nation first won three consecutive races should hold the trophy in perpetuity. Only seaplanes were eligible. The first two races were held in Monaco in 1913 and 1914. The races were not held during the war years, so the next contest was in England in 1919. From 1919 to 1925, the races were held annually, the country which won the previous year's race playing host for the next race. Generally, in order to take advantage of the availability of equipment, crews, and observers, attempts on the absolute world speed record were made after each Schneider Trophy race.

The United States won the races in 1924 and 1925 with a Curtiss biplane and captured the world speed record in 1925 at 245 mph. The American teams were funded by the Navy. Even though there was considerable interest in the races, they seem somehow to have never quite touched the mainstream of American consciousness. In reporting the 1926 race, won by an Italian monoplane, the *New York Times* correspondent recorded with some dismay that the spectators were so entranced by the high-speed aircraft "that announcement of the quarter results of the Army–Notre Dame and other big Eastern football games failed to bring a murmur from the crowd."[30] Because of presidential and congressional opposition to such frivolous expenditure, the United States entered none of the races after 1926. From 1925 to 1931, the races could be organized only every other year, since the cost of building the very sophisticated aircraft by then necessary to be competitive and of assembling and training the required crews was indeed becoming prohibitive.

The Schneider Trophy races and the record attempts probably did more to engender, or at least demonstrate, rapid progress in maximum aircraft performance than any other activity between the two world wars. The world speed record in 1913 stood at 126 mph, in 1919 at 162 mph; by 1928 it had been nearly doubled, to 318 mph. The races and the speed record attempts consistently elicited the cleanest and most powerful aircraft that industry could produce. Yet those aircraft were increasingly difficult to fly and their engines so highly stressed that they were barely good for the duration of the race. By the 1927 Schneider Trophy race, the monoplane, with its superior streamlining and lower induced drag, had clearly come to predominate. But just as clearly, the airplane itself and the naturally aspirated engine with which it was powered were approaching the limits of their capabilities. Further speed increases would be bought dearly in terms of time, money, and engineering effort.

The Origins of the Turbojet Revolution

Representative of the effort required to get the necessary performance were the British preparations for the 1929 and 1931 races. R. J. Mitchell of Supermarine had designed the winning aircraft in the 1927 race, the S.5, which was powered by an unsupercharged but highly modified Napier Lion engine of 875 h.p. Mitchell realized that for the 1929 race a cleaner plane and, above all, more power would be imperative. The National Physical Laboratory aided Mitchell in the design of the S.6 for the 1929 race. Since the Lion had reached the limit of its development, L.F.R. Fell, who had left the Air Ministry to join Rolls-Royce in 1927, persuaded Sir Henry Royce to authorize development of a racing engine. The result was the highly supercharged R engine, which produced 1,850 h.p. for the 1929 event. In the speed record attempt following the Schneider Trophy race, which the British won, the S.6 raised the record to 357 mph. For the 1931 race, a slightly refined version of the S.6, the S.6B, was fitted with an R engine boosted to 2,330 h.p. The numerous, ingenious expedients resorted to in that developmental program were discussed in Chapter 5. The British won the 1931 race, and thus the Schneider Trophy outright. For the subsequent speed record attempt, in which the absolute speed record was raised to 407 mph, the R engine was boosted to produce 2,650 h.p. Its life expectancy was only an hour.[31]

Massive and remarkable though this progress was, it generated, in addition to the natural elation on the part of participants and amazement on the part of the public, a certain apprehension, an unease, among some members of the aeronautical community. As early as January of 1928, P. A. Ralli, who had designed the metal propeller for the Supermarine S.5, told the Royal Aeronautical Society in a lecture:

> It would seem that, adhering to the present disposition of airscrew, engine, and aircraft structure, only minor advances are possible. Such, for instance, would be improvements in blade shape and sections as regards adaptability to high velocities near the speed of sound. The requirements are exacting, inasmuch as a very appreciable increase in thrust horse-power has but a trifling influence on actual speed.
>
> I suggest, however, without desiring to make any positive assertion on the subject, that advances of a really important order will more probably be the result of a deliberate departure from the present arrangements, whereby some of the disabilities under which the airscrews are now working would be partially or totally removed.[32]

Ralli was among the foremost practitioners of conventional aero-engine technology, yet he shared the same sense of imminent change that would lead others to the turbojet.

H. R. Ricardo, the piston engine expert who helped solve the problem of detonation (knock), shared Ralli's feeling. In a 1930 lecture to the

Supermarine S 6 B, 1931. (Courtesy of the Smithsonian Institution.)

Royal Aeronautical Society, Ricardo remarked of the piston engine: "What further scope is left for increase of power, reduction of weight, or improvement in economy can now be of a small order only, and, by analogy from the past, it would seem that the time is almost ripe for some innovation more radical than mere detail improvement."[33] (As noted in Chapter 5, Ricardo's ideas of what constituted radical innovation turned out to be highly conservative.) Nevertheless, detail improvement would occupy the conventional aeronautical community almost completely and would in a decade slowly bring service aircraft up to the standards of performance set by the Schneider Cup machines of 1931.

Although England won the Schneider Trophy permanently in 1931, that did not signal the end of indirect competition. The Italian team was not prepared for the 1931 races and did not appear. That apparently hurt Mussolini's feelings, and he ordered a special unit of his air force to raise the world speed record to 600 mph. They never made it, but a Macchi seaplane powered by a twenty-four-cylinder (two V-12s in tandem) Fiat engine of 2,850 h.p. driving contrarotating propellers did raise the record to 424 mph in 1933 and to 440 mph in 1934.[34] The Italian record stood until March 1939, when a Heinkel He 100 jumped it to 463 mph. Less than a month later a Messerschmitt Me 209 raised the record to 469 mph, which would stand for piston-engined aircraft for thirty years. The Schneider Cup races and the speed record attempts demonstrated the extreme of development possible for the piston engine–propeller combination as well as the limitations of that system..

Until 1939, the world speed record was held by a seaplane, which carried two floats nearly as large as the fuselage of the airplane itself.

Heinkel He 100, 1939. (Courtesy of the Smithsonian Institution.)

Seaplanes were used for the Schneider Cup races and for the speed record attempts up to 1934. In order to get the best attainable streamlining, the wing loading (weight per unit of wing area) of the aircraft had to be, for the time, extremely high. High-lift devices such as flaps and slots, and retractable landing gear, were all undeveloped. Variable-pitch propellers were likewise undeveloped, and this necessitated installing on the aircraft fixed-pitch propellers suitable for the very highest speeds, propellers that were extremely inefficient at low speeds. These factors combined to give the high-speed seaplanes very high landing speeds (130 mph) and landing and takeoff distances longer than any runways then existent. In the years from 1934 to 1939, as noted, much technological progress occurred: retractable landing gear, flaps, and variable pitch propellers were all developed. It is perhaps significant that the two German aircraft that broke the Italian record of 1934 were landplanes—and that their Diamler-Benz engines were pulling a good 1,000 h.p. less than the Fiat, which gives some indication of the reduction in drag on the German landplanes.

High speed was only one dimension of aircraft performance, however, and the Schneider Cup races and world speed record attempts represented only one aspect of aeronautical competition during the interwar years. American withdrawal from international competition did not signal an end to American conpetitiveness. American landplanes competed in the domestic Thomson Trophy races. Americans held the world speed record for landplanes for most of the period from 1932 until 1938. The names of some American aviators are still household words: Charles Lindbergh, who flew nonstop from New York to Paris in

1927, and Amelia Earhart, who vanished in the Pacific on an attempted around-the-world flight in 1937. Others are still recognized. Wiley Post set a solo around-the-world record of 7 days, 18 hours, 49 ½ minutes in 1933, and died in a crash in Alaska in 1935 with his passenger, Will Rogers. Howard Hughes set a whole series of records. He piloted his Hughes Special monoplane to a landplane record of 352 mph in 1935. He set two official transcontinental records: Los Angeles to New York in 9 hours, 26 minutes (1936) and 7 hours, 28 minutes (1937). In 1938, he flew a specially modified Lockheed 14 (a commercial transport model) around the world in 3 days, 19 hours, 14 minutes, for an average speed in the air of 206.7 mph. When World War II erupted, he had just bought a custom-built Boeing 307 Stratoliner to use in another around-the-world speed record attempt. During the war, he flew an early version of the Lockheed Constellation from Los Angeles to Washington in 6 hours, 58 minutes, for an average speed of 355 mph.

Other names and other accomplishments, spectacular in their time, are long since forgotten: the feats of de Pinedo, Chamberlain, Mears and Collyer, Boardman and Polando, Panghorn and Griffin.[35] Some aviation record attempts had overt and sinister political purposes. K. E. Bailes has recently shown that the Soviet long-distance-record flights, which resulted in several nonstop distance records during 1936 and 1937, were used deliberately by the Stalin regime to distract attention from the purges and to cover up the poor overall quality of Soviet aircraft (the record-breaking single-engine ANT-25 was in fact technically backward).[36] Numerous German exploits were similarly contrived for propaganda.

In general, sharp national differences in participation in international aviation competition are clear. After 1926, only Great Britain, Italy, and Germany build specially designed aircraft capable of the very highest speeds attainable with piston engine and propeller. American landplane racers were much more clearly based on contemporary commercial practice and were substantially slower than the ultimate European machines: Hughes's 352-mph record was almost 90 mph under the 440 mph Italian absolute record set the year before. The vast bulk of American aviation achievement exploited American commercial transport aircraft to set long-distance flight and long-distance speed records and to make flights to and from exotic places. American participation in international competition after 1926 was directly keyed to commercial aviation development. As Hughes remarked after his 1938 around-the-world flight; "It was in no way a stunt. It was the carrying out of a careful plan. We are in no way supermen. Any one of the airline pilots of this nation with any of the trained army or navy navigators and competent radio engineers in any of our modern passenger transports could have

Howard Hughes's Lockheed L-14 (1938). (Courtesy of the Smithsonian Institution.)

done the same thing."[37] Other nations, notably the Soviet Union and France, made what efforts they could to set distance or altitude records with their relatively backward aircraft. What is important for the turbojet revolution is that ultimate high-performance aircraft were the exclusive province of Britain, Italy, and Germany from 1926 to 1939.

Overall, significant and pervasive national differences persisted within the world aeronautical community throughout the years between 1920 and 1940. Those differences were manifest in distinctive national patterns in the pursuit and utilization of aerodynamic science. High-speed aerodynamics was theoretically developed in Germany and to a lesser extent in England and was all but ignored in the United States. Theoretical and empirical investigations of turbomachinery phenomena were carried out in Germany, Switzerland, and Great Britain, but not in the United States. Conversely, the United States carried out the world's most comprehensive and fruitful empirical investigations of subsonic aircraft behavior. These national choices in the pursuit of aerodynamic science were reflected in choices of experimental technology, especially wind tunnels.

National Patterns in Scientific Knowledge

The United States, by private initiative and lavish government support, built the world's largest and best commercial aviation system. Other nations, although capable of building high-performance transport aircraft, pursued other objectives. The United States withdrew from international high-speed competition after 1926. Britain, Italy, and Germany built the world's fastest aircraft from 1929 to 1939. American achievement in competitive aviation was based completely on the development of commercial aircraft, on normal technology.

It would not be implausible to argue that American tardiness in recognizing the presumptive anomaly upon which the turbojet was based resulted from American deficiency in high-speed aerodynamics, from American fixation with commercial aviation, from American isolationist withdrawal from international high-speed competition. The simple fact is that the nations which exhibited the obverse of the American pattern, England and Germany, produced the first turbojets. Yet to argue that one thing did not happen because something else did not happen becomes tenuous and bespeaks a naive faith in unilinear causal sequence. A more subtle explanation may be more nearly accurate and have similar overt historical consequences.

The national patterns in the pursuit and utilization of aerodynamic science observed here may reflect fundamentally differentiated cultural traditions. No later than 1900 Germany certainly had an unequalled tradition of mathematical and theoretical excellence in science and also had developed a deliberately close relationship between science and industry. Britain shared a similar if more empirical and less mathematically rigorous tradition in science. In contrast, the United States still was possessed of a scientific tradition extreme in its empiricism and utilitarianism.[38] Just as specific values within the French scientific and technological communities seem to have shaped the behavior of Fourneyron and his successors, just as cultural values in general seem to influence if not determine the direction both of scientific acitivity in the large[39] and of particular theoretical developments,[40] so broad cultural values seem to have acted both as a vicarious selector for general scientific activity in aeronautics and as a more direct determinant of utilization patterns for that science.

Sharp differences likewise characterized European and American technology. American technology tended to be functional, cheap, rugged, and energy-intensive. The quest for technical excellence or extraordinary performance at the expense of utility or reasonable cost was quite foreign to most American technological practice. European craftsmen and manufacturers were more comfortable with small-volume, very expensive, and often not very practical technological *tours de force*. The basic values of national cultures were reflected in national scientific and technological subcommunities. Membership in the great international invisible college of scientific or technological practitioners was mediated by national cultural influences.

The ultimate historical explanation for the particular spatiotemporal configuration of the turbojet revolution may lie at this rather remote level of cultural differentiation. Certainly Americans' failure to recognize promptly the presumptive anomaly upon which the turbojet revolution was based is inseparable from the total cultural matrix of American technological practice, so well defined by American choices in aerodynamic science and in aeronautical practice. Timely recognition of that presumptive anomaly by a handful of Europeans is similarly couched in specific national milieux.

7

The Turbojet Revolution

A handful of men created the turbojet revolution. Those men overthrew the tradition of practice centered around the piston engine and propeller and set in its place a new normal technology founded on the turbojet. The presumptive anomaly upon which those men predicated their actions derived solely and directly from advances in aerodynamics and comprised the conjunction of radical assumptions about total airframe performance with radical assumptions about internal combustion gas turbine component efficiencies. Those men saw what mature subsonic aerodynamics and the first insights of supersonic aerodynamics implied: that but for the propeller the new stressed-skin, well-streamlined airframes were perfectly capable of flying at near-sonic speeds. Those men saw that the application of mature subsonic theory to gas turbine component design could produce internal combustion gas turbines of unprecedented power, efficiency, simplicity, and lightness. Those men saw that such a gas turbine, when used for reaction propulsion at high speeds, could be propulsively efficient even though thermally less efficient than existent piston aero-engines or proposed gas turbines used for driving propellers.

The protagonists of the turbojet revolution were uniformly aliens to the normal aero-engine community. No turbojet was produced independently by a conventional aero-engine firm. The creators of the turbojet shared but one common bond: each was either a part of or exposed to the leading edge of aerodynamic advance, and each separately deduced from that science the radical performance assumptions that led him to the turbojet. Each reasoned through a slightly different path to the common conclusion. The turbojet was the creation of these few men who through the prism of science saw technological reality differently.

Yet that those protagonists were aliens to the conventional aero-engine community does not imply that they were strangers. Each was deeply committed to some facet of aeronautics. Each was the product of a uniquely Western scientific and technological culture. However totally and passionately committed to his own ideas, however certain that new performance parameters were necessary and proper to judge fairly the product of his inspiration, each of these men recognized the authority and power of community-held norms of testability. Each would recognize the urgent need to redefine critical performance parameters to suit his own purposes. Yet each would work to overthrow only a technological tradition, not to violate the norms of a scientific or technological culture.

Turbojets emerged independently from the work of four men: Frank Whittle in England, and Hans von Ohain, Herbert Wagner, and Helmut Schelp in Germany. Only Whittle and von Ohain directed their efforts toward a turbojet from the beginning. Wagner initiated a more general inquiry that led directly to the turbojet, while Schelp, an engineer in the German Air Ministry, separately arrived at the turbojet conclusion and promoted a government-supported turbojet program. These four men were the protagonists of the turbojet revolution. They used prior technology; they required certain preconditions. But theirs, personally and individually, was the turbojet.

Significantly, the words *invention* and *patent* appear rarely here. The turbojet was created, and that creation bears little but superficial resemblance to alleged precursors. Admittedly, reaction propulsion was discussed as early as the seventeenth century,[1] and numerous devices were patented during the First World War. Indeed, in 1921, a Frenchman named Charles Guillaume patented a complete axial-flow turbojet in very nearly its modern form. There was one small difference: Guillaume's patent drawings, in addition to showing the expected compressor, combustion chamber, and turbine, also show, protruding from the front of the engine, a very large manual starting crank.[2] One wonders how aeronautical engineers would have streamlined that. Guillaume's concept, in short, although of the same configuration as a turbojet, could not have been farther from the valid scientific assumptions that made the turbojet a practicable possibility. To say that he "invented" the turbojet in any meaningful sense is absurd. Such would seem to be true of many antique patents of modern devices. What matters here is the actual creation of the turbojet—and of a technological revolution.

The first man to design a turbojet in accordance with sound aerodynamic insight was Frank Whittle.[3] Whittle was the eldest son of a

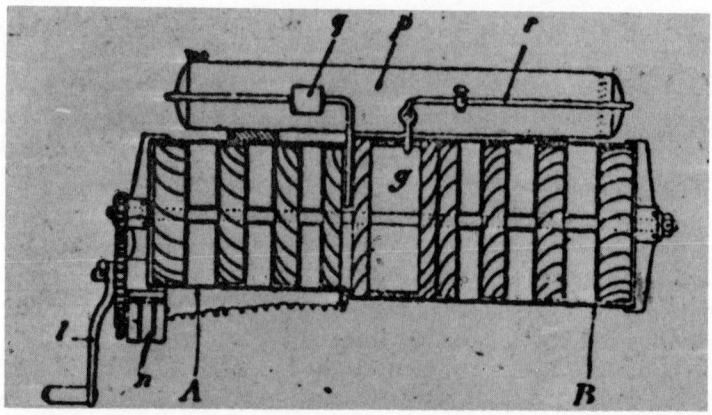

Guillaume patent drawing of a turbojet, 1921. (From Gohlke, "Thermal-Air Jet-Propulsion," *Aircraft Engineering* [London] 14 [1942]: 33.)

Midlands machinist and amateur inventor who had purchased a small engineering firm, the Leamington Valve and Piston Ring Company, when the younger Whittle was nine years old. Frank Whittle thus grew up in a shop in which he learned early to use machine tools and, because of his father's influence, in an atmosphere in which inventiveness was encouraged. Although from a working-class family, Whittle managed to secure a scholarship to a local secondary school. His academic record was not spectacular, but what he lacked in the classroom he seems to have made up by reading voraciously, especially on scientific subjects, including aeronautics. After some difficulty caused by his small stature, which he allegedly overcame by a combination of exercise and diet, Whittle joined the Royal Air Force (RAF) in 1923 and was admitted to the apprentices' (ground mechanics) school at Cranwell. He was trained as a rigger for metal-frame airplanes, which were new at the time. From his childhood, Whittle had been an avid builder of model airplanes, and he continued his hobby at Cranwell. After three years as an apprentice, Whittle obtained one of only five cadetships at the RAF college, also at Cranwell.[4]

The two-year curriculum at the cadet college was designed to prepare flight officers for the RAF, and included engineering subjects as well as flying instruction. In his last semester at Cranwell, in 1928, Whittle wrote his required term thesis, entitled "Future Developments in Aircraft Design," on high-speed, high-altitude flight. A version of this thesis was published in the *RAF Cadet College Magazine* in the fall of 1928 under the title "Speculation."[5] The origin of Whittle's interest in high-speed, high-altitude flight can be traced to informal discussions of rockets and other advanced flight systems. Whittle began his "Speculation" by noting:

Frank Whittle at Cranwell in 1925. (From Frank Whittle, *Jet* [London: Frederick Muller, 1953].)

> I was once asked by an optimistic sub-editor of this magazine for an account of how I intended to reach the moon. I was naturally a little shaken at first, as I have never contemplated leaving this homely planet, but, thinking that I might write a little light fiction, I promised; only to find that I cannot rise to the level of Verne or Wells. It, however, caused my thoughts to soar above the tropopanze (for the benefit of those who have never been initiated to the mysteries of meteorology, the tropopanze is that altitude above which the temperature of the atmosphere remains constant), and the following speculation is the result. (p. 106)

Whittle "came to the conclusion that if very high speeds were to be combined with long range, it would be necessary to fly at very great heights where the low air density would greatly reduce resistance in proportion to speed."[6] Whittle independently derived, from aerodynamic 'first principles," a still-air-range formula essentially the same as the later well-known Breguet formulation.

The Turbojet Revolution

For high-speed, high-altitude flight, Whittle rejected the conventional piston engine and propeller: airscrew speed was limited both by engine speed and by the onset of compressibility shock on the airscrew blades, and piston-engine power was sharply limited by supercharger efficiency (pp. 107–8).[7] He noted the possibility of rocket propulsion: "We have heard much recently about the rocket-driven car, and proposals for an aeroplane to be driven on the rocket principle" (p. 108). But after careful analysis of likely propulsive efficiency, Whittle concluded that rocket propulsion would be hopelessly inefficient at the aircraft speed he was then considering, 300 mph.

Consideration of reaction propulsion, however, directed Whittle's attention to the use of reaction in "a rotating nozzle where high linear speeds are possible," which, of course, was the principle of the turbine. Whittle was considering a turboprop engine for aircraft speeds of up to 300 mph and altitudes as high as 115,000 feet at a time when contemporary service aircraft were capable of 150 mph at 15,000 feet. Whittle concluded that the turboprop was most promising:

> The advantage of such a power unit may be stated as follows: —
>
> (1) The only limit to the compression ratio is the maximum temperature which the heating chamber may stand.
> (2) Power is not dependent on r.p.m., as in the case of the petrol engine.
> (3) The work which may be done by 1 lb. of air increases with altitude, and partly compensates for the smaller quantity of air available.
> (4) Supercharging does not appear to be necessary.
> (5) Rotors of different diameters may be used as gearing.
> The main disadvantage as far as air work is concerned is the gyroscopic effect of the rotors, but on reviewing the points for and against it seems as though the air turbine is the aero engine of the future. (p. 110)

Whittle graduated from Cranwell in July 1928, second in his class, and received an award in aeronautical sciences. He was posted to a fighter squadron at Hornchurch, and then, toward the end of 1929, to Wittering for a flying instructor's course. In the year and a half after he left Cranwell, Whittle continued to ponder the problem of high-speed, high-altitude flight. In order to avoid the problems of propeller blade-tip compressibility foreseen in his 1928 paper, he considered a jet propulsion system in which a conventional piston engine powered a low-pressure blower. Both the blower and the engine would be located in a duct, and fuel could be burned in the flow-stream aft of the engine to increase thrust. Whittle later discovered that this system had been patented by Harris in 1917. Moreover, about the same time Whittle was considering it, the ducted fan was being developed by Secondo Campini in Italy. (As noted below, an airplane with a Campini propulsion system was flown in 1940, but was not successful.) Whittle concluded directly from his calculations, however, that the piston-engined jet propulsion scheme was far too heavy and not really superior to the piston engine–propeller combination, and he rejected it.

By early 1929, then, Whittle seems to have been in somewhat of a quandary. Initially, he had sought high speed not through airframe streamlining but rather through lessened drag at very high altitudes. For flight under such conditions, he required a light, reasonably efficient propulsion system which was altitude compensating (as a turbocharged system or a gas turbine was) and which dispensed with the propeller (because of insufficient propeller "bite" in less dense air, and because of the earlier on-set of blade-tip compressibility at high altitudes). A pure rocket could meet neither of the first two conditions, a piston engine with ducted fan none but the last, and a turbo-prop all but the last.

In the meantime, Whittle learned of the conclusions of B. M. Jones's study of aircraft streamlining. Prior to his design of the turbojet, Whittle had not seen Jones's original Report and Memorandum (R&M), but he had most probably either read or heard of Jones's Royal Aeronautical Society lecture based on it.[8] From the age of fourteen, Whittle had been an eager reader of aeronautical material, and he attended lectures whenever possible, so it is highly likely he did know of Jones's work. It is also virtually certain that Whittle took an active interest in the results of the September 1929 Schneider Cup contest and the subsequent raising of the world speed record to 357 mph—and that he recognized the implications of such progress for other high-speed aircraft.

By late 1929, Whittle clearly had deduced from sound aerodynamic and thermodynamic precepts both the possibility of high-speed flight at high altitudes and the possibility of high cycle efficiency from a gas turbine at those altitudes; he certainly saw the implications of well-streamlined airframes for attaining very high speeds. At the end of 1919, Whittle "invented" the turbojet. In his words,

> While I was at Whittering, it suddenly occurred to me to substitute a turbine for the piston engine [in the ducted fan system]. This change meant that the compressor would have to have a much higher pressure ratio than the one I had visualized for the piston-engined scheme. In short, I was back to the gas turbine, but this time of a type which produced a propelling jet instead of driving a propeller. Once the idea had taken shape, it seemed rather odd that I had taken so long to arrive at a concept which had become very obvious and of extraordinary simplicity. My calculations satisfied me that it was far superior to my earlier proposals.[9]

The turbojet was born. In order to get adequate efficiencies, Whittle assumed speeds of 500 mph (for propulsive efficiency) and altitudes of

35,000 feet (for low ambient temperature). He also assumed component efficiencies far above current practice.

Two things that Whittle was not then aware of were the Stern report and Griffith's work on axial compressors. Neither did he know of Guillaume's 1921 patent nor of the work on axial compressors in Germany or Switzerland. He was, however, well acquainted with the arguments against internal combustion gas turbines current after the turn of the century, especially those regarding the problem of temperature limitation. Yet in addition to a basic faith that compressor and turbine component efficiencies could be significantly improved by better design, Whittle realized that there were three other factors favorable to reaction propulsion by turbojet "not present in other applications" of the gas turbine: "1. The fact that the low temperatures at high altitudes made possible a greater ratio of positive to negative work for a given maximum cycle temperature. 2. A certain proportion of the compression could be obtained at high efficiency by the ram effect of forward speed, thereby raising the average efficiency of the whole compression process. 3. The expansion taking place in the turbine element of such a motor was only that which was necessary to drive the compressor; and therefore, only part of the expansion process was subject to turbine losses."[10] In short, although Whittle did not have at his command scientific insight as comprehensive as that of some later turbojet proponents, he did deduce completely valid assumptions from what he did know.

But Whittle, like all visionaries, had a great deal of difficulty making others see what he saw so clearly. One of the first men with whom Whittle discussed his idea was W.E.P. Johnson, one of his flying instructors, who had had experience as a patent agent. Johnson and Whittle explained the turbojet to their commanding officer, and he, in turn, arranged for Whittle to see W. L. Tweedie, "a technical officer in the Directorate of Scientific Research" of the Air Ministry. After an initial interview, Tweedie took Whittle to see A. A. Griffith. Griffith, although a believer in the gas turbine for propeller drive, thought Whittle's assumptions overly optimistic; and the Air Ministry, probably reflecting Griffith's opinion, informed Whittle that since his design was basically a gas turbine, its development was considered impracticable because of limitations on materials' temperatures and stresses.[11]

Following rejection by the Air Ministry, Whittle applied for a patent on his engine on 16 January 1930. The patent was subsequently granted. Whittle's patent application is indicative of his thought at that time: it shows a lightly constructed turbojet with two axial intake fans followed by a single-sided, single-stage centrifugal compressor; the compressor feeds to a number of straight-through combustion chambers which exhaust through a two-stage axial turbine and thence out through

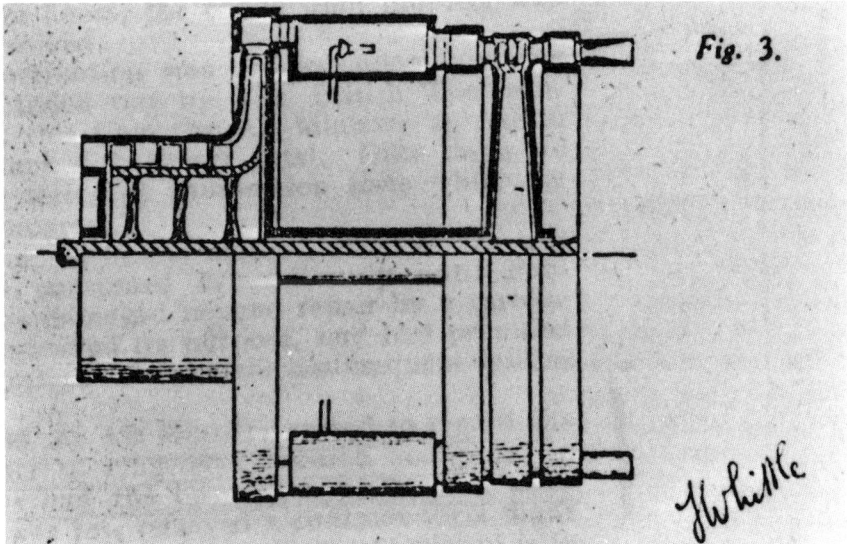

Original Whittle turbojet patent drawing, 1930. (From F. Whittle, "The Early History of the Whittle Jet-Propulsion Gas Turbine-I," *The Aeroplane* [London], Oct. 19, 1945.)

multiple exhaust nozzles. Whittle's 1930 design does not indicate a clear awareness of the importance of high mass-flow to the turbojet's propulsive efficiency.[12] Nevertheless, it may be assumed, on the basis of Whittle's careful analysis of the deficient propulsive efficiency of a pure rocket system in his 1928 paper, that he was in fact fully cognizant of the need for high mass-flow in his turbojet.

After securing his patent and being once again informed of the Air Ministry's lack of interest, Whittle futilely approached a number of commercial firms, several through Johnson's contacts. Late in 1930, for example, the British Thomson-Houston Company (B.T.-H.), manufacturer of steam turbine and centrifugal industrial air compressors, declined to support Whittle because of lack of money (the company estimated it would cost £60,000 to develop the engine), because of the uncertainty of the whole undertaking, and because aero-engines were out of its field. Armstrong Siddeley, which made piston aero-engines and was affiliated with a variety of other heavy industrial manufacturing firms, rejected the Whittle design in 1931 mainly because of materials' temperature limitations. Also in 1931, Bristol Aircraft, which was pioneering sleeve-valve radial piston engines, refused to try to develop the Whittle engine. The Bristol engineer who studied the Whittle proposal, Frank Owner, reported to his board that "the combination of a centrifugal type of supercharger of the best efficiency we knew of, matched to a turbine with a limiting temperature which we believed would run without failure, would not in fact, 'pull the skin off a rice pudding.'"[13] Whittle in 1930 was only twenty-three, and suffered from a considerable "lack of presence."[14] As is indicated by the tone of the Bristol remark, he and his ideas were not always treated with the greatest respect.

Whittle first learned of Griffith's work on axial compressors when someone at the Air Ministry handed him a copy of Griffith's report to read while he was waiting for an appointment. He found Griffith's analysis "over his head" because he "had not then received the necessary degree of engineering training."[15] Whittle first read the Stern report when the Air Ministry sent him a copy to support its contention that the gas turbine was impracticable. Whittle's reaction to the Stern report was that of a classic protagonist of scientific or technological revolution responding in Mitroff's counter-normative mode: "I rejected Stern's views because it seemed to me that he confined himself to industrial forms of multi-stage compressors and turbines and came out with weights then usual in heavy engineering practice for such components. From the earliest days I believed that the future for aircraft power units lay with small high speed components very different from the sort of thing visualized by Stern."[16] Whittle himself later realized that the initial reaction to his proposal was not completely unreasonable. But he (and most who have studied the matter since)[17] felt that even though his engine was probably not practicable in 1929, a research program into its possibilities should have been funded.

As it was, Whittle was unable to interest anyone in his engine. He realized that he "would need to convince people of the value of [his] compressor proposals before [he] could hope to get them to accept the much more comprehensive scheme for a complete engine." In 1930–31, therefore, he wrote a paper published in the *Journal of the Royal Aeronautical Society* on the centrifugal compressor.[18] Whittle's turbojet would require a centrifugal compressor which developed a pressure ratio of 4 : 1 at an efficiency of at least 75 percent, while the best existent compressor as late as 1931, the supercharger of the R engine in the S.6B, developed a ratio of only 2 : 1 at 62 percent efficiency. Whittle's paper, which argued that much higher pressure ratios and efficiencies were possible, constituted a significant contribution in its own right and demonstrated both a complete understanding of centrifugal compressors and a familiarity with the best available theoretical engineering sources—Aurel Stodola and William Kearton.[19] The article also disclosed details of a B.T.-H. centrifugal high-speed compressor used for conveying clay along compressed air pipes. Still, Whittle was unable to get backing for his turbojet.

After four years' service, flying officers in the RAF were required to choose a specialty in which to receive further training. Whittle quite naturally chose aeronautical engineering, and in 1932 was posted to the Officers' Engineering Course at Henlow. On his entrance examinations to Henlow, Whittle scored an "aggregate 98" percent, which is indicative of the degree to which he had succeeded in educating himself. Because of his high score, he was permitted to complete the normal two-year course in eighteen months, which he did with distinction. Then, although the Air Ministry had discontinued the program, Whittle petitioned for special consideration for assignment to the Mechanical Sciences Tripos at Cambridge (roughly equivalent to a couple of years' graduate work in mechanical engineering in the United States). Whittle was duly posted to Cambridge in the summer of 1934.

Cambridge was the turning point for Whittle personally and for his engine. Because of his earlier training at Henlow, Whittle completed the normal three-year course at Cambridge in two years. Through remarkable efforts, especially considering that this was the period in which he finally was able to start development of his engine, Whittle managed a First (honors) in the Tripos. During his time at Cambridge he worked with one of the most outstanding aerodynamicists in England, B. M. Jones, and assisted him in the Aeronautical Laboratory at Cambridge in developing the Pitot traverse method of measuring aerofoil drag.

So far as his turbojet was concerned, however, Whittle's first months at Cambridge had been dismal. His original 1930 patent came due for renewal in January of 1935, for a fee of £5. Whittle had about given up hope of ever seeing his engine built. The Air ministry declined to pay the renewal fee, and because of "medical expenses arising out of an illness of his elder son and the birth of his younger," Whittle let the patent lapse. But then in May 1935, Whittle received a letter from R. Dudley Williams, a fellow cadet at Cranwell with whom Whittle had flown as a test pilot at Felixstowe during 1931-32, and with whom he had discussed his engine. Williams had retired from the RAF, but he, in company with J.C.B. Tinling, another former RAF officer whom Whittle did not know, thought he could arrange financing for the Whittle engine. Williams' and Tinling's activities ultimately drew the interest of M. L. Bramson, who led, in turn, to the investment banking firm of Falk and Partners. One of the Falk partners was L. L. Whyte, who, although by profession a banker, held a Natural Sciences degree from Cambridge. After Whittle had designed and patented all sorts of strange, supposed "improvements" on his engine to get some sort of patent protection (the lapsed original patent could not be renewed), a very modestly funded company (anticipated expenses were only £50,000), Power Jets, Ltd., was formed to build the Whittle engine. Thus, by an odd string of coincidences, Whittle at last would get a chance to develop his engine.

The Turbojet Revolution

Power Jets obviously did not have the money to set up its own production facilities, so Whittle, already acquainted with British Thomson-Houston, contracted with them to fabricate the main components of the first engine. The first contract with B.T.-H., for shop drawings, was let in March 1936, and marks the beginning of actual development of the Whittle engine. Since B.T.-H. was going to fabricate the engine rotor (the compressor impeller, the rotor shaft, and the turbine disc and turbine), and since Whittle had to approve personally all the drawings and any modifications, he and Power Jets set up light housekeeping in the B.T.-H. shop at Rugby. The first engine was not intended for flight, but was built as a prototype.

The only likely commercial application Power Jets could think of was for a fast, high-altitude trans-Atlantic mailplane, which was not then so ridiculous as it might sound now. Numerous aircraft manufacturers (such as Boeing and Heinkel) had good markets for special mailplanes. The mid-1930s also saw the Germans refueling flying boats from ocean liners in mid-Atlantic and catapulting Blohm and Voss floatplanes off the bows of Atlantic liners to cut a day or so off trans-Atlantic mail service. The Short Aircraft Company in England even experimented with a composite aircraft, consisting of a large flying boat carrying a smaller floatplane piggyback. The smaller plane was to be launched in midair over the Atlantic to complete the crossing. A small, extremely fast, high-altitude plane with long range—precisely the desiderata Whittle had first explored in 1928—was thus not altogether a commercial implausibility.

The first Whittle engine was designed for an assumed speed of 500 mph at 70,000 feet. It was to produce a usable thrust of almost 1,400 pounds. It had a double-sided centrifugal compressor 19 inches in diameter with a designed efficiency of 80 percent at a pressure ratio of $4:1$. The double-sided impeller was used to get adequate air-flow volume without a large diameter. No diffuser blades were used, reliance being placed in the "vortex space" diffuser in accordance with Whittle's conclusions in the 1931 paper on centrifugal superchargers. On the first engine, a single combustion chamber fed into an axial-flow turbine through a nozzleless scroll turbine inlet. The turbine was only 16 inches in diameter, and from it Whittle had to get the equivalent of just over 3,000 h.p. to run the compressor—more than the net power then produced by any piston engine.

For the first engine, Whittle felt well qualified through both his scholastic and his personal studies to design the compressor and the turbine, but he was rather at a loss in regard to the third major component, the combustion system. To get help with his combustion chamber, Whittle went to the British Industries Fair at Castle Bromwich in Feb-

Original Whittle W.U. turbojet. (From General Electric, *Aircraft Gas Turbine Engineering Conference* [West Lynn, Mass.: G.E., 1945].)

ruary 1936. He visited the displays of several firms which specialized in combustion equipment and, without revealing anything of his turbojet, outlined his combustion problems. He wrote later: "For the most part I met with blank astonishment and was told that I was asking for a combustion intensity at least twenty times greater than had ever before been achieved."[20] Finally, Whittle persuaded Laidlaw, Drew and Company to aid in the development of the combustion system, and experiments were begun at the B.T.-H. shop at Rugby. The combustion problem would continue to plague Whittle into early 1940, cause him a delay of as much as eighteen months, and bring him to say, "I cannot possibly give an adequate picture of that heartbreaking period."[21]

Because Power Jets could not afford the massive plant necessary to test each component of the engine separately, it was forced to build the complete engine and then test it as a whole. Such a procedure promised quick and fairly cheap results, but would make the isolation of specific deficiencies almost impossible. The engine was first run on 12 April, 1937, and testing continued at Rugby until August of 1937.

Whittle, meanwhile, was completing his tenure at Cambridge. His final examinations came in the spring of 1936, and normally he would have been posted back to a service assignment. But he managed to wrangle from the Air Ministry a graduate year at Cambridge for "research," which would consist mainly of work on the engine. When the graduate year was up in 1937, the Air Ministry placed Whittle on the special duty list and assigned him to Power Jets. Thus, although the Air

Ministry did initially refuse any direct support for the turbojet project, it did provide indirectly the most precious of all resources, Frank Whittle.

In March 1936, Whittle had met Henry Tizard at a Cambridge Air Squadron Dinner. Tizard, chairman of the Engine Sub-Committee of the Aeronautical Research Council (ARC), was probably the most influential scientist in the Air Ministry. Whittle met with Tizard several times later that year. Tizard had supported Griffith's work in 1926-30; and in late 1936, largely at Tizard's urging, the Engine Sub-Committee reopened the whole issue of gas turbine power plants. One result of the Engine Sub-Committee's deliberations was authorization for the Royal Aircraft Establishment (RAE) to pick up the gas turbine investigation abandoned by Griffith in 1930. The committee also asked Griffith to write an evaluation of the Whittle engine, which Griffith did and submitted in February 1937.

Griffith was known for his personal reserve as well as for his sense of engineering elegance,[22] but his evaluation of Whittle's ambitious and immodest design seems positively icy. After suggesting that Whittle was overly optimistic about possible centrifugal compressor efficiencies (Griffith felt 65 percent more likely than Whittle's estimated 80 percent), he proceeded to compare the expected performance of the Whittle engine with that of a conventional piston engine and propeller. Griffith concluded:

> In its present form, the proposed jet propulsion system cannot compete with a conventional power plant in any case where economical flight is demanded (e.g., the transport of the maximum percentage of useful load over a given distance). It is of value only for special purposes, such as the attainment of high speed or high altitude for a short time, in cases where take-off requirements are not stringent.
>
> In order that the proposed system may become a competitor in the field of economical flight, a large improvement, of the order of at least 50-100 percent, must be made in the ratio of take-off thrust to power plant weight.[23]

Quite unconciously, Griffith seems to have adopted then current Air Ministry tactical doctrine as engineering principle. The increased speed and altitude capabilities of conventionally powered aircraft apparent during the early 1930s were thought to have made interception by defending aircraft and therefore the air defense of Great Britian impossible: in Stanley Baldwin's grim 1932 phrase, "The bomber will always get through."[24] The only plausible defense was deterence, the ability to bomb back, which translated into a requirement for aircraft with good load and range capabilities. Ironically, Griffith was submitting his report to a subcommittee chaired by Tizard, who since 1934 had also been intimately involved in England's ultra-secret development of radar, which just then, in the spring of 1937, was proving its value in bomber interception in what came to be known as the Biggin Hill experiments.[25] For Tizard, perhaps more than for any other man in England, the proposition of a very fast, high altitude fighter with a rapid rate of climb, even at the expense of sharply limited load and endurance, was not absurd.

The net result of the engine sub-committee's deliberations was a small Air Ministry contract to Power Jets for information on the first series of engine tests and minor support for a second series. The total value of the first contract was to turn out to be only £1,900.[26] Although receiving sporadic but somewhat larger Ministry support, Power Jets would remain in financial straits until the fall of 1938, and in fact very nearly went under during the early summer of 1938.

Power Jets moved from Rugby in August 1938, at the request of B.T.-H., because Power Jets' rather volatile activities tended to disrupt plant production. (Plant personnel generally thought the Power Jets people to be working on a new and highly dangerous type of flamethrower.) The Power Jets team therefore moved to an abandoned foundry at Ladywood, leased from B.T.-H. There, in April of 1938, testing began again on the engine, which had been considerably modified. The rebuilt engine employed a vaned diffuser with separate passages, a redesigned combustion chamber, a new turbine inlet, and a redesigned turbine. The new turbine was of free-vortex design. The blades were constructed with "twist" to compensate for differential radial velocity and pressure along the diameter of the blade. Whittle had independently arrived at the free-vortex blade design before 1936, as had Griffith at the RAE in 1929. Griffith's analysis of the new blades, however, was contained in his still-secret 1929 paper on turboprops. B.T.-H. had built the turbine of the first engine according to steam turbine practice, and Whittle had to argue quite vigorously to convince them that the new blades offered better efficiency. Eventually, B.T.-H. even adopted free-vortex blading for its steam turbines. The higher turbine efficiency which derived from the superior blade design of the British turbojets accounted in large measure for the lower fuel consumption enjoyed later by the British engines over their German counterparts, which used turbines of conventional steam turbine design throughout the war. Testing of the rebuilt Whittle engine continued until 5 June 1938, when the turbine disintegrated and extensively damaged the engine.

The same engine was once more reconstructed, not because of any great faith in it but because Power Jets could not afford completely new components. With the third reconstruction, the Whittle engine took on the basic form it would keep throughout the war. In order to use as many of the older components as possible, the new version used ten reverse-flow "can" combustion chambers. The same compressor impeller, slightly modified, was used, along with basically the same diffuser; the combustion chambers were new, as was the turbine inlet and turbine. Testing of the third version began in October 1938 and continued until February 1941, when it too was finally destroyed by a turbine failure.[27]

The Air Ministry had agreed to finance the third reconstruction of the engine, and received continuous reports on progress. Tizard and Hayne Constant, who was running the RAE turboprop program, evinced mounting confidence in Whittle's engine. On 30 June, 1939, David R. Pye, director of scientific research at the Air Ministry, visited Ladywood and witnessed a run-up of the engine to 16,000 rpm. Pye, who had been vociferously skeptical, was now convinced that Whittle "had the basis of an aero-engine," and indeed became so enthusiastic that Whittle later recalled,

> Later that day, when I drove Pye to Rugby station, I had the curious experience of having him recite to me all the advantages of the engine. His manner of doing so was almost as though he were trying to convert a sceptic. I was tactful enough not to point out that he was preaching to the first of all converts.[28]

Two weeks later Power Jets received a contract for a flight engine, the W.1., Gloster Aircraft received a contract for an experimental airframe, specification E.28/39. By mutual consent of the Air Ministry and the RAE group, the RAE turboprop project was redirected toward development of a turbojet shortly after war broke out in September 1939. At the end of January 1940, Tizard and Air Vice-Marshal Tedder came to Power Jets to witness a demonstration of the test engine. Afterward, Tizard's comment was simply: "A demonstration which does not break down in my presence is a production job."[29] Gloster got a contract for a service interceptor, the E.9/40. Because no single turbojet then produced enough thrust to power a fully armed fighter carrying enough fuel to have a useful endurance, the E.9/40 was to have two wing-mounted engines.

At least fifteen months before the first turbojet aircraft in England would fly, the turbojet revolution had occurred. Probably the most accurate date for the revolution in England would be September 1939, for it was then that the turbojet ceased to be regarded as a matter of long-term research, and it was then that not only the Power Jets program but also the RAE project was committed to rapid service development of the turbojet.

Commitment to the revolutionary device that was the turbojet in no way meant that all problems were solved or even that solutions to all problems were in sight. As late as the end of 1940, the Power Jets engines were still highly temperamental. Until December of that year,

Power Jets had only one test engine, the third reconstruction of the original engine. Total test hours were still fewer than 100. Total expenses were about £50,000, half of which had come from the Air Ministry. Neither the combustion problem nor the problem of turbine blade failure were solved. In the fall of 1940, I. Lubbock of the Asiatic Petroleum Company (Shell) and a team working with him, who had advised Power Jets before, designed and tested a combustion chamber of their own that ultimately eliminated the combustion system problem. Not until the beginning of 1942, however, did Power Jets adopt a very expensive new nickel-chromium alloy, called Nimonic 80, for making turbine blades that substantially solved the problem of turbine blade failure.

The first test flight of the Gloster E.28/39 took place on 15 May, 1941. The E.28/39 was an all-metal midwing monoplane. It had a circular-section semimonocoque fuselage with an annular air intake and a retractable tricycle landing gear. The W.1 engine which powered the E.28/39 in its test program developed a maximum of 1,000 pounds thrust at 17,000 rpm. It weighed only 623 pounds installed and had a specific fuel consumption (sfc, pounds of fuel per pound of thrust per hour) of 1.4. The ultimate maximum test speed of the E.28/39 was 370 mph at 25,000 feet, which was superior to that of the contemporary Spitfire mark. The most serious problem to manifest itself in tests of the E.28/39 was frictional losses in the long intake and exhaust ducts. By May 1941, Power Jets still had only two engines with which to experiment, but by the end of 1941 these engines had accumulated nearly 500 hours of running time. Great as progress was, many long and tedious months of development still lay ahead. The whole problem of mass prodcution on a scale suitable for service use had yet to be solved.

In April 1940 the Air Ministry had chosen the Rover Company, manufacturer of motorcars, to undertake quantity production of Whittle engines under subcontract to Power Jets. Both personal and engineering design differences plagued the Power Jets–Rover effort, however, and the lack of cooperation caused not only delays in delivery of production engines, but also created a great deal of dissension and recrimination between Power Jets, Rover, and the Ministry of Aircraft Production (MAP), which had replaced the Air Ministry proper in control of aircraft and aero-engine production in June 1940. As a result, in April of 1943, Rolls-Royce took over production of the Whittle engines from Rover. The W.2.B (from Rover) and the W.2/500 (the development mark then current at Power Jets) became the Royce Welland I that went into service in the Gloster Meteor I (as the E.9/40 came to be called) in July 1944. The Welland I initially produced 1,600 pounds of thrust at 16,750 rpm for a complete installed weight of 860 pounds, a diameter of 43 inches, and a specific fuel consumption of 1.12. The Welland I had a rated service life between overhauls of 180 hours.

Whittle's work was naturally held in the utmost secrecy throughout the early war years. Although his patents, both of 1930 and 1935, were freely published, the Air Ministry was rather perturbed when complete reproductions of them appeared in an article on reaction propulsion in the German magazine *Flugsport* in early 1939. The omniscient *New York Times* reported some details of Whittle's work on 20 July, 1939, but erroneously described his engine as "rocket propulsion" (the distinction between a true rocket and the turbojet had not yet been made in the United States) and also wrongly reported that flights had already occurred. Some cooperation between Power Jets and the RAE took place early on, but it was haphazard. To reduce duplication and to speed progress, H. Roxbee Cox of the MAP created a Gas Turbine Collaboration Committee, upon which were represented all British organizations engaged in turbojet work, in October 1941. Whittle's engine was also revealed to representatives of the United States Army Air Corps in July 1941.

Secrecy had its usual odd effects. Whittle attended a lecture on high-speed flight and compressibility given at the Royal Aeronautical Society in November 1937 by C.N.H. Lock, the outstanding ballistician from the National Physical Laboratory (NPL). In the discussion period, Whittle asked about the effects of compressibility on a streamlined cylinder having a "kind of Venturi hole in the center." Lock replied that the thin cylinder would "presumably have no drag at all, but he did not quite see where the pilot was to be placed or how such a construction was to be of service."[30]

Frank Whittle individually and independently deduced assumptions about aircraft speed and altitude performance and about possible gas turbine component efficiencies that could and did make the turbojet successful. Engines similar in basic form to the original flight engine of 1941, which was in turn fundamentally the same as the third reconstruction of the first engine of 1938, would continue in service use until well into the 1950s. Whittle created a turbojet and an aeronautical revolution.

The work of Hans von Ohain in Germany was completely separate from and done in complete ignorance of any of the projects in England. Yet in many respects von Ohain's undertaking remarkably parallels Whittle's. Von Ohain was the son of a career Army officer who entered business after the 1918 debacle. The younger von Ohain received the usual upper-class "Humanistic Gymnasium" secondary education, which emphasized classical Greek and Latin as well as mathematics and physics.

Von Ohain's first, privately funded, demonstration turbojet, Bartels and Becker's garage, Goettingen, 1935. (From Hans von Ohain, "The Evolution and Future of Aeropropulsion Systems," *The Jet Age* [Washington, D.C.: National Air and Space Museum, 1979].)

He went on to Göttingen University in the early 1930s, where he stayed to pursue doctoral study in general physics. Von Ohain received his doctorate in physics in early 1936.[31] His principal professor, to whom he was a teaching and research assistant, was Robert W. Pohl, head of the Science Institute and a physicist noted for his ability to make clear basic concepts.[32] At Göttingen, von Ohain took courses in aerodynamics but did not belong to the Aerodynamische Versuchanstalt (AVA).[33]

Von Ohain's interest in jet propulsion derived directly from assumptions about aircraft performance deduced from aerodynamic insight. On the basis of what was possible in terms of airframe streamlining and structures, von Ohain felt that speeds much higher than 300 mph ought to be attainable. He noticed that aircraft performance had risen significantly each year until the early 1930s, when progress appeared to slow. Since he knew from aerodynamics that airframes were capable of much better performance, he reasoned that the lag in actual performance must be due to the piston engine's and propeller's incapacity for further radical improvement. Von Ohain therefore sought "new and more elegant methods for aircraft propulsion."[34]

Von Ohain was convinced that some "continuous aerothermodynamic propulsion process could be inherently more powerful, smoother,

The Turbojet Revolution

lighter, and more compatible with the aerovehicle than a propeller-piston engine."[35] Von Ohain recalled:

> In the Fall of 1933, my thoughts began to focus on a steady aerothermodynamic flow process in which the energy for compressing the fresh air would be extracted from the expanding exhaust gas. Such a steady flow process promised a far greater air volume handling capability than that of a reciprocating engine and consequently a much greater power concentration and power-to-weight ratio. Also, the air ducted into such a system could be decelerated prior to reaching any Mach number-sensitive engine component. Both of these characteristics are of greatest significance for a high-speed propulsion system.[36]

From the outset of his inquiry, then, von Ohain wanted to attain high Mach-number speeds by replacing both the propeller and the conventional reciprocating engine.

Initially, von Ohain considered an ejector system similar to that of Melot, but after careful analysis concluded that such devices were likely to be inherently inefficient. Instead, he turned to a turbojet:

> Searching for an extremely lightweight, compact and simple configuration having a minimum development rise, I chose a radial outflow compressor rotor back-to-back with a radial inflow turbine rotor. This configuration also promised correct matching simply by providing equal outer diameters for the straight radial outflow compressor rotor and the straight radial inflow turbine rotor. I was aware of the possibility of employing axial flow compressors and turbines, and I considered an axial flow configuration as very desirable for future developments from a standpoint of small frontal area, but as too complex and expensive for the beginning. In particular, stage matching of a multistage axial flow compressor and matching of axial flow compressor and turbine without component test facilities appeared to me too risky.[37]

Von Ohain's reasoning should not be mistaken for a perception of functional failure in the piston engine and propeller system. Piston-engined, propeller-driven aircraft were performing badly only in comparison to a theoretical ideal specified by the insights of aerodynamics. Hence, von Ohain perceived presumptive anomaly, not functional failure. The aesthetic element in his quest for a new system is likewise revealing. Von Ohain had no knowledge of either Whittle's or Guillaume's patents, but he was fully aware of Betz and Encke's work on axial compressors at the AVA.[38] From assumptions derived from theoretical aerodynamics, from the speed records, and from the work of Betz and Encke, von Ohain too arrived at the turbojet conclusion.

Von Ohain undertook preliminary design studies for his turbojet during 1934; he assumed a compression ratio of 3:1, a turbine inlet temperature of 1200 to 1400°F, and an aircraft speed of 500 mph. According to his calculations, the greater fuel consumption of the turbojet would be compensated by its much lighter weight compared to a piston engine and propeller and would make it especially suitable for high-performance aircraft such as fighters.[39]

Von Ohain began patent proceedings and, at the end of 1934, resolved to have a small demonstration engine built at his own expense. He was well acquainted with the head mechanic and machinist, Max Hahn, in an automobile repair shop (Bartels and Becker) in Göttingen, and during 1935 induced him to undertake fabrication of the turbojet engine. Total cost came to just over 1000 marks, or the equivalent of perhaps $10,000. Although tests during 1935 showed the initial model promising, the liquid gasoline combustors did not function properly and the engine never really ran. Von Ohain recognized that full scale development was beyond his personal means and that he would have to seek external support.

Von Ohain managed to convince Professor Pohl of the potential of his turbojet, and Pohl agreed to write him a letter of recommendation to any company he chose. Von Ohain correctly sensed that conventional aero-engine manufacturers would likely be averse to such a radical innovation, and decided to approach Ernst Heinkel, whom he knew to be interested in ultra-high-performance aircraft. Pohl, it turned out, knew Heinkel, and in February of 1936 wrote him describing von Ohain's ideas.

Heinkel was indeed a particularly apt person to approach. He had built aircraft for the German Air Force during the First World War and in secret after Versailles. He had attended the 1927 Schneider Trophy contest in Venice and reported: "Face to face with those beautiful fast machines, I was seized with an intense desire to build similar ones myself and perhaps design the fastest of them all."[40] He became "intoxicated" with speed and plotted how he could build a faster plane: "I saw, of course, that the floats must disappear in order to reduce wind resistance. I had to build fast land planes with a moderate landing speed, the least possible wind resistance, and the smoothest possible surfaces."[41] It was the end of 1932 before Heinkel's dream began to be fulfilled. By then he had built his own wind tunnel and begun his own experimental program. At the end of 1932, he produced for Lufthansa a beautiful four-passenger mailplane, the He 70. The He 70 was an extremely clean monoplane with retractable landing gear, and when it went into service with Lufthansa in 1933, its top speed of 230 mph made it faster than any service military aircraft in the world. It naturally caused something of a stir, and was so good, in fact, that Rolls-Royce purchased one to serve as a flying test bed for their new Kestrel V engine, the forerunner of the Merlins of war fame.

The Turbojet Revolution

Things did not go quite as well for Heinkel after the He 70. His fighter design based on He 70, the He 112, lost the 1935 competition for selecting a new Luftwaffe interceptor to a plane manufactured by a small Bavarian firm which theretofore had specialized in sport planes and gliders: the winning design was Willy Messerschmitt's Bf 109. The loss to Messerschmitt seems to have piqued Heinkel's vanity, and he became more determined than ever to build the world's fastest airplanes. In 1935 Heinkel was working on an article entitled "An Inquiry into Engine Development" in which he argued that 500 mph was the maximum speed at which a propeller-driven aircraft could be expected to fly. Heinkel was also aware, as were most highly placed designers in Germany, of the great inferiority of German high-performance piston engines which had resulted from the fifteen-year development hiatus imposed by the Versailles Treaty.

In November of 1935, Heinkel met Wernher von Braun, one of a small group of enthusiasts who had formed the core of the Verein für Raumschiffahrt (VfR, the German Rocket Society). Von Braun told Heinkel that one of his rocket motors should be able to propel an aircraft. Heinkel gave von Braun an He 112 fuselage with which to experiment, financial aid, and technical personnel. After blowing up two test fuselages, von Braun and his group finally flew an He 112 on rocket power alone in 1937. Later, a more highly developed Walther motor, which had been developed with Air Ministry support for assisting the takeoff of conventional aircraft, was substituted for the von Braun engine. The program culminated in construction of the Walther-powered He 176, the world's first pure rocket plane. Von Braun, of course, moved on to the V-2 project at Peenemunde. The point is this: when Heinkel received Pohl's letter, he was already favorably disposed, both intellectually and personally, toward radical, high-performance propulsion systems that dispensed with the propeller.

In response to Pohl's letter, Heinkel invited von Ohain to come to see him in March 1936. Heinkel also arranged a meeting for von Ohain with his top engineers who, although disturbed by the turbojet's projected high fuel consumption, did acknowledge its potential for high-speed propulsion. Heinkel's two superb airframe designers, Siegfried and Walter Guenther, stressed the importance of high-thrust per-unit-frontal-area, but were equally cognizant of the need to eliminate the propeller in very high speed aircraft.[42] As a result of the meetings, Heinkel hired von Ohain, then only twenty-four, and Max Hahn, and set them up in a small shop at the Heinkel works at Marienehe. Work began on a new demonstration engine in April 1936, just a month after Whittle began work on his engine, and the Heinkel-von Ohain engine first ran in late February or early March 1937, about the same time the Whittle engine first

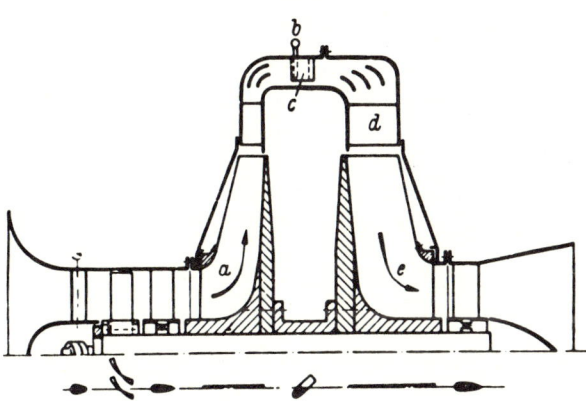

Von Ohain's hydrogen-fueled He S-1 demonstration turbojet for Heinkel, 1936. (From Hans von Ohain, "The Evolution and Future of Aeropropulsion Systems," *The Jet Age* [Washington D.C.: National Air and Space Museum, 1979].)

ran. The new von Ohain engine employed a single-sided centrifugal compressor, a single very simple combustion chamber, and a radial inflow turbine. It was made almost entirely from mild sheet steel fabricated in the Marienehe shop, and in order to avoid the combustion problems encountered in 1935, hydrogen was used as a fuel. The use of hydrogen fuel in the test engine proved to be a master stroke. Hydrogen burns easily and, being a gas at normal pressures, presented no problems of atomization or vaporization. As a result, other engine components could be developed at the same time that the combustion problem for normal fuels was attacked. Both von Ohain and Whittle suffered the constraint of having no adequate test facilities for components and thus had to develop their engines whole. Von Ohain's choice of hydrogen as an experimental fuel largely accounted for the nearly two-year development lead he had built up by late 1939.[43]

Von Ohain's crude demonstration engine produced 550 pounds thrust, and Heinkel, impatient and eager, immediately authorized construction of a flight engine and of an experimental airframe, the He 178. The total cost of the first engine was only about $50,000. In order to save time and avoid the inherent difficulties bound to be encountered in any completely new design, von Ohain kept the first flight engine essentially similar to the demonstration unit: centrifugal compressor (but preceded by an axial intake fan), single combustion chamber, and radial inflow turbine. Von Ohain, like Whittle, continued to be troubled by combustion problems and visited an industrial fair at Leipzig to try to secure the aid of a burner-maker. Unlike Whittle, von Ohain could find no one willing to help. Von Ohain, Max Hahn, and Heinkel engineer Wilhelm Gundermann had to tackle the problem alone, but by early 1938 they had evolved a satisfactory gasoline combustor configuration. The first flight engine, tested in 1938, did not, however, produce the designed thrust (between 1,760 and 1,980 pounds) because of insufficient compressor and combustor size. The engine was therefore redesigned with larger and slightly different components as the HeS-3B.[44]

After some test flights in a conventionally powered He 118, the HeS-3B engine first flew in the He 178 on 27 August, 1939, four days before the outbreak of the Second World War. It was the world's first flight on turbojet power alone. The HeS-3B gave 1,100 pounds static thrust at 13,000 rpm for 795 pounds weight and a specific fuel consumption of 1.6, which compared without embarrassment to the W.1 of eighteen months later. The He 178 was a very small high-wing monoplane with a metal semimonocoque fuselage incorporating an annular air intake. It had plywood wings and a retractable tailwheel type undercarriage. The He 178 attained a maximum speed of only 250 mph, and the engine's combustion chamber overheated severely, but the practicability of turbojet-powered flight was irrefutably proven. Later, the HeS-6, a modified HeS-3B with thrust raised to 1,300 pounds of weight to 925 pounds, was installed in the He 178, but because of airframe deficiencies the speed of the plane remained low. Heinkel decided to abandon the He 178 entirely and to proceed to development of a prototype service fighter, what was to become the He 280. Because of the large losses experienced in the intake and exhaust ducting of the He 178 (as in the E.28/39), and because no single turbojet was sufficiently powerful, the new Heinkel plane was designed with two turbojets mounted in pods under the wings.

Heinkel had kept his whole program secret even from the German Air Ministry (RLM, the Reichsluftfahrt Ministerium) until early 1938. He accepted no government support for development until late 1939, largely by his own choice, since he wanted to maintain unfettered control over his own enterprise. Even after 1939, Heinkel did much independently. The RLM began to discourage long-term development in September 1939, and in mid-1940 placed a flat prohibition on all research or development projects that could not be ready for production within a year, by which time the war was supposed to be won. The He 280 too was therefore designed and built in secret from the RLM.

The He 280 was designed by Robert Lusser, Heinkel's technical director, who had been Messerschmitt's chief of development until 1939. The He 280 was an all-metal low-wing monoplane. It was the first German aircraft to have a tricycle undercart and the first aircraft anywhere to have an emergency ejection seat for the pilot. The He 280 was powered by two von Ohain HeS-8 turbojets. The HeS-8 was a progressive development of the HeS-6, with a similar axial-fan intake blower followed by a centrifugal compressor, a new, smaller diameter, straight-through annular combustion chamber, and a radial inflow turbine. It developed 1,100 pounds thrust (1,300 pounds in the HeS-8A) at 13,000 rpm. The HeS-8 weighed 836 pounds and had a diameter of only 30 ½ inches. The He 280 made its first pure jet flight, with HeS-6's, on 5 April, 1941— forty days before the experimental E.28/39 would fly. The HeS-8, which later received the RLM designation 001., marked the culmination of von Ohain's independent turbojet development. The next engines he designed would be a part of a larger RLM program.

Von Ohain, then, was the second man to design a scientifically engineered turbojet. Like Whittle, he deduced from aerodynamics the critical assumptions: first that aircraft could fly faster, fast enough for jet propulsion to be efficient, if the propeller were eliminated; second, that the propeller had to be abandoned if speeds approaching that of sound were to be attained; and third, that component efficiencies within a turbojet could be raised to acceptable levels. Von Ohain progressed more rapidly than Whittle, especially in terms of actual flights, principally because he had chosen hydrogen as an experimental fuel and because he had at his disposal the tremendous (compared to Power Jets) financial resources of Heinkel Aircraft and, usually, the unrestrained support of Heinkel himself. Furthermore, as a purely private project, the first von Ohain engine was subjected to none of the rigorous governmental type or certification tests which the Whittle engine had to pass before being cleared for its first test flight.

The third man to initiate a turbojet program was Herbert Wagner, head of airframe development at the Junkers Aircraft Company. Wagner received his secondary education at the Austrian Imperial and Royal Naval Academy during 1914-17, before serving as a junior ensign in the Austrian Navy in 1917-18.[45] In the library of the officers' club of the Austro-Hungarian Navy, Wagner read a technical journal article about steam turbines which so impressed him that he resolved to leave the navy and become an engineer. Following the political collapse of Austria-Hungary in 1918, Wagner studied mechanical engineering at the Technical University of Graz (Austria). He then went to the Technical University of Berlin to study naval engineering, receiving his master's degree in 1922. He continued work at Berlin during 1923-24, serving as teaching assistant to the professor of steam turbines and propellers in the department of mechanical engineering. Wagner received his doctorate from the Technical University of Berlin in 1924, the title of his thesis being "On the Arising of Lift on Aerofoils."

From 1924 to 1927, Wagner was a designer for the Rohrbach Metal

Aeroplane Company, which pioneered the development of all-metal flying boats. Wagner became dissatisfied with the latticework framing system used to support the hull skins of Rohrbach's aircraft. As a result, he undertook a theoretical examination of stress-carrying thin metal sheets, which led him, in 1925, to the theory of the diagonal-tension field beam to describe the behavior of stressed-skin structures.[46] As noted in Chapter 5, Wagner's beam theory provided the analytical foundation for the revolution in aircraft structures during the 1930s. From 1927 to 1930, Wagner was a professor of aeronautics at the Technical University of Danzig, and from 1930 to 1938 at the Technical University of Berlin. He was responsible for new design approaches to cantilevered wings and did other valuable work in aircraft structural stress analysis.[47] Significantly, during the 1920s and 1930s, the Deutsche Versuchsanstalt für Luftfortforschung (DVL)—the official German aerodynamic institute, analogous to the U.S. NACA—and the Technical University of Berlin were virtual sister institutions, with some students and professors doing work in both.[48] Wagner was on leave from Berlin after 1935 to work for Junkers Aircraft.

Wagner thus had three critical elements in his background by 1936: he was familiar with turbine machinery; he knew intimately, indeed contributed greatly to, what the new streamlined monoplanes could achieve in performance; and he was fully versed in the latest aerodynamics. Wagner was interested in high-altitude flight, and later had a special high-altitude test plane built at Junkers. In common with Whittle and von Ohain, Wagner concluded that airframes were capable of speeds beyond those conventional engines could produce.

Wagner first considered a turboprop in late 1934, while still a professor at Berlin. He was familiar with early French gas turbine experiments but with nothing later. He theoretically investigated "whether the additional reaction of the exhaust gases might make a gas turbine an acceptable engine for driving the propeller of an aircraft."[49] Wagner, then, was the first person investigating a turboprop to realize that the energy not extracted by the turbine to turn the compressor or the propeller could be at least partially reclaimed in the form of residual thrust. The greater the aircraft speed the more propulsively efficient that residual thrust would be. After making "quite elaborate" design drawings of a prospective engine, Wagner concluded that a gas turbine could be very promising if materials to withstand blade temperatures of 850° C could be found. Wagner considered only axial-flow components and was aware of the benefits of ram-effect on compressor efficiency. At that time, Wagner knew neither of Betz's and Encke's axial compressor investigations nor of any other turbojet or turboprop project. He did not discover that others were interested in the same problems until he ordered a patent search after beginning his own project at Junkers.

When Wagner joined Junkers Aircraft at Dessau in 1935, his instructions from Heinrich Koppenberg, chairman of the board and president of both the Junkers Aircraft and the Junkers Engine companies, were "to lead, in aircraft development, the company in a few years to the peak position."[50] At the time Junkers Aircraft and Junkers Engine were separate corporate entities, sharing only a common chief executive and board of directors; although the two firms were merged in July 1936, the aircraft and engine divisions continued to operate independently. Wagner felt, with some justification, that Otto Mader, head of engine development at Junkers Engine, was too conservative in his approach to engine design. Mader, having been singed once working with Hugo Junkers on the free-piston turbine engine, and with a tremendous amount of work to do to get the Junkers Jumo piston engines at all competitive with foreign designs, was perhaps understandably reluctant to diffuse his efforts. Wagner's low opinion of Mader's willingness to innovate, however, was apparently shared by Koppenberg, and in late 1935 Wagner got permission to conduct an inquiry into unorthodox propulsion systems, especially the turboprop, at Junkers Aircraft without the interference of Junkers Engine. Koppenberg provided funds as well as a building at the engine factory in Magdeburg.[51]

To run the operation in Magdeburg, Wagner brought his assistant from the Berlin laboratory, Max Adolf Müller, who had fulfilled all requirements for his doctorate except mathematics. Although Wagner himself thought the turboprop most promising, Müller was encouraged to analyze a variety of systems: turboprop, turbojet, and diesel-driven ducted fan. Wagner, having studied the residual thrust from the turboprop, apparently did not attach great novelty to the turbojet. He seems instead to have regarded it as one end of a continuum, from pure jet thrust to maximum shaft output, for the use of the power developed by a gas turbine. By early 1937, however, systematic studies done jointly by Wagner and Müller "showed that, without propeller, a very light engine would result, and that an acceptable efficiency could be attained with gas temperatures that were easily low enough for then available blade materials." Wagner had had "intensive studies made of fast aircraft" as part of his airframe development program at Junkers, and had "become aware of the problems of high-speed propellers."[52]

Wagner had become head of all aircraft development at Junkers Aircraft in 1937, and by mid-1937, Wagner and Müller had definitely settled on the turbojet as "the shortest path to high aircraft speeds,"[53] although investigations of the other systems continued less actively. The first Junkers turbojet was built at Magdeburg under Müller's direction. It had a five-stage axial compressor, a single annular combustion chamber, and a two-stage turbine. The axial compressor was apparently chosen because it offered straight-through airflow. The compressor itself—a remarkable design by a Junkers aerodynamicist, Rudolph Freidrich that owed nothing to the work of Betz and Encke—managed to get a pressure ratio of 3 : 1 out of only five stages by using 50 percent reaction blading.[54] The whole engine was only 24.3 inches in diameter. The Wagner-Müller engine was under bench test before the end of 1938, and the Junkers people were successful in solving the combustion problem. Yet, the highly loaded if efficient compressor made the whole engine temperamental and unable to run under its own power as late as May 1939. The Junkers group encountered the severe component-matching problems foreseen by von Ohain. In order to get materials priority, Wagner had revealed the turbojet and other projects to the RLM in mid-1938, but the programs were supported exclusively from Junkers Aircraft Division funds until May 1939. At that time, the RLM intervened, and the Junkers projects became, after some dispute, a part of the larger German turbojet program.

Once again, a man of extraordinary aerodynamic insight, a man reputed to be one of the most brilliant aeronautical engineers in Germany,[55] had deduced from aerodynamic theory the radical performance assumptions that led to the turbojet. No later than mid-1937, Wagner and Müller, both fully versed in the latest theoretical aerodynamics, had become, like Whittle and von Ohain before them, turbojet practitioners solely on the basis of presumptive anomaly deduced from science.

The fourth man to reach the turbojet conclusion was Helmut Schelp. Schelp also was a German, but had earned his master's degree at the Stevens Institute of Technology, Hoboken, New Jersey, in 1936. He returned to Germany and joined the RLM, where he participated in a new RLM program that led to the title of "Flugbaumeister." The object of the program was to create engineers "who would be neither highly specialized theorists nor practical handbook engineers, but men broadly trained in all aspects of aviation and aeronautical engineering, and who it was expected would soon make their way into responsible positions in industry and government."[56] Schelp's course of study entailed spending six months assigned to Diamler-Benz and six months at the Deutsche Versuchsanstalt für Luftfahrt (DVL), Berlin-Adlershof.

DVL research prior to 1937 had indicated that 480 mph was the limit for propeller-driven aircraft.[57] As mentioned, the He 70 had appeared in 1933, and in July 1937 the Bf 109 exhibited its then-stunning performance at the Zurich International Flying Meeting. Both aircraft gave exceptional performances with unexceptional engines. The question therefore arose within the DVL of what, in view of the new airframes, the limits to aircraft performance really were. It was this problem which Schelp was given as a part of his DVL study. Schelp's analysis indicated that the limit would probably be set by compressibility, which, according to "existing wind-tunnel data," became obtrusive about 0.82 Mach, or just over 600 mph. The question then became how to bridge the gap between the speed at which the propeller became prohibitively inefficient and the compressibility limit to airframe speed.[58]

Schelp, uniquely among turbojet pioneers, was thoroughly familiar with early work on gas turbines, especially that of Armengaud and Lemale, and also with the work on various reaction propulsion systems during the First World War. He realized that for flight at high Mach numbers, two things were necessary: a power plant with a significantly lower power-to-weight ratio than current engines to overcome increased high speed drag, and some means of dispensing with the propeller. He concluded that the system most likely to offer the desired characteristics would be some form of reaction propulsion not requiring a reciprocating engine. Schelp therefore carried out a systematic analysis of all reaction propulsion engine types, including ram-jets and pulse-jets, finally concluding that a turbojet using an axial-flow compressor (because of its small diameter and therefore low drag) offered the best solution.[59] Thus by mid-1937, Schelp, quite independently of Whittle, von Ohain, or the Wagner group, also deduced the turbojet conclusion from aerodynamic first principles.

In August 1937 Schelp was assigned to (and virtually constituted) a special jet propulsion office within the Research Division of the Technical Office of the RLM. Schelp's section supervised work on the Schmidt pulse-jet (which would power the V-1) and a bifuel rocket, RLM support for which had begun in the early 1930s. Schelp, however, could not get Research Division funds for turbojet research. In August 1938 Schelp by chance met Hans A. Mauch, who in April 1938 had come to head the Sondertriebwerke (Special Propulsion Systems) section under the Power Plant Group of the Development Division of the Technical Office. Mauch had charge of RLM-sponsored rocket development, specifically the Walther rocket engines. He had accidentally heard of Heinkel's turbojet project and had visited the Heinkel works and seen the He 178; he also had heard of Wagner's engine and of the projects under Schelp's control. Schelp explained the whole theoretical rationale of the turbojet to Mauch. Impressed, Mauch persuaded Schelp to leave the Research Division and come to work for him. About the same time, Schelp learned of the highly promising work of Betz and Encke on axial compressors.

The Turbojet Revolution

The Heinkel He 178, the world's first aircraft to fly purely on turbojet power, 1939. (Courtesy of the Smithsonian Institution.)

Although Mauch and Schelp felt that the time was ripe for turbojet development, they also felt that only experienced aero-engine manufacturers could develop and produce a practical engine. Therefore, they refused to support either the Heinkel or the Junkers Aircraft projects. In the fall of 1938, Mauch and Schelp visited all four major German aero-engine manufacturers—Diamler-Benz, Junkers Motors, B.M.W., and Bramo (which would merge with B.M.W. in 1939)—to interest them in turbojet development. Initially, Fritz Nallinger, head of development at Diamler-Benz, refused outright to undertake work on reaction propulsion, a not unreasonable response, considering Diamler-Benz's piston engine commitments. Mader at Junkers Engines accepted a study contract, but unenthusiastically. Helmut Sachse at B.M.W., with that company's background in turbosuperchargers, accepted more willingly, and gave the project to Kurt Lohrner, his head of research. Bramo, which feared withdrawal of government support for its piston engines, also accepted a reaction propulsion study contract, and placed the task in the hands of its research director, Hermann Oestrich. Thus, by early 1939, in addition to the projects at Heinkel and Junkers Aircraft, every major German aero-engine manufacturer except Diamler-Benz was engaged in some sort of reaction propulsion program.

The Germans had not neglected airframe development. Heinkel, of course, was building his own test plane, the He 178, without government support. In the fall of 1938, Hans M. Antz joined the Airframe Development Group of the Development Division of the Technical Office. Antz had followed the same "Flugbaumeister" course as Schelp, and the two were well acquainted. Soon after joining the Airframe Group, Antz approached Robert Lusser, then still chief of development at Messerschmitt, to make preliminary investigations of airframe requirements for turbojet propulsion. In the interim, Antz and Schelp thoroughly

Gloster E. 28/39, first British jet aircraft to fly. (G.E., Gas Turbine Engineer, Conference, 1945.)

discussed the problem. They knew of the difficulties Heinkel expected with the long intake and exhaust of the He 178, and they knew of the high-efficiency axial compressor designs of the Aerodynamische Versuchsanstalt (AVA). Schelp computed specific weight (thrust-per-pound-engine-weight) and specific dimension (thrust-per-unit-diameter) estimates based on the AVA compressor figures. Schelp and Antz concluded that twin, small-diameter axial-flow turbojets mounted in individual pods would most likely provide the basis for a practical fighter. In the late fall of 1938, Antz gave Messerschmitt a contract for a turbojet-powered interceptor. Schelp's figures were provided as a basis for design, but no aircraft configuration was specified. The fighter was to have a speed of 528 mph and an endurance of one hour. Woldemar Voigt, whom Messerschmitt had named project engineer, directed design studies of a number of different configurations. The studies indicated a design with two wing-mounted engines would be the best solution. Messerschmitt calculated that to meet the performance required he would need about 1,500 pounds of thrust from each engine.[60] The favored design, which received the Messerschmitt project designation P.1065, became the Me 262.

In Germany as in England, before any turbojet-powered aircraft had flown, the turbojet revolution had already occurred. Preparations for the production of turbojet combat aircraft were underway. Solely on the basis of presumptive anomaly deduced from theoretical aerodynamics, three Germans had joined Frank Whittle in creating a technological revolution.

8

The New Normal Technology

The turbojet revolution initiated by Whittle in England and by von Ohain, Wagner, and Schelp in Germany was the beginning of a new aeronautical age. That technological revolution was a *fait accompli* by the end of 1939: the most advanced aeronautical practice in England and in Germany was by that date firmly committed to development of the turbojet. Yet initial commitment marks only the beginning of a new normal technology—not its success or its community-wide adoption. The Second World War saw the triumph of the turbojet: the development for service use of the engines themselves and the conversion of the world aeronautical community to the newly established tradition. Germany exploited the turbojet most fully and most quickly. But the revolution spread rapidly through the English aeronautical community as well, and from there literally flew the Atlantic to the United States.

During the last desperate years of the war, the Germans put more different types and a larger number of turbojet-powered aircraft into service than any other nation. Germany also probably had the most comprehensive program for advanced turbojet development, both engine and airframe, of any of the powers. Given Germany's military situation, the very magnitude of the German programs may have been counterproductive. But what the Germans did produce represents remarkable technical achievement, especially considering wartime privation and near-total lack of strategic materials, which, as noted below, sharply distorted the quality and performance of their systems.

Ironically, the two original German turbojet projects—that of von Ohain and that of Wagner—miscarried. The He 280, Heinkel's private venture, did fly with von Ohain HeS-8 engines in April 1941, as mentioned. Altogether, eight He 280 airframes and at least fifteen HeS-8 and HeS-8A (001) engines were built. Some HeS-8A engines were rated at 1,540 pounds thrust, but a more common figure was 1,300 pounds. The engine was still not dependable, largely because of diffuser problems. The 001 was finally abandoned in favor of a more advanced project no later than mid-1942.

Meanwhile, in May 1939, Heinkel had taken over the Müller engine projects from Junkers Aircraft. When Schelp interceded in the turbojet programs in late 1938, he wanted the Junkers Aircraft projects transferred to Junkers Engine. Wagner was in the midst of a dispute with the president of the combined Junkers Company about the kind of aircraft to be developed and left Junkers in 1939 to go to Henschel to work on guided-missile development. Neither Wagner's staff nor Müller nor any of his people wanted to work under Otto Mader at Junkers Engine. A number of Wagner's staff returned to the University of Berlin, and Müller and most of his staff moved, over the protests of the Junkers board, to Heinkel.[1]

Müller brought with him the Wagner-Müller engine, which still would not run under its own power, plus hardware for a ducted-fan engine driven by a sixteen-cylinder high-speed two-stroke diesel and for a ramjet, all of which Heinkel contemplated developing. Although pursuing a number of lines of development simultaneously to see which is most promising makes good technological sense (it was RLM policy until 1941), it makes sense only if adequate resources and talent are available. Heinkel had neither: he had no experience in, nor staff for, engine production, and with the war emergency, new technical people, especially competent engineers, were impossible to hire. Müller ultimately left Heinkel in May 1942, apparently because of personal differences, and the ducted-fan and other peripheral projects were then dropped.

The Wagner-Müller turbojet was given the German Air Ministry (RLM) designation 006 at Heinkel. The main development difficulty with the 006, with its very highly loaded compressor, remained that of matching the compressor and turbine characteristics. Development work on the 006 did continue at Heinkel, however, and by the end of 1942, the 006's basic problems had been solved. It was developing almost 1,900 pounds thrust for 857 pounds weight, the best thrust-for-weight ratio of any German engine built during the war.[2]

But the 006 was abandoned too. By mid-1941, the B.M.W. and Junkers Engine Division turbojets were making promising progress. It was Schelp's opinion that the "reaction" blades of the 006 compressor were inherently less efficient than the Aerodynamische Versuchsanstalt (AVA) pure-impulse blading (he was wrong) and that the diameter of the 006 was too small to offer great development potential. Furthermore, in view of the progress at Junkers and B.M.W., he wanted to press on with development of more powerful, more efficient second-generation engines. He therefore persuaded Heinkel to drop work on both the 001 and the 006 and to devote all his efforts to an advanced engine, the 011. As part of the inducement to Heinkel to take up the advanced design, Schelp arranged for him to acquire the Hirth Motor Company, which made small piston engines for training planes, and thus gave Heinkel, for the first time, engine production facilities.

The 011 was designed by von Ohain, and development work on it began in September 1942. The 011 was a very large engine, just over 13 feet long with a diameter of 42.4 inches, and weighed 2,090 pounds. It was designed for a thrust of 2,860 pounds, with supposed potential for an eventual 3,520 pounds; it was designed to run at 11,000 rpm and to give a specific fuel consumption of 1.31. It had a unique compressor, with an initial "diagonal" stage (a combination of axial and centrifugal flow) followed by three straight axial stages. It had a large straight-through annular combustion chamber and a two-stage axial-flow turbine. A turboprop variation of the same engine carried the RLM designation 021.[3] By the end of the war, numerous tests of the 011 had been made, including some flights in a piston-engined Ju 88 flying test bed, but no 011 had flown under its own power. Preparations for production were underway, but production had not commenced when the war ended.[4]

The cancellation of the 001 and the 006 in favor of the 011 left the He 280 without an engine. The He 280 was extremely fast, having attained a reported 578 mph at 19,685 feet with two HeS-8A engines. In mock combat against a Fw 190A it proved markedly superior. But at high speeds the He 280 suffered from severe tail flutter (uncontrollable, rapid oscillation), and its fuel capacity was relatively small, having been planned on the basis of expected fuel consumption figures for the 001 and 006 that turned out to be grossly optimistic. As a result, with available engines, the He 280 had a uselessly short endurance. Heinkel himself attributed the He 280's rejection to RLM prejudice against him, and considering some of the more improbable enterprises supported by RLM, there was perhaps something to Heinkel's suspicion.[5] Whatever the reason, the He 280 was cancelled, and neither the original von Ohain nor the original Wagner engine went into production.

Again ironically, the only German turbojet to achieve mass production was the Junkers Jumo 004, designed and built by the Junkers Engine Division. Junkers Engine had finally accepted a development contract for a turbojet in the summer of 1939. Anselm Franz, who had headed supercharger development at Junkers Motors and whom Mader had selected to conduct preliminary turbojet investigations when approached by Schelp and Antz in 1938, was placed in charge of the new Junkers Engine turbojet project. Franz examined the Wagner-Müller engine, and decided that its design was too far advanced to be capable of easy or rapid development. He favored building a simple, unsophisticated engine first in order to get the necessary experience for more advanced designs. He therefore refused to take over the Wagner-Müller engine (which eventually went to Heinkel) and began a completely different engine, the 004, designed above all else to run: "The guiding principle behind every decision in the design of the 004 was to make the choice which would lead with the greatest certainty and speed to an operating engine, even though this entailed some sacrifice in performance. This conservatism is shown, for example, in the fact that the ratios both of weight to thrust and of diameter to thrust were appreciably greater in the 004 than in the Müller turbojet."[6] The complete 004 was first run in October 1940, and its more serious development problems had been solved by January 1942.

The 004 had an eight-stage axial compressor designed by Encke of the AVA. The single-stage axial turbine was designed in cooperation with Allgemeine Elektrizatasgesellschaft (A.E.G.), the steam turbine manufacturers. The compressor produced a pressure rise of 3.1:1 at an efficiency of 78 percent. Six "can" combustion chambers were used to simplify development. The 004B, the definitive production version, generated 1,980 pounds thrust at 8,700 rpm for a weight of 1,590 pounds. The 004B was 12 feet 8 inches long and 31.5 inches in diameter, and had a specific fuel consumption of 1.4.[7] It entered pilot production in mid-1943 and was ordered into full-scale production near the end of that year. About 6,000 004's were built before the end of the war.

The 004 was a remarkable technical achievement in the minimal use of strategic materials. The first experimental 004 engines had solid turbine buckets, but later engines employed hollow, internally air-cooled buckets that permitted the use of a manganese alloy that could be fabricated simply by folding and welding sheet stock. The combustion chambers were mild sheet steel anodized with aluminum (which, with uneven internal temperature distribution, resulted in their having a service life of only about 25 hours). As a result of extraordinary design effort, a complete 004B required no nickel at all and used less than 5 pounds of chromium.[8]

In addition to the 004, Junkers had responsibility for the 012, a very large axial-flow turbojet of 6,000 pounds thrust which was to weigh 4,400 pounds, and for the 022, a turboprop development of the 012. Neither the 012 nor the 022 had proceeded beyond the initial design stages when the war ended.

The New Normal Technology

B.M.W.'s turbojet, the 003, was the second most widely used German jet engine. The ultimate B.M.W. program was really a composite of three different projects. When originally approached by Schelp in 1938, Kurt Loehner at B.M.W. had proposed an experimental engine with a two-stage centrifugal compressor. Herman Oestrich at Bramo, in contrast, initially favored a ducted-fan system. Meanwhile, Helmut Weinrich, who had been developing a contrarotating gas turbine for the German Navy, had his project assigned to Bramo by the RLM. Tests at the AVA proved the ducted fan impractical, and work on the centrifugal unit at B.M.W. stopped at the outbreak of the war so that B.M.W. could devote all its efforts to developing its radial piston engines. Bramo had received a development contract for a contrarotating turbojet on the Weinrich pattern, given the RLM designation 002, before Bramo and B.M.W. merged in mid-1939. But B.M.W. quickly recognized the difficulties likely to be encountered in any engine as complicated as the contrarotating unit, and in the autumn of 1939 applied for and got Ministry support for a preliminary straight-through axial engine, ultimately the 003. By 1942, B.M.W. had discovered just how difficult development even of the straight-through turbojet could be, and the 002 was dropped completely in favor of the 003.

The 003 was first run in early 1940. It was designed for higher efficiency and, especially, lower weight than the Junkers Jumo 004. A great deal of difficulty was encountered with the first test engines, necessitating much component redesign. The new 003's were not ready until early 1941. Even then, two more years of experimentation and development were necessary before pilot production could begin. Production of service engines began in the second quarter of 1944, but production levels never reached those of the 004.

The definitive version of the B.M.W. turbojet was the 003A-2. It developed 1,760 pounds thrust at 9,500 rpm and weighed 1,250 pounds. Its specific fuel consumption was 1.47; it had a length of just over 10 feet and a diameter of only 27 inches. In place of the original prototype's six-stage compressor designed by Encke at AVA, the production 003 employed a B.M.W.-designed seven-stage compressor having NACA blade profiles and producing a 3.1:1 pressure ratio. The 003 had an annular combustion chamber and an internally air-cooled single-stage axial turbine, developed with the help of Brown Boveri, Mannheim, and Maschinenfabrik Augsburg-Nürnberg (M.A.N.). In its final form, the 003 had a considerably longer life (about 100 hours) than the 004, required fewer man-hours to build, and was easier to disassemble for maintenance. The 003, like the 004, required little use of strategic metals. Brown Boveri designed an advanced compressor for the 003, the model with it being designated 003C, but it was not ready for production before the war's end. B.M.W., like Junkers, also had a contract for a much larger engine, the 018, of 7,500 pounds thrust and 5,100 pounds weight, but only components had been built by 1945.

Diamler-Benz (D.-B.) did the only other major German work on gas turbine aero-engines. D.-B. had refused even to study the turbojet in 1938, but by late 1939 realized that its competitors, Junkers and B.M.W., were completely serious about the new engines. Diamler-Benz therefore assigned an engineer, Karl Leist, who had come to them from research on superchargers at the DVL, to work on a very complicated contra-rotating turbofan engine, the D-B 007. Work on the 007 continued with minimum effort until the fall of 1943, when the impossibly slow progress at D.-B. caused the RLM to cancel the 007 altogether and order them to develop the 021 turboprop version of the Heinkel 011 turbojet. No D.-B. gas turbine engine flew during the war. The D.-B. 600 series piston engines were far and away the most important military aircraft engine for the Luftwaffe. Because of the development lag resulting from the Versailles prohibition of high-performance aero-engine production, German piston engines were inferior, in terms of power per cubic inch, displacement, weight, and specific fuel consumption, to foreign engines, and would remain inferior throughout the war. It is therefore understandable that D.-B. did not divert resources from piston engine development.

Although the engines built by von Ohain and by Wagner's group never saw production, they did provide part of the technological base, and, above all, critical verification at the critical time, for Schelp's Ministry-initiated turbojet program. The independent work of von Ohain and Wagner, interacting with Schelp's own independent study, created the turbojet revolution in Germany long before any turbojet-powered aircraft had flown, and ultimately led, through the established aero-engine manufacturers, to service turbojet engines.

In England, Whittle's work had much the same effect. As mentioned, the Air Ministry—in the persons of Tizard of the Aeronautical Research Council (ARC) and Pye, the Air Ministry's director of scientific research—was convinced by the beginning of 1939 that Whittle had a practical engine. By the outbreak of war, the ministry realized that the turbojet was the only gas turbine aero-engine likely to be developed in a reasonable length of time. It therefore ordered other gas turbine projects redirected to the turbojet to supplement the efforts at Power Jets.

As a result of that decision, the second British turbojet project grew out of Griffith's work at the Royal Aircraft Establishment (RAE). In July 1936, as a part of general rearmament, Tizard's Engine Sub-Committee of the ARC had authorized the RAE to proceed with the turboprop investigations originally proposed by Griffith in 1929. Construction was begun on an eight-stage straight-through axial-flow test compressor designed by Griffith and Hayne Constant, but it was not completed until early 1938. Unfortunately, as a result of a faulty oil seal, the experimental compressor stripped its blading 30 seconds into its first run.[9] The compressor was reconstructed, this time according to Ackeret's design precepts: Griffith himself had visited Switzerland in 1937 to study Brown Boveri's work, so the RAE had full knowledge of the latest Swiss practice. The rebuilt compressor was run again in October 1938, and was used for experiments until August 1940, when it was destroyed in a German air raid. In the interim, C. A. Parsons and Company had built another, similar test compressor to RAE blade specifications.

A nine-stage contrarotating, contraflow compressor and turbine experimental unit designed according to Griffith's original ideas was also authorized by the RAE in 1938 and constructed by Armstrong-Siddeley in 1939. Each stage of the unit consisted of concentric compressor and turbine sections; each stage rotated independently, which meant labyrinth seals between each separate disc. According to Griffith's concepts, pressure at each turbine and compressor stage should have been proximately equal, thus minimizing leakage losses. Practice, however, was different:

> At the time the first turbo-compressor was constructed there was insufficient knowledge of blade design available to make it possible to attain in fact the high performance predicted by Griffith theories. As a result the turbine blades produced less power and the compressor blades absorbed more power than predicted. In consequence the wheels could not be run up to their design speed, and gross pressure differences developed between the compressor and turbine annuli. These pressure differences resulted in large leakage flows which further depreciated the machine's performance.[10]

Griffith left the RAE in June 1939 to continue development of his ideas at Rolls Royce. There he designed an immensely complicated contraflow ducted-fan turbojet with fourteen high-pressure and six low-pressure stages together with an entirely novel combustion system. Only the fourteen-stage high-pressure section was actually built, and when tested during 1942–44 proved susceptible to the same problems as the earlier RAE test unit; the whole project was finally abandoned in 1944. In the closing months of the war, however, Griffith's work did lay the foundation for Rolls Royce's extraordinarily successful postwar Avon and Conway series straight axial-flow turbojets.[11]

Meanwhile, Constant had presented a comprehensive paper on turboprop engines to the Engine Sub-Committee in March 1937. At the same meeting at which the ARC recommended limited support for

Whittle's turbojet work, it authorized Constant at the RAE to proceed with work on a complete turboprop engine (in addition to the test components mentioned above). Design of the RAE engine thus began in April 1937. To gain access to production experience and facilities, the RAE asked for the collaboration of the Metropolitan-Vickers Electrical Company, a manufacturer of steam turbines. The RAE and Metro-Vick designed a very complicated two-stage turboprop with separate high- and low-pressure compressors powered by separate high- and low-pressure turbines. In this design approach, the RAE and Metro-Vick were following standard steam turbine practice, which commonly employed separate high-, medium-, and low-pressure turbines to extract the maximum energy from the expansive power of steam. Only the high-pressure compressor and turbine, together labeled the B.10, were actually constructed. Various components were under test from December 1939, and the complete high-pressure unit from October 1940. Simultaneously, in late 1938, the RAE began design of a third axial compressor, also based on the data from Ackeret. The new unit, known as the D.11, was intended for a 2,000-h.p. turboprop which was to have straight-through airflow with no contrarotating provisions. The D.11 compressor was built during 1940 and first tested in 1941. During the same period, two other similar experimental compressors were built for the RAE by other private firms.

Griffith and the RAE, like other early investigators of the aircraft gas turbine, were handicapped by their fixation with shaft output and with performance parameters—aircraft speed and engine weight and specific fuel consumption—defined by conventional piston-engined aircraft. As a result, they failed to see the turbojet's capacity to utilize excess air ingestion to increase mass-flow and reduce exhaust velocity and temperature, which would obviate materials temperature limitations and, more critically, would boost propulsive efficiency, especially at high aircraft speeds. They thus did not recognize that a turbojet could be propulsively efficient with individual component efficiencies (compressor and turbine) well below those necessary for a practicable turboprop.[12]

Significantly, the axial compressor design produced by the slow, methodical, "research" oriented approach of Griffith and the RAE, like the design that would emerge from Eastman Jacob's equally protracted work at NACA, was notably more efficient than contemporary German compressors, although the German units may have been intentionally compromised for ease of production. The RAE-NACA approach to axial compressors nevertheless would seem to represent an alternative program for exploring radical technological alternatives. Where the successful protagonists of the turbojet revolution immediately grasped the full technological implications of only partially corroborated aerodynamic insight, the RAE and NACA programs sought, rather, to develop a more complete scientific understanding of turbomachinery phenomena before making any commitments to radical technological systems. These differences between the RAE-NACA and the successful turbojet programs suggest a continuum of responses to new basic theoretical information. At one end, the implications of theory for radical technological systems may be quickly realized and new technological development projects initiated immediately. Such an approach incurs considerable risks associated with applying incompletely developed or corroborated theory. Alternatively, better scientific information can be bought at the cost of further research, thus offering the ultimate but postponed prospect of lower risks in technological development. Yet ironically, in this case quick, decisive success in the RAE or NACA turbocompressor experiments probably would have led to early development of a turboprop rather than a turbojet.

Events, however, quickly overtook all the RAE investigations. By the time the Second World War began in September 1939, the RAE had done enough work that both it and the Air Ministry realized how complicated, difficult, and time-consuming building any turboprop was going to be. They therefore decided, wisely, to build a simple, straight-through axial-flow turbojet first and to postpone turboprop development. Preliminary design work was done by Constant at the RAE, and Metro-Vick began detailed design of the engine that was to become the F.2 turbojet in July 1940. The F.2 was first run in December 1941, and first flew under its own power (in a specially modified Meteor) in November 1943. The original F.2 had a nine-stage axial compressor, an annular combustion chamber, and a two-stage turbine. It offered 1,800 pounds of thrust at 7,300 rpm for 1,510 pounds weight and a specific fuel consumption of 1.14. Although further development of the F.2 continued (the F.2/4), it was not put into production during the war. It was nevertheless the most highly developed British axial-flow turbojet. Had England found herself in straits similar to those of Germany and had she been willing to accept the same sacrifices in efficiency and especially in reliability that the Germans did, the F.2 could have been produced for service use.[13]

The only other gas turbine engine to reach a high state of development in England during the war was the de Havilland Goblin. In January 1941, Tizard concluded that in addition to the Whittle engines and the RAE–Metro-Vick engines, which would obviously be a long time in development, it would be advisable to begin another, separate line of engine development as a backup. He therefore invited de Havilland, a manufacturer of both airframes and engines (mostly small trainer engines), and the brilliant independent aero-engine designer Frank B.

Halford to design and build a turbojet. Tizard gave Halford—who had a long line of successful piston engines to his credit, most notably the Napier Sabre—complete design freedom for the new engine. Halford was also given complete access to the work at Power Jets and at the RAE.

De Havilland and Halford decided that a single-engined fighter should offer significant savings in construction costs and complexity. But no one turbojet then under development would produce enough thrust to power a combat aircraft. Halford therefore undertook the design of a larger engine with a minimum thrust of 3,000 pounds. Since in a single-engined fighter with a fuselage-mounted power plant, outside engine diameter was not so critical, Halford, drawing on his own experience with centrifugal superchargers and his knowledge of Whittle's work, chose a large, single-sided centrifugal compressor. Because emphasis was on getting a production engine as quickly as possible and because most previous work had been done on "can" combustion chambers (they were the easiest to experiment with), Halford employed sixteen straight-through "can" combustors. He used a single-stage axial-flow turbine. Altogether, Halford's Goblin, as it was called, was a remarkably clean, elegant design.

Construction of the Goblin began in August 1941, and the engine first ran in April 1942. Because of de Havilland's other commitments, the airframe for which the Goblin was intended, the D.H. 100 Vampire, was delayed. Derated versions of the Goblin were therefore flown in a modified Meteor in March 1943—which, because of delays in the delivery of Whittle engines, was also the Meteor's first flight. The Goblin I, with a thrust of 2,700 pounds, was ready for production in January 1945, and the Goblin II, uprated to 3,000 pounds thrust, in July 1945. The Goblin II produced its 3,000 pounds thrust at 10,200 rpm and weighed only 1,500 pounds. Its specific fuel consumption was 1.14. It was 50 inches in diameter and had an overall length of 8 feet 4 inches.[14] The de Havilland-Halford engines were in production at the end of the war, as was the D.H. 100, but neither the planes nor the engines had entered squadron service.

There were two other gas turbine projects in England. Because only one firm, Metropolitan-Vickers, was at work on an axial-flow turbojet, Armstrong-Siddeley was given a contract to develop another in November 1942. Its engine, designated the ASX, was a very large, 2,800-pound-thrust unit with a fourteen-stage axial compressor and a two-stage turbine. It bagan tests in April 1943. Because the ASX duplicated the performance of the D.H. Goblin, however, it was adapted to turboprop configuration as the ASP, which, in April 1945, gave 3,600 horsepower on test. No Armstrong-Siddeley gas turbine was in production during the war.

The New Normal Technology

Finally, the Bristol Aeroplane Company approached the Ministry of Aircraft Production concerning a gas turbine project in the spring of 1943. Bristol's interest in gas turbines had been delayed because it had been fully occupied in getting into production its revolutionary sleeve-valve radial piston engines, urgently needed by the RAF. After some foot-dragging by the MAP, Bristol finally received a development contract for a 2,000-h.p. turboprop in early 1944. It was not completed before the war ended.

All British turbojet, as distinct from turboprop, programs therefore stemmed directly from Whittle's work. Uniformly, they began after he had shown the way, after he had demonstrated an operable engine. Yet there is some question how far Whittle would have gotten in gleaning Air Ministry support had not Griffith's proposals introduced the whole idea of a light-weight, efficient gas turbine aero-engine. Certainly neither Griffith's own turboprop nor the turbojet that grew out of the axial-flow work Griffith spawned at the RAE saw war service, but Griffith is so intimately connected with the turbojet revolution that he does deserve considerable credit in its creation. Griffith was the first man in any country to recognize that the high-efficiency components he had designed meant that a gas turbine aero-engine could be practicable. Yet whatever accolades Griffith might deserve, it was Frank Whittle who created the first English turbojet.

The most important U.S. turbojet was a direct copy of the Whittle engine. H. H. ("Hap") Arnold, chief of staff of the U.S. Army Air Corps, visited England in March 1941. A liaison officer, A. J. Lyon, and a G.E. turbocharger service representative, D. R. Shoults, informed him of the Whittle engine and of its nearness to flight readiness. Arnold, though a nontechnical officer, immediately grasped the revolutionary tactical importance of the turbojet. When he returned to the United States in April, he set about arranging for domestic production of the Whittle engine. Official negotiations were opened with the British Ministry of Aircraft Production in July 1941, and Major D. J. Kiern was sent to England to learn everything possible about the Whittle turbojet. Arnold sat down with a catalogue of U.S. government suppliers to pick the one most qualified to build the Whittle engine—and chose the turbosupercharger division of General Electric, West Lynn, Massachusetts, which had been nursed along by Sanford Moss.

Ironically, G.E. independently had come perilously close to the turbojet twice in the preceding twelve years. Glenn Warren, as noted above, had followed Sanford Moss in studying the internal combustion gas turbine at G.E. early in the 1920s. Warren had remained at Schenectady after Moss was transferred to West Lynn, but Warren had been diverted during the late 1920s to work on flow in steam turbine nozzles. In some large turbines, steam velocities in the nozzles exceeded the local speed of sound, which set up compressibility shock waves and resulted in sharp losses in efficiency. Investigation of these problems led Warren in 1929 to propose research into an advanced ram-jet aircraft propulsion system.[15] Warren's analysis of supersonic flow in convergent-divergent nozzles convinced him that a ram-jet powering an aircraft at a speed of 1,350 mph at an altitude of 80,000 feet would be thermally and propulsively efficient. Warren recognized that exceptionally high speed was imperative for the ram-jet to be efficient and that very high altitude was necessary to reduce aircraft drag.

Warren's preliminary internal G.E. memorandum included a number of sophisticated provisions. A movable cone in the ram-jet's exhaust would vary the nozzle size and shape in accordance with varying speed and load conditions. "Starting jets," or rockets, would take the aircraft quickly up to cruising speed and altitude. The rockets would employ liquid oxygen both as an oxidant and as a coolant for the combustion chambers. Warren noted the need for passenger cabin "supercharging" at such high altitudes, and acknowledged that cabin cooling also might be needed, despite the −60° F outside temperature, since air friction at such high speeds, even at 80,000 feet, might well heat the surface of the aircraft to 240° F. Warren visualized designs with separate wing-mounted propulsion units as well as with the ram-jet duct built into the aircraft fuselage.

The only serious problems Warren foresaw for his design related to aircraft aerodynamics at supersonic speeds. He noted:

> The real problem is going to be in the aerodynamics of the airplane at these speeds—that is, above the velocity of sound. So far as the writer knows, there is little if any data relative to this factor. It is a well-known fact that near the velocity of sound the efficiency of an aerofoil drops quite low. It is the writer's opinion, based on theory alone and unsupported as yet with tests, that at speeds appreciably above the velocity of sound the lift and efficiency of an aerofoil is again good, but that the lift is in a direction opposite to that at velocities below the velocity of sound.[16]

Warren reached this novel conclusion through analogy with supersonic flow in nozzles; he clearly at this point had no familiarity with European developments in supersonic aerodynamic theory—specifically, he had no knowledge of Ackeret's 1925 paper on supersonic lift on a two-dimensional aerofoil.

The New Normal Technology

G.E. took little notice of Warren's ideas, and nothing came of his proposal; he went on to become general manager of G.E.'s steam turbine division. Looking back in 1952, Warren observed:

> Although I have always been very much interested in the fact that I apparently thought as early as I did of the mechanism which is now generally called the "ramjet," I have been very chagrined at myself that, with all of my turbine and gas turbine background, I did not conceive at that time of the application of a compressor and turbine to this device which greatly increases its efficiency and makes it a practical machine for propulsion at 100 to 700 m.p.h., as is now done by the turbojet.[17]

Given his knowledgeability and infatuation with gas turbines, Warren probably would not have been long delayed in discovering the turbojet had he been permitted to indulge his ram-jet fantasy in 1929.

Moss's turbosupercharger group at West Lynn came even closer to the turbojet. During the early 1930s, the turbochargers were frequently tested by hooking up a simple combustion chamber between the compressor and turbine: G.E. was inadvertently running a turbojet. Some of the younger engineers wanted to build a true turbojet, but Moss, having been over the bumpy road of gas turbine development once, rejected any such program: he thought any gas turbine impractical.[18] In connection with its turbocharger work, G.E. also held Air Corps contracts from 1931 on for research on high-temperature turbine nozzles and blades, for which G.E. constructed a special hot gas test stand. These programs stimulated frequent informal discussions between G.E. engineers and the personnel at Wright Field (the Army Air Corps test establishment) concerning turboprops, but the consensus remained that component efficiencies and high-temperature materials were both inadequate.

But during the summer of 1937, Moss heard of Whittle's work (apparently) and of that of Ljungstrom in Sweden through contacts at British Thompson-Houston (B.T.-H.), and wrote to a colleague at Schenectady that fall:

> Mr. A. R. Smith has just returned from England and he and his assistant, Mr. R. J. Gaughey, saw an exhaust gas turbine at the BTH factory for airplane service.... At any rate, we would like to visit Mr. A. R. Smith and get from him a complete description of the gas turbine apparatus he saw.... We have the idea that the apparatus involved a centrifugal compressor with ramming intake driven by a gas turbine wheel which furnishes the jet for jet propulsion, or that only part of the gases for the combustion chamber drive the turbine wheel to drive the compressor and the rest of the gases furnish the jet direct.... We have in mind going to Wright Field with a similar proposition, and want to know all we can about the British outfit before we start.... I do not doubt that I can get a lot of information when I am in England myself

next summer. However, we would like to present some sort of proposition to Wright Field now.[19]

G.E. did conduct a number of internal studies of gas turbine aeroengines, turboprop and turbojet, but no formal turbojet proposal was submitted to the Air Corps prior to initiation of the Whittle project. Moreover, prior to September 1939, when a young Schenectady engineer, Dale Streid, first considered jet propulsion for speeds greater than 450 mph,[20] no G.E. study or proposal assumed aircraft speeds greater than 300 mph,[21] at which, of course, any turbojet would be propulsively inefficient.

By January 1940, however, Sanford Moss had completed a comprehensive theoretical comparison of gas turbines used either to turn a conventional propeller or to provide a reaction jet. Moss based his calculations on centrifugal compressors with pressure ratios of from 6 to 18 to 1 at an efficiency of 75 percent and on a turbine with an efficiency of 85 percent. Propeller efficiencies were assumed to range from 87 percent at 300 mph to 65 percent at 600 mph. Moss deduced hypothetical overall thermal efficiency curves for the two types of power plant at aircraft speeds of 300 to 600 mph and at altitudes from sea level to 40,000 feet. Moss concluded, "that at all altitudes the overall efficiency of the jet propulsion is higher than the conventional propeller propulsion for speeds above 500 miles per hour."[22] Thus G.E. was clearly moving in the direction of turbojet development by late 1939 or early 1940. Yet Moss's fixation with thermal efficiency and his assumptions of very high compressor compression ratios suggest that he still had not freed himself from traditional notions of what characteristics (especially with regard to fuel consumption) an aircraft powerplant should have.

Whether General Electric's persistent myopia reflects the broader American cultural values discussed in Chapter 6, some deficiency characteristic of large, diversified corporations, or some bias peculiar to G.E. remains problematic. G.E. teetered on the edge of the turbojet revolution throughout the 1930s. At least superficially, the critical factors preventing its going over appear to have been Moss's hard-learned conservativism and, even more importantly, G.E.'s isolation from the theoretical high-speed aerodynamics which might have fostered radical airframe performance assumptions.

Whatever the debates within G.E. might have concluded, however, the evidence of the Whittle engines, especially after the first flight in May 1941, was incontrovertible, and G.E. eagerly accepted the Army contract to develop and produce Whittle type engines. The Army formally let contracts to G.E. for engines and to Bell Aircraft for an experimental airframe of roughly the same specifications as the British E.9/40 in September 1941. Kiern smuggled the W.1X engine and drawings of the W.2B out of England under U.S. diplomatic seals in October 1941. After a Keystone Cops squabble with the U.S. Customs Service at Washington National Airport, the engine and the drawings were turned over to G.E. The Army gave G.E. full design authority. The first G.E. engines were designated, for security, "I," "H" being the then-current turbosupercharger series. The first G.E. engine ran on 18 March 1942. Slightly modified engines—I-A's, which were basically Whittle engines with some G.E. changes—first flew in the Bell XP-59 in October 1942, well before either the production Whittle engines or the Meteor itself had flown. The first G.E. production engine, run in February 1943, was the I-14, with 1,400 pounds thrust; it was rapidly succeeded by the I-16, with 1,600 pounds thrust. American progress in development was indeed remarkable. "In July 1943 a fully-armed P-59A was flown with these modified I-16's at 46,700 feet. The Welland had first flown in the Meteor in June 1943, and was still having trouble with surging at about 25,000 feet."[23]

G.E. developed two other engines of its own design during the war. One, the I-40 (later J-33), was a direct outgrowth of the Whittle engines. At the beginning of 1943, the Army asked G.E. to study the possibility of building a much larger turbojet (ultimately of 4,000 pounds thrust). The I-40, begun in June 1943 and first run in January 1944, was the result. The I-40 first flew in June 1944 in a Lockheed airframe, the XP-80, originally designed for the de Havilland Goblin engine that was to be manufactured under license by Allis-Chalmers in the United States.

The other G.E. turbojet, the TG-180 or J-35, came about in a more circular fashion. American intelligence got the first reports of German jet propulsion activity in late 1940. As a result, Arnold requested NACA to investigate the entire field. Vannevar Bush, NACA chairman, appointed a Special Committee on Jet Propulsion under William R. Durand. The Special Committee included rockets as well as turbojets and turboprops in its study, the distinction between the systems not yet having been fully grasped. Westinghouse, Allis-Chalmers, and the Steam Turbine Division of G.E. at Schenectady were among those belonging to the Special Committee. All were asked to submit studies of possible unorthodox engines. None were told then, or at any time before 1943, of Whittle's work or of that of West Lynn. Even the Schenectady Division of G.E. never knew officially of the turbojet activities at the turbosupercharger division.

G.E.-Schenectady's study proposed on axial-flow turboprop, the TG-100, for which an Army contract was granted on 8 December 1941. The TG-100 had a fourteen-stage axial compressor (based essentially on published Brown Boveri designs), nine "can" combustion chambers, and a single-stage axial turbine.[24] Development of the TG-100 proceeded, though slowly, through 1942 and the early part of 1943. When the Army demanded a more powerful turbojet in mid-1943, the TG-180 turbojet was designed, based on the TG-100. The TG-180 had an eleven-stage axial compressor derived in part from Eastman Jacobs' studies at NACA, but was otherwise similar to the TG-100. Development of neither engine was complete at the end of the war.[25]

Two other projects emerged from the Special Committee's activities. Westinghouse concluded that the gas turbine could best be utilized for reaction propulsion, and on 8 December 1941 received a Navy contract for a small (1,000-pound-thrust) booster turbojet for conventionally powered fighters. The Westinghouse engine, the 19A, also was designed in accordance with Jacobs' axial-flow compressor results. It first ran in March 1943 and first flew, as a booster, in a Navy Corsair fighter in January 1945. Allis-Chalmers received a contract for a design study of a very complicated ducted-fan turbine engine in February 1942. This engine was still in the study stage in mid-1943 when it was dropped in favor of Allis-Chalmers' production of the de Havilland Goblin engine. The G.E. I-40, however, proved so successful that the Goblin was not needed, and its production was canceled.

Several other gas turbine projects emerged in the United States, although none was ultimately successful. Probably the most sophisticated was sponsored by Lockheed Aircraft. Clarence Johnson, the brilliant Lockheed airframe designer who would later be responsible for the P-80, U-2, F-104, YF-11, and SR-71, first began to consider the problems of compressibility and of supersonic flight, together with suitable power plants, in the mid-1930s during the initial design studies for Lockheed's twin-piston-engined P-38 Lightning. Indeed, Johnson even approached von Karman and Clark Millikan at the California Institute of Technology concerning possible solutions to compressibility problems. The extremely well streamlined P-38 certainly confirmed Johnson's apprehensions: when power dived, it did encounter compressibility flutter.[26]

Johnson's concern with compressibility and supersonic speed, combined with experience with the P-38, led by the fall of 1940 to the initiation of what was to become Lockheed's extraordinary L-133/L-1000 proposal, officially submitted to the Army Air Corps in March 1942.[27] The L-133/L-1000 was to be a fully integrated aircraft-and-turbojet propulsion system designed to attain 625 mph at 50,000 feet. The L-133 airframe was to be of Canard (tail-first) configuration to achieve lower drag, with very thin section wings incorporating boundary layer removal to minimize compressibility effects and drag, and was to be constructed entirely of stainless steel in order to withstand the high skin temperatures due to friction expected to result from such speeds. Wind tunnel tests indicated that such a design would not be inherently stable aerodynamically, so an artificial stabilization system ("fly-by-wire") was provided, employing small reaction "jets" (using air bled from the engine compressor) in place of conventional control surfaces (as in the X-15).

The L-1000 engine was to be equally extraordinary. The engine itself was designed by Nathan C. Price, an engineer who worked for Lockheed. Price had begun his career as a steam turbine specialist and during the 1930s had designed a number of light, high-speed reciprocating steam engines with Velox-type boilers, and even one light steam-turbine aircraft engine. Price initially tried to design an engine fully competitive in fuel economy with conventional piston engines, so the L-1000 was large, heavy, and complicated, at first having an axial followed by a reciprocating compressor. By mid-1943, the prospective L-1000 was simpler but still complex: a sixteen-stage axial inlet compressor followed by a second sixteen-stage axial high-pressure compressor (total compression ratio, 50:1), with an annular combustion chamber followed by a single impulse-stage plus three reaction-stage turbine. Estimated efficiencies were 85.5 percent for the inlet compressor, 90 percent for the high-pressure compressor, 80 percent for the turbine, and 94 percent for the propulsion nozzle. Afterburing was anticipated. The extremely high compression ratio was intended "to attain high internal cycle efficiency." Specific fuel consumption (pounds of fuel burned per pound of thrust produced per hour) was expected to be 1.5 at the full 5,100 pounds thrust at sea level, but only 0.60 at the cruising thrust of 750 pounds at 50,000 feet. Total weight was pegged at only 1,235 pounds, with a diameter of 25 inches.[28]

The Army declined support of the entire L-133/L-1000 project, but did provide modest support for the revised L-1000 engine from mid-1943 until the end of the war, by which time the L-1000, so advanced in so many ways, was still incomplete. As noted below, the Army Air Crops did turn to Lockheed for the more conventional P-80 aircraft. Lockheed president Robert Gross recalled the Army's attitude about the L-133/L-1000 more succinctly: "The Army was very stern with us. It told us our double responsibility lay in turning out the P-38 and B-17, which the military regarded as the two great workhorses of the war."[29]

Johnson and Price at Lockheed clearly made the same basic, scientifically valid performance assumptions as Whittle and the Germans, although several years later. Certainly the L-133/L-1000 does represent the most advanced aerodynamic thinking in the United States before the European-originated turbojet revolution had its effect.

Only one other turbojet was designed in the United States. In 1937, a

Swiss engineer named Rudolph Birmann (who had written a dissertation on gas turbines under Stodola at Zurich in 1922), the De Laval Steam Turbine Company of the United States, and some private investors formed a small firm to manufacture turbochargers for the Navy. After preliminary studies that began in December 1940, their company, Turbo Engineering Corporation, received a Navy contract for a booster engine similar to the Westinghouse 19 in October 1942. Owing to the company's turbocharger commitments, however, the T.E.C. turbojet was soon dropped.

There were two other indigenous turboprop projects in the United States. A Czech engineer who had acquired a taste for gas turbines in Europe, Vladimir H. Pavlecka, was among the original founders of Northrop Aviation in 1939. He convinced John K. Northrop that a turboprop could be built that was competitive in efficiency with conventional engines. Northrop got a joint Army-Navy development contract in June 1941, but the engine was still incomplete at the end of the war. Finally, the earliest U.S. gas turbine project deserves mention. In 1934, R. E. Lasley, who had formerly worked for the Allis-Chalmers steam turbine division, designed and constructed a turbine engine. He offered it to both the Army and Navy, but both turned it down. Lasley had designed a very complex engine in a futile attempt to match conventional engine fuel economy at conventional speeds. His engine was abandoned in 1934.[30]

In addition to gas turbine projects, two other unconventional propulsion systems received support in the United States. After 1940, Pratt and Whitney, in cooperation with Andrew Kalitinsky of the Massachusetts Institute of Technology, began work on a very large 4,000- to 5,000-h.p. free-piston turboprop as a purely private venture. The goal of the project was an engine of superior fuel economy for very long ranges. Throughout the war the project was confined to component research, and was abandoned shortly after the war ended.

NACA imported the second unconventional system to the United States. Eastman N. Jacobs, the NACA aerodynamicist who had attended the Volta High Speed Conference in 1935, had seen models and read descriptions of the Campini ducted-fan system. Although, as noted below, the Campini system ultimately would prove unsuccessful, Jacobs thought it promising and began theoretical studies of it in 1939 and construction of a test unit in 1940. Several experimental units were ultimately built at Langley Field, but by 1943 it was apparent that the Campini system had no hope of competing with the turbojet, and work was halted.[31] It is indeed ironic that Jacobs, who was doing first-rate work on axial compressors apparently intended eventually for either turbosuperchargers or turboprops, should have simultaneously begun work on the Campini jet propulsion system. Jacobs clearly grasped both the radical assumptions about gas turbine component efficiency and the radical assumptions about airframe performance which together implied a turbojet, but he seems never to have put the two sets of ideas together. The only turbojets in production in the United States at the end of the Second World War, then, were those derived directly from the Whittle engine: the G.E. I-16 and I-40.

Several groups in other nations undertook turbojet or turboprop projects either before or during the Second World War. None was successful in building a production engine before 1945. Probably the most advanced were the French projects. Three groups in France had begun work on turbojets by late 1940. The first was Société Rateau, the steam turbine company. Société Rateau had manufactured turbosuperchargers as well as steam turbines, and in 1939 two officers of the company, René Anxionnaz and Roger Imbert, obtained French patents on a turbojet with a multistage axial compressor, multiple reverse-flow combustion chambers, and an axial turbine. Development began on the engine in 1939 and continued in secret during the German occupation. Actual component work began in 1942, but the complete engine was not tested until after the war. The Rateau turbojet was built in ignorance of foreign efforts, and, given the conditions under which it was built, it is not surprising that its performance was not comparable to that of foreign models.[32]

The other French turbojet project belonged to Société Turbomeca. Turbomeca was formed in 1938 by Joseph Szydlowski and M. Planiol to manufacture turbosuperchargers for Hispano Suiza piston engines. Sydlowski and Planiol had done considerable work on high-pressure superchargers, and in 1941, in secret, they designed a small turbojet. Nothing could be done, however, until after the Liberation.[33] The third French project was developed by Electro-Mechanique, the French licensee of Brown Boveri. Under the direction of P. Destival, a subsidiary of Electro-Mechanique began development of a large axial-flow turboprop in 1941, based in part on Brown Boveri patents. Like the other French engines, it was not completed until after the war, too late to lead the revolution.[34]

The Soviet Union claims to have done limited development work on ram-jets, rockets, and some turbojets during the 1930s.[35] At least three different Soviet turbojets had reportedly been tested by the end of 1940, but despite some official statements to the contrary, the Soviets do not seem to have developed an indigenous engine that was even remotely airworthy.[36] The Soviet Union did not acquire turbojets capable of being used as primary propulsion units until they captured their first German

Jumo 004's and B.M.W. 003's. The Red Army occupied many of the most important German research and industrial centers, which had been moved east and south to escape bombing from England. The Soviet Union thus fell heir to some of the most advanced German work. Furthermore, in 1947, Britain sold the Russians twenty-five Rolls-Royce Nenes and thirty Derwents, both derivatives of the Welland. Purchase of the Rolls-Royce engines cut about two years off the Soviets' development time, and turbojets based on the imported engines appeared in the MIG 15 over Korea. The Japanese, in contrast to the Russians, had no domestic turbojet program, although they had done research on axial compressors. During the war, Japan did take licenses on both the German Walther rocket engines and the 004 turbojet. A Japanese copy of the Me 262 flew on 6 August 1945, but was ditched in the ocean when its rocket assist units cut out and its pilot thought the turbojets had quit.[37]

Work on four other reaction propulsion systems formed a backdrop to the development of the turbojet. Even though none of the other systems employed a gas turbine, each was at least partially based on the same radical assumptions about airframe performance as the first turbojets. The most fully developed of the other alternative systems was the ducted propeller design of Secondo Campini. Campini was an Italian engineer who in the late 1920s recognized both that airframes ought to be capable of much greater speeds and that the propeller posed the major obstacle to attaining those speeds. As a result, in 1930, Campini patented a ducted fan system comprising an annular fuselage enclosing a multistage fan (propeller) driven by a conventional piston engine, aft of which fuel could be introduced and ignited to increase thrust. From the early 1930s, Campini received financial backing from the Caproni Aircraft Company, and in 1940 a Campini-powered aircraft did fly. Nevertheless, the Campini system was simply too big, heavy, and inefficient for the thrust it produced. The Caprioni airframe in which the Campini power plant was installed was twice as big and twice as heavy as the first experimental turbojet aircraft. Accordingly, the performance of the Campini system was disappointing, and work on it was discontinued.[38]

Two of the other systems were ram-jets similar to that proposed by Warren. The first was proposed by René Leduc in 1933–34, also for attaining supersonic speeds. It consisted of an annular fuselage which contained a pod for the pilot and which served as the duct for the ram-jet. A model of the Leduc design was displayed at the Paris Air Show in 1938, but little development was done during the war.[39] Better progress was made by Eugene Sänger. Sänger had become interested in liquid-fueled rocket motors for suborbital flight during the late 1920s, and had experimented with rockets at the Technical University of Vienna during 1933–34 before being recruited in 1936 for a new Luftwaffe experimental establishment (in Germany) devoted to rocket development. Beginning in 1939, Sänger designed ram-jets for near-sonic-speed interceptors. In the later stages of the war, numerous tests of Sänger ram-jets were made, with the ram-jet mounted either on a truck or above a Dornier Do 217 conventionally powered bomber which served as a flying test bed. Sänger even made some tests with pulverized coal as fuel, owing to Germany's acute shortage of petroleum in 1944–45. Although Sänger's work was probably more advanced than that of either Campini or Leduc by 1945, no aircraft powered solely by a Sänger ram-jet had flown.[40]

The fourth reaction propulsion system used for aircraft was the pure rocket. The rocket engines developed by Helmut Walther at Kiel, beginning in 1933, for use as takeoff assistance units for conventionally powered aircraft were adapted for use in a high-speed rocket interceptor, the Me 163. The Me 163 airframe, initially designated DFS 194, was the creation of a brilliant glider designer, Alexander Lippisch, who had been associated with the Rhön-Rossitten Gesellschaft, Germany's preeminent glider society, and who also had directed Fritz von Opel's nonsensical rocket-powered flight experiments in 1928. Work on the rocket-propelled plane, then intended purely for research, began in 1938 at the Deutsche Forschunganstalt für Segelflug (DFS, German Research Institute for Sailplanes, successor organization to Rhön-Rossitten), and was transferred to Messerschmitt in 1939. An experimental Me 163 attained 623 mph in October 1941. Such astonishing performance caused the RLM to order development of the ME 163 as a service interceptor; and redesign began under Lippisch's direction in December 1941. Operational units were first formed in June 1944. Although moderately successful in combat against Allied bomber formations, the Me 163 had a very short powered endurance (12 minutes) and was extremely dangerous to its pilot because of its hypergolic fuels (hydrogen peroxide and hydrazine hydrate), so dangerous, in fact, that none of the Allied countries permitted their test pilots to fly it after the war.[41] Thus the only really practicable reaction propulsion engine for aircraft developed during or after the war proved to be the turbojet.

Intrinsic to the turbojet revolution in all countries was the evolution of testing techniques and equipment appropriate to the new propulsion system. Conventional dynamometers and propeller test-stands had no relevance to the turbojet; even units for measuring output differed—

pounds thrust rather than horsepower. The importance of high altitude for turbojet performance and the huge air volume ingested by the turbojet likewise rendered the altitude testing installations used for piston engines obsolete. The difficulties that all the turbojet pioneers had in finding (or not finding) component test facilities has already been noted. Once it was clear that the turbojet revolution would not just go away, the necessary facilities were built, albeit at great cost. The work of Rateau and others did provide some guide to the fundamental design of compressor and turbine test cells. Testing of complete turbojets, however, also required accurate thrust measurement; precise monitoring of temperatures and pressures at numerous points throughout the engine; exact recording of engine rpm, fuel flow, air flow, and oil consumption; regulation of ambient temperature and pressure; and, obviously, protection of test personnel from exploding or disintegrating engines.[42] As with all of the radical new technologies surveyed here, the creation of the turbojet demanded the creation of new testing technologies that in themselves were major technological achievements. The scale of test equipment could be awesome. When Whittle visited the United States for the first time in mid-1942, he was amazed to find General Electric not only preparing for production of 1,000 turbojet engines a month, but also "planning a full-scale test plant for the individual testing of compressors, turbines, etc., on a far bigger scale than anything we [the British] had visualized, and for which the cost was estimated to be about three million dollars, i.e., about twelve times as much as we [Power Jets] were spending on our full-scale test plant at Whetstone."[43] The new normal technology demanded radical alteration and extension of the traditions of technological testability central to prime mover development since Fourneyron.

Finally, a study of the turbojet revolution would be incomplete without some consideration of the aircraft built to take the new engines. The turbojet-powered airframes of the Second World War were really more significant for what they were not than for what they were: with one exception, all were of conventional configuration; all, without exception, were of conventional construction. In general, the turbojet airframes verified the performance assumptions made by the men who first imagined the turbojet: the first jet aircraft represented the perfection of aluminum stressed-skin construction and of subsonic streamlining techniques. They made fullest use of all the interwar developments in leading-edge slots, trailing-edge flaps, retractable undercarts, even dragless radiator ducting. The tremendous power made available to aircraft designers by the turbojet did inspire supersonic aerodynamic research

Messerschmitt Me 262 A, 1943. (Courtesy of the Smithsonian Institution.)

during the war, especially research on swept wings. After the war, new aerofoils, new construction methods, indeed a new aerodynamics, would use the turbojet's power to attain ever higher performance. But all that is beyond the turbojet revolution itself. The first turbojet airframes, while embodying a revolution, were quite conventional.

As mentioned, the Germans produced more and a greater variety of turbojet aircraft than any other belligerent. The German Messerschmitt Me 262 was produced in larger numbers and employed operationally in larger numbers than any other World War II turbojet aircraft. It was the first turbojet fighter to enter service anywhere. It was also probably the best turbojet aircraft of the war. The Me 262 is widely acknowledged to have been aerodynamically superior to any Anglo-American aircraft as late as 1946–47.[44] In service configuration the Me 262 had a maximum speed of 540 mph at 20,000 feet, a sea level rate of climb of 3,937 feet per minute, an average endurance of one and a quarter hours, and a maximum service ceiling of 37,565 feet. The only real flaws in the Me 262's performance were in takeoff and landing. It carried a very high wing loading of 66 pounds per square foot, had a takeoff speed of 190 mph, and required a takeoff run of at least 3,200 feet into a 15-mph wind. Landing speed was correspondingly high, just under 200 mph. Late in the war the Me 262 was fitted with rocket assist units that shortened takeoff run to about 2,000 feet.[45]

The Me 262 was a low-wing, all-metal monoplane. Its fuselage was a five-section semimonocoque structure. The wing was built on a composite single "I-beam" spar and had a leading-edge sweepback of 18.5

degrees. The wing was designed with sweepback for center-of-gravity reasons; at the time the Me 262 was conceived, the benefits of wing sweep in delaying Mach effects were not yet verified. Nevertheless, the wing sweep of the Me 262 did delay high-speed compressibility problems. The leading edges of the Me 262 wings carried rolled sheet steel slots that moved on tracks in the wings. The slots opened automatically at high angles of attack to prevent airflow breakaway and preserve lift. The tail structure and control surfaces of the Me 262 were of conventional configuration and construction. The first through fourth prototypes used a tailwheel-type undercarriage, but all prototypes after the fifth and all production Me 262's had a tricycle landing gear to facilitate takeoff and to reduce runway damage from jet blast. Production Me 262's were powered by two Junkers Jumo 004B turbojets carried in pods under either wing. No high-strength alloys were used in construction of the Me 262: it was built exclusively of aluminum alloy and some steel. It was successfully designed for ease of fabrication, and many Me 262's were built by slave labor. The basic single-seat production version of the Me 262 mounted four Mk 108 30mm. cannon in the nose. Some Me 262's were equipped to carry two 550-pound bombs on shackles under the fuselage, and some were equipped with underwing pylons for twenty-four or forty-eight R4M air-to-air rockets for the bomber interception role. Several Me 262's were experimentally modified to carry one 50mm. cannon in the nose. A two-seat trainer, and a two-seat radar-carrying night fighter version, the Me 262B, were built in small numbers. Some Me 262's were also equipped with cameras for photo reconnaisance missions.

The Me 262 first flew, with a 700 h.p. Jumo 210 piston engine in the nose and empty turbine housings, on 4 April 1941. In November, still with the prop engine in the nose, the Me 262 flew with two prototype B.M.W. 003's. Next, two Heinkel 001's were installed but provided inadequate power for the Me 262, which was considerably larger than the He 280. The first pure turbojet flight of the Me 262 occurred on 18 July 1942, with two preproduction Jumo 004's. Testing continued through the remainder of 1942 and into 1943. With a modicum of foresight, the Me 262 could have been ready for quantity production in the second half of 1943. But Reichsmarshal Göring and Erhard Milch, head of the Technical Office, had at the time awarded priority to bomber production, and throughout 1942 and early 1943 blocked production of the Me 262. Hitler also forbade Me 262 production until August 1943, when the decision was finally made to build it in quantity. Production was delayed by heavy Allied bombings in February and April 1944. Then, in May 1944, Hitler ordered that all available Me 262's be converted to "Blitzbombers"—fast, low-altitude fighter-bombers, a role for which the

Me 262, with its turbojets efficient only at high altitudes, was completely unsuited. It was the end of 1944 before Hitler reversed himself and the Me 262 was available in quantity for deployment as a fighter. By then it was too late. The opportunity thrown away at the end of 1942 could not be redeemed.[46] Altogether, about 1,400 Me 262's were built. Only about 200 actually reached squadron service, and many of those were squandered on low-level bombing missions.[47]

Nevertheless, the Me 262's which did fly reconnaisance or bomber interception missions proved stunningly successful. It seems likely that had the Me 262 been deployed quickly and in quantity, it would have changed the course of the war. In all probability, the Me 262 could have crippled the Allied bomber offensive over Germany and could have denied the Allies the absolute local air superiority requisite to the Normandy invasion. The Allies had no airplane during the war to equal the Me 262, and it would have certainly been the beginning of 1945, at the earliest, before they could have deployed a turbojet fighter with sufficient range to challenge the Me 262 over the Continent.

The only other German turbojet aircraft to enter service in significant numbers was the Arado Ar 234 bomber-reconnaissance plane. The Ar 234 was built in several versions. It was designed in 1941, first flew in mid-1943, and entered production in early 1944. The Ar 234A used a launching trolley and a skid landing system; the Ar 234B and Ar 234C had conventional nosewheel-type undercarts. The Ar 234 was a high-wing, single-seat, all-metal monoplane, somewhat larger than the Me 262, but still quite small by bomber standards. The most common variant, the Ar 234B, had two Jumo 004 turbojets slung in pods under the straight wings. Some Ar 234B's used B.M.W. 003's. The Ar 234B had a maximum speed of 461 mph at 20,000 feet, a range of roughly 1,000 miles, and could carry a maximum bomb load of 3,300 pounds. A later version, the Ar 234C, employed four B.M.W. 003 engines housed in twin pods and could attain 542 mph at 20,000 feet. Just over 200 Ar 234B's were built, but probably at most 100 reached operational units. Very few Ar 234C's saw service. The Ar 234 was the world's first turbojet-powered bomber.

The last turbojet-powered aircraft to be issued to Luftwaffe operational squadrons was the Heinkel He 162 Volksjager ("People's Fighter," literally "People's Hunter"). It is ironic that after having his He 112 passed over in favor of the Bf 109, his very promising He 100 stifled, and his He 280 aborted, Heinkel should have the RLM turn to him in desperation in the fall of 1944. The He 162 was designed to be produced with an absolute minimum of strategic materials and labor and to be flown by an unskilled pilot trained to fly in a glider and to shoot in a swing. It was the embodiment of Germany's final realization that she must have superior numbers of superior fighters if she were to survive. Design of the He 162 began on 8 September 1944, and the first prototype flew on 6 December of the same year: from drawings to flight took only sixty-nine days. The He 162 was a high-wing monoplane. The wing was a one-piece bonded-plywood structure. The fuselage was semi-monocoque aluminum except for the nose and the doors for the tricycle landing gear, which were wood. The twin tail unit was of composite construction. The He 162 was powered by a single B.M.W. 003 turbojet mounted above the wing, and the plane carried an armament of two Mk 108 30mm. cannon. It had a maximum speed of 522 mph at 20,000 feet and an endurance of almost an hour. A production of 4,000 He 162's per month was planned for mid-1945. At the end of the war, only 116 had been completed, and certainly fewer than a score had been issued to operational units. But partially completed He 162's were found scattered all over Germany.

The Germans produced three other turbojet-powered aircraft, but none got beyond the prototype stage. The most unconventional was the Horten-Gotha Go 229. Two brothers, Reimar and Walter Horten, had designed a series of all-wing gliders before the war. In 1942 they began design of an all-wing (tailless) turbojet fighter. Construction and development, under the Hortens' direction, was undertaken by Gotha. With two Jumo 004C engines of 2,200 pounds thrust each, the all-metal Go 229 was expected to attain 640 mph at 22,000 feet. Flight tests were about to begin when the Gotha works were occupied by Allied ground units.

The Junkers Ju 287 was the world's first turbojet-powered heavy bomber. By the time design of the Ju 287 began in 1943, the Germans had discovered the beneficial Mach number effects of swept wings. They had also discovered the very poor low-speed handling characteristics of backward swept wings. Dr. Hans Wocke of Junkers therefore designed the Ju 287 with wings swept forward 25 degrees, which provided delay of compressibility without sacrificing low-speed stability. The Ju 287 was originally intended to have four Heinkel 011 turbojets, but difficulties in that engine's development led to redesign of the Ju 287 to take six B.M.W. 003's, one on either side of the nose and two in twin pods under each wing. With the six 003's, a speed of 537 mph at 17,000 feet was anticipated. By the end of the war, only a turbojet-powered (four 004's) test aircraft with the new wing had flown.

The final German turbojet aircraft was a dive bomber, the Henschel Hs 132. The Hs 132, begun in early 1944, was similar in configuration to the He 162, only in the Hs 132, the pilot lay prone. Like the He 162, the Hs 132 had a plywood-skinned wing, an aluminum semimonocoque fuselage, and a twin tail. With a single B.M.W. 003 turbojet mounted

Junkers Jumo 004. (Courtesy of the Smithsonian Institution.)

above the fuselage, the Hs 132 was expected to attain 485 mph at 20,000 feet.[48]

The Germans also had a whole array of advanced supersonic projects underway when hostilities ended. Both Messerschmitt (P.1111) and Focke-Wulf (T.A. 183) had swept-wing fighters under development. Lippisch had begun design of a tailless supersonic fighter (P.11) and was investigating a very advanced delta-wing design (P.13).[49] Some projects were far enough along that experimental aircraft were actually under construction. At war's end, however, none had flown. At the B.M.W. facility near Munich the Germans had the finest high-altitude, high-speed engine test-plant in the world. With the exception of high-temperature alloys, for which they did not have the rare metals, the Germans by 1945 had a lead of from two to four years in turbojet engine and aircraft technology.

The English had two turbojet-powered aircraft flying at the end of the war. Only one, the Gloster E.9/40 Meteor, was operational. The Meteor had first flown with Whittle engines in July 1943, although, as noted, it had flown in March of that year with prototype Goblins. The Meteor was an all-metal, low-wing, single-seat monoplane with a four-section monocoque fuselage, straight, tapered wings, and a tricycle undercart. A turbojet was mounted in a pod through the center of each wing. The Meteor carried an armament of four 20mm. cannon. The first Meteor I's, only twenty of which were built, were rushed into service in July 1944 against the German V.1 flying bombs sent against England. The Meteor I employed Rolls-Royce (Whittle design) Welland I engines

Gloster F.9/40 "Meteor." (Courtesy of the Smithsonian Institution.)

of 1,700 pounds thrust and had a maximum speed of 410 mph at 30,000 feet. The next version of the Meteor to enter service was the Meteor III, which went operational on the Continent in mid-April 1945. With uprated 2,000-pound-thrust Derwent I engines, the Meteor III had a speed of 493 mph at 30,000 feet and a maximum range of 1,340 miles. Something on the order of a hundred Meteors actually saw service during the war, but there is no record of their ever having encountered any German turbojet aircraft. The Meteor was a complex airplane, and while inferior to the Me 262 aerodynamically and therefore in performance, it was developed to much higher degrees of finish and reliability. Such painstaking development and manufacture was of course made possible by England's relatively secure air position after 1942.

The other English turbojet fighter was the de Havilland E.6/41, the D.H. 100 Vampire. The Vampire was an all-metal, single-seat, single-engine, straight-wing monoplane. To avoid the losses entailed by long intake and exhaust ducting, the D.H. Goblin engine was positioned immediately behind the pilot, with air intakes in each wing root, and the fuselage was bobbed off just aft of the engine. The vertical and horizontal stabilizers were therefore mounted on twin booms extending back from the wings in a fashion similar to the Lockheed P-38. The Vampire first flew in September 1943, but the first production aircraft did not fly until 20 April 1945, three weeks before the end of the war in Europe. The Vampire was thus not in service until after the war was over. The Vampire, with a 3,100-pound-thrust Goblin II engine, could attain 531 mph at 17,500 feet.

The United States had three airworthy pure turbojet aircraft by the end of the war. None were operational, although the earliest of the three could have been. The first U.S. turbojet-powered aircraft was the Bell P-59A Airacomet. Design of the P-59A began in September 1941, and

The New Normal Technology

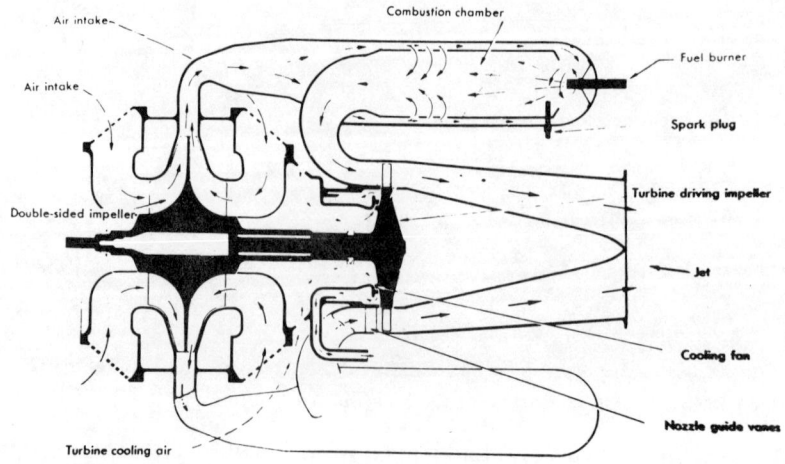

Air flow path in Whittle–Rolls Royce Welland. (From General Electric, *Aircraft Gas Turbine Engineering Conference* [West Lynn, Mass.: G.E., 1945].)

the plane first flew on 1 October 1942 with G.E. prototype engines. The P-59A was an all-metal, midwing monoplane with two turbojets mounted beneath the wing roots next to the fuselage. The P-59A and the G.E. I-16 engines for it were fully developed by July 1943. The maximum speed of the P-59A was 414 mph. By mid-1943, however, American piston-engined P-51's and P-47's were attaining higher speeds with much better fuel economy than the P-59A, so it was relegated to the training role. The performance of the P-59A was comparable to that of the Meteor I, but the P-59A had a lighter wing loading and was therefore more maneuverable.

The other U.S. Army Air Corps jet, the Lockheed XP-80, was initially designed to take the D.H. Goblin engine. The XP-80 was designed, built and flown with a British-supplied Goblin in 143 days, the first flight occurring on 9 January 1944. Because of the great promise of the G.E. I-40 engine, the XP-80 was substantially modified to become the XP-80A, and first flew with the new I-40 in June 1944. The XP-80A was ordered into production in August 1944, but had not reached operational squadrons by the end of the war, although two early production models were sent to England, and two to Italy, very late in the war to raise the morale of American pilots who had encountered German jet aircraft. The Lockheed XP-80A was a low-wing, single-seat, all-metal fighter powered by one G.E. I-40 (J-33) 4,000-pound-thrust turbojet mounted in the after section of its semimonocoque fuselage. Air intakes were located just forward of the wing roots; the tail unit was of conventional configuration with the jet exhaust beneath it. The maximum

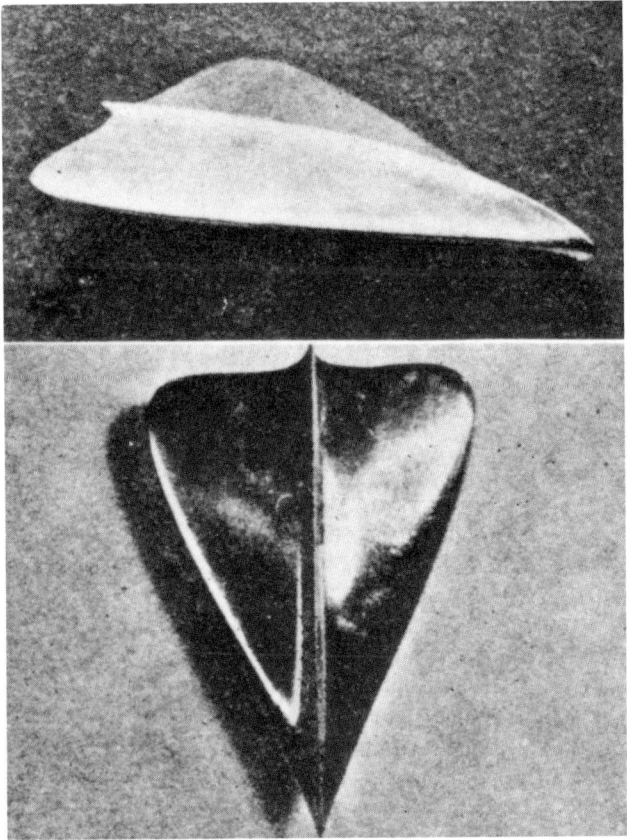

Wind tunnel model of Lippisch P. 13 supersonic delta wing design, 1944. (From R. Smelt, "A Critical Review of German Research on High-Speed Air Flow," *Jl. RAeS.* 50 [1946]: 899–934.)

speed of the XP-80A was 550 mph, but that speed was dependent upon a specially polished and waxed airplane surface. The third U.S. turbojet aircraft was a twin-jet fighter built under a Navy contract let in January 1943. The McDonnel FD-1, the original "Phantom," was powered by two of the small 1,300-pound-thrust Westinghouse 19B turbojets, and had flown in January 1945. It was not in production at the end of the war.

In England and in Germany the same basic, radical assumptions about aircraft speed and about possible gas turbine component efficiencies led different individuals to the turbojet engine. From their work the turbojet revolution spread throughout the world aeronautical community. Yet not all turbojet engines or aircraft were the same. Each country

Bell P-59 A "Airacomet," 1943. (Courtesy of the Smithsonian Institution.)

required of the turbojet slightly different characteristics based on that country's particular strategic and industrial situation. England and the United States emphasized safety, reliability, and fuel economy, which, combined, would provide at least the possibility of adequate flight endurance. By 1943, the strategic problem facing the air staffs in the United States and Great Britian was not asserting control over their own skies but maintaining it over Germany. Similarly, to the United States and Britain, costs in terms of labor, machine tools, or rare metals were of little moment, and the Allied engines used all three lavishly.

The Germans, with different strategic requirements, approached development of their turbojets differently. By late 1943, the primary problem facing the Luftwaffe was blocking the massive Allied bomber offensive over Germany. The RLM therefore was forced to seek the most rapid possible production of usable engines. Owing to Germany's military overextension, the engines could use hardly any rare-metal alloys and had to be built with a minimum of relatively unskilled, often "foreign" (slave) labor. The absence of rare metals in the Jumo 004 and the B.M.W. 003 has already been noted. The 004, moreover, required only 700 man-hours to build and the 003, only 500, compared to 3,000 to 5,000 for a conventional D-B 600–series piston engine. In order to eliminate rare alloys and to reduce production costs, the Germans accepted large performance penalties in engine efficiency and in service life—penalties the Allies did not have to accept.[50] But range was not as

Lockheed P 80A. (G.E., 1945.)

vital to the Germans, who desperately needed interceptors to fight over their own territory. And, "although the life of the turbojets was only a very small fraction of the normal life of a conventional engine, it was a very large fraction of the average life of German conventional engines in fighters on combat missions."[51]

Airframe-design philosophies also differed. Both the Me 262 and the He 280 were designed for a maximum speed of at least 528 mph—just below the velocity at which theoretical investigation indicated compressibility would become a serious problem. Indeed, the Me 262 was redlined for 658 mph, about 0.85 Mach. In contrast, the first two Allied jet aircraft, the Meteor and the P-59A, were designed for maximum speeds of about 400 mph—just slightly faster than then-current (1942) piston-engined fighters. As a result, the Meteor and P-59A were outclassed by Allied piston-engined fighters by mid-1944, while the Me 262 was markedly superior to all Allied fighter aircraft.

In the years following World War II the new normal technological tradition would produce turbojets of ever greater thrust, efficiency, and reliability. Those achievements would derive from continuing progress in the subsystems which constituted the turbojet: greater experience and refined theory would yield more efficient compressors and turbines, and new alloys and production processes would permit ever-higher cycle temperatures. Design innovations, such as bypass and turbofan engines and the use of afterburning, would provide much greater propulsive efficiency or extraordinary supersonic performance.

Airframes, which had coevolved so effectively with piston engines and which had defined an aeronautical environment ideally suited to the first turbojets, would in turn be revolutionized by the turbojet. The immense thrust available from very lightweight turbojets in effect defined a new technological niche—a new supersonic environment—to which conventional subsonic airframes were totally unsuited. Entirely new wing shapes and profiles, together with new structural approaches and new materials, would be necessary to pierce the sound barrier and then to fly comfortably beyond it.[52] New turbojets designed for transsonic speeds would co-evolve with new supersonic airframes. In time, also, turbojet-powered aircraft would revolutionize commercial air transport as thoroughly as had the first modern piston-engined transports in the 1930s.[53]

The turbojets would bring revolution to science no less than to technology. A new supersonic aerodynamics would be created, only to give way to aerothermodynamics as increasingly powerful turbojets pushed aircraft to speeds at which the generation of heat on the surface of the aircraft became a major factor in airflow behavior. Eventually, turbojet-powered aircraft would reach speeds at which magnetothermodynamic considerations would become paramount: temperatures would become so great that air would dissociate into charged submolecular ions.[54]

All that lay in the future. Even by 1945, the turbojet revolution had fulfilled the promise its progenitors had visualized. A renewed and redefined community of practitioners under the leadership of new members both in old firms (Pratt and Whitney, Rolls-Royce) and in new firms (General Electric), and minus some old members (most notably, Curtiss-Wright),[55] were pursuing with awesome success a new normal technology.

9

Conclusion

The turbojet revolution depended upon prior technology, upon a two-centuries-long evolution of turbine systems: water turbines, turbine pumps, steam turbines, turbo-air compressors, internal combustion gas turbines, and turbosuperchargers. It required the prior successful co-evolution of piston aero-engines and steamlined airframes. Yet the turbojet itself was a simple, linear extrapolation of no prior technology. The turbojet is a holistic system with novel properties no predecessor system ever contemplated exhibiting. The turbojet would overthrow and very nearly extinguish the technological tradition founded upon piston engines and propellers, and while fully exploiting the production technology associated with steam turbines, the turbojet would revolutionize their design as well.

The turbojet revolution also depended upon rigorously maintained traditions of technological testability, upon the valves of "scientific" technology, and upon the culture from which those traditions and values derived. The protagonists of the turbojet revolution, no less than their adversaries who were committed to normal technology, fully understood and fully subscribed to community-defined requirements of testability and replicability. The turbojet pioneers, no less than their opponents, recognized the necessity of adapting their ideas to meet total-systems interface constraints imposed by the whole community of aeronautical practitioners. Yet if all the protagonists of technological revolution portrayed here fully acknowledged the sovereignty of community norms, they also responded to strong counternormative traditions within those same communities. The men who created the turbojet would test their engines with merciless rigor and report their success, or their failure, with unequivocal candor while at the same time challenging the most sacred performance parameters and most fundamental assumptions of normal technology.

The turbojet revolution not only overthrew a tradition of practice, but also sharply redefined, on its own terms, a relevant community of practitioners. The revolution was the progeny of men uniformly outside the conventional aero-engine community. Yet not one of the original creators of the turbojet personally saw his engine through to production. The turbojet revolution became a community revolution fomented by outsiders.

The creation of the first turbojets, while clearly linked to prior turbine technology, to previous aeronautical practice, and to the traditions and values of scientific technology, depended most critically upon radical assumptions about airframe and gas turbine performance deduced from contemporary advances in theoretical aerodynamics. The perception of presumptive anomaly derived from science demarcates successful from unsuccessful proponents of revolutionary alternative aircraft propulsion systems. Only the complete recognition of presumptive anomaly separates the four men who created turbojets from the two men at General Electric who came so close or from the proponents of turboprops or other reaction propulsion systems.

Yet the specific search processes of even the successful protagonists of revolution differed. Whittle and von Ohain appear to have followed an essentially random, iterative program, exploring one alternative at a time, letting each idea lead them on to the next idea, in a manner more or less in accord with introspectionist descriptions of inventive processes. Wagner and Schelp, in contrast, conducted an overtly structured, parallel, virtually morphological search which considered and evaluated by *a priori* criteria a wide range of alternative propulsion systems.[1] These perceived differences in search processes may reflect fundamental underlying cognitive differences. Alternatively, these apparent differences may result only from Whittle's and von Ohain's working alone and thinking to themselves, while Wagner and Schelp deployed the resources of large organizations in a formal, interpersonal search process. In either case, however, clear and accurate perception of presumptive anomaly both sharply reduced the very large search space otherwise presented by possible alternative propulsion systems and rigorously structured the search process itself.

Nevertheless, the technological universe vicariously discovered through presumptive anomaly ultimately would be directly explored. The immaculately streamlined, world-speed-record-setting He 100 and Me 209 when flat-out and level, and the early Mark British Spitfires and American P-38's and P-39's when power dived, all directly encountered the propeller compressibility and airframe performance limits described in aerodynamic theory. Similarly, the late prewar Brown Boveri compressors and turbines designed in accordance with aerofoil theory precisely corroborated radical assumptions about component efficiencies of internal combustion gas turbines. In each case the world discovered vicariously in scientific theory proved to exist in technological fact.

This logical and expected relationship between vicarious and direct exploration holds an important implication for the alleged inevitability of invention. The pursuit of conventional technology—of piston engine and propeller practice, of steam turbine and internal combustion gas turbine practice—clearly leads to direct encounter with precisely those circumstances underlying the turbojet revolution and defined in presumptive anomaly. In this sense only—in the sense that valid vicarious exploration discovers the same world as direct exploration—the turbojet revolution was inevitable. Yet this tautological inevitability says nothing about the specific aspects of nature or the specific elements of technology which in fact formed the foundation for the turbojet revolution. While science may behave as if there were but "one world to discover," technology is both equifinal and equipotential. That is, within the broad and changing structure of the laws of nature, technology develops via the meanderings of technological co-evolution: development of current technology defines new ecological niches for further technological development. Technology itself defines both the substance of further variation and the criteria of retention. Technology thus hypothetically can go in several directions equally well, or alternatively, can reach the same end point from diverse starting points through diverse developmental paths. This nonteleological system is determinant, then, only in the very limited sense noted above. A mere pious incantation of this "inevitability" yields no historical understanding of the turbojet revolution as it occurred.

In a similar sense, examination of the turbojet revolution solely from the perspective of economic factors or of the patent system yields little historical insight. Expanded military budgets prior to World War II clearly account for the financial resources devoted to turbojet development. Yet without presumptive anomaly, without prior conceptualization and development of the turbojet, no plausible amount of money spent on aeronautics for any purpose would seem likely to have led quickly to the turbojet. Science, not rearmament, created the turbojet; air ministries merely bought what was offered. Rearmament gave the ministries the surplus funds to invest in radical, marginal, or superficial projects not feasible in times of severe budgetary constraints. But for science, however, those funds would not have produced a turbojet engine. The patent system likewise neither portrays nor accounts for the turbojet revolution. To a vastly greater extent than recounted here,

Conclusion

patents and proprietary rights were of critical and understandable concern to the protagonists of the turbojet revolution. Yet as the examples of Guillaume and Whittle so pointedly illustrate, patents are hardly a reliable guide to the historical reality of that revolution.

As suggested in Chapter 6, the historical explanation for the particular spatial and temporal configuration of the turbojet revolution may lie at the fairly abstruse level of national cultural differences. The United States lagged in theoretical development of high-speed aerodynamics and in the provision of high-speed research facilities, and neglected theoretical investigation of turbomachinery phenomena. It dropped out of international high-speed competition after 1926, and had no aircraft capable of approaching 400 mph until late 1939. Instead, the United States developed the world's best commercial transports and the world's best and largest commercial aviation system, and devoted its aerodynamic research efforts almost exclusively to development-oriented problems. European, especially German and British, aerodynamicists and aeronautical practitioners exhibited obverse patterns in the pursuit and utilization of scientific knowledge. Thus, the information field in which American practitioners worked, in sharp contrast to that of the Europeans, was clearly biased against recognition of the theoretically founded presumptive anomaly so critical to the turbojet revolution. That bias, in turn, would seem to reflect fundamental differences in the broad cultural values underlying American and European technological traditions.

The model for technological change presented here, and the perspectives deriving from its application to the origins of the turbojet revolution, may have wider relevance. A consideration of the evolution of traditions of technological testability would seem likely to inform discussion of much modern technology, at least since the late eighteenth century. Similarly, more careful attention to the role of systems interface constraints and the evolution of holistic technological systems should provide a fuller and richer context for comparing the development of various modern technologies. Examination of the process of technological co-evolution should yield greater understanding of the direction of technological change and of the choice constraints imposed upon decisionmakers. The emerging and intensifying relationship between scientific theory and technological practice, mediated by community traditions of testability, seems central to the development of sophisticated technologies since Watt. Finally, the mechanism of presumptive anomaly–induced technological change, so fundamental to the turbojet revolution, also appears to be fundamental to a whole array of other modern technological revolutions.

These insights, in combination with much greater attention to facets of technological change barely touched upon here—the nature of normal technological development, the practice of "forced invention" in response to "log-jams" in technological systems hierarchies, the special roles of development of and innovation in materials and fabrication techniques—may lead both to a more comprehensive and a better-integrated understanding of technological change.

Still, when the epistemology has been dissected and the values and cultures examined, when the traditions of practice and the insights of science have all been properly analyzed, what is left are men. One of the most striking facets of the origins of the turbojet revolution is the remarkable continuity of men and organizations—the recurrence of names like Francis, Parsons, Rateau, and Moss; of companies like Brown Boveri, British Thompson-Houston, or General Electric; of academic figures like Burdin, Hesse, Stodola, Prandtl, or Ackeret; of institutions like the Ecole des Mines, Zurich, the AVA, or the RAE. This persistence implies neither that "history is a seamless web" nor that new inventions are merely "new combinations of old ideas," but rather that information fields are powerful and, most of all, enthusiasms contagious. Finally, the excitement and passion, the elegance and pride bordering on arrogance barely restrained by the mask of scientific objectivity, that imbues men like Griffith and Whittle and Wagner and Johnson reveal in technology a familiar and human dimension. They are impressive men.

And it is in the insight and determination of these individual protagonists that the final historical cause of the turbojet revolution is embodied. That insight and that determination were born of something more than reason or avarice. In June of 1922 the British Institution of Mechanical Engineers met jointly in Paris with the Société des Ingenieurs Civils de France—for the first time since the summer of 1914. The day before the meeting, the members had, in the words of the president of the Institution, "passed through the devastated regions, and it was well that they should do so before coming to the Meeting, because it brought home to them, in a way they could never have realized, the terrible effects of the War upon France."[2] With that prelude, Auguste Rateau read his great paper on turbosupercharging and high-speed, high-altitude flight.[3] Comprehensive discussion of Rateau's paper was postponed until a special general meeting of the Institution back in London in October. In that discussion, William Reavell, Rateau's virulent antagonist in the turbocompressor debates fifteen years before, commented on Rateau's paper:

> When he read through the Author's [Rateau's] conclusions, he could not help drawing another "conclusion" of his own. A good many of the members would remember meeting the Author in Paris. He was not in his early youth, and he had already been the Author of many fine contributions to science; but the conclusion that he (Mr. Reavell) came to was that the Author was still

one of the young men who saw visions of the future and not one of the old men who dreamed dreams of the past. The Author reminded him of his famous compatriot, Jules Verne, whose works all the members read with such avidity in their youth. They had seen Jules Verne's stories come true. They all enjoyed reading about the "Nautilus" in "Twenty Thousand Leagues under the Sea," and some of the members during the War had a great deal to do with voyages under the sea. And now another of France's famous sons, not in a story, but in a scientific treatise, had shown the way in which their children, or perhaps their children's children perchance some day might move with incredible speed in the stratosphere.[4]

In 1930, the year Frank Whittle patented the first true turbojet, the American Society of Mechanical Engineers struck a medal to commemorate the society's fiftieth anniversary. That medal was inscribed simply, "What Is Not Yet, May Be."[5]

A Technical Appendix: Some Essentials of the Aeronautic Environment and of Aero-Engines

Specialized communities of practitioners evolve their own internal languages which serve them as highly efficient communications media. For outsiders, however, beneficiaries neither of formal preparation nor of participation in the technological evolution which the language describes, such an argot is frequently intimidating. Yet the language is anything but meaningless: it tags important, often fundamental theoretical concepts or empirical facts. This technical appendix is intended to provide a rudimentary guide to essential concepts and nomenclature in aeronautical science and in mechanical engineering relevant to the turbojet revolution. It is not meant to be a sophisticated protrayal of those sciences; it is meant only to convey enough information to prevent the reader's being confused by, say, engine type designations, as this author initially was.

AERONAUTICS

All aircraft operate well within the earth's atmosphere—that is, somewhat below an altitude of 100,000 feet. As is commonly known, the earth's atmosphere is composed of a mixture of gases which declines in density, pressure, and temperature as altitude increases. The density of the atmosphere at 44,000 feet, for example, is only one-fifth that at sea level; and at 59,000 feet, oxygen concentration is not sufficient to support even the burning of a candle.[1] Atmospheric air is viscous—that is, has internal cohesion—and therefore offers frictional resistance to the passage of an object through it. For velocities relative to an object not approaching the speed of sound (less than 200 mph), air generally obeys the hydrodynamic laws which describe the behavior of an incompressible fluid. For velocities approaching that of sound (which decreases with altitude), the earth's atmosphere behaves as a compressible fluid. As fluid speeds, relative to the aircraft, increase above 350 mph, the simplifying assumption of air incompressibility becomes increasingly invalid. (Although the speed of sound at sea level is about 740 mph, some parts, such as wings or propeller tips, of aircraft flying at less than half that speed will, due to aerodynamic circumstances, experience flow rates in the transonic regime.) Still, the commonly used phrase "sonic barrier" or "sound barrier," which is associated with compressibility, is a misconstruction. Although early encounters between aircraft and near-sonic velocities did often produce violent results, no real "barrier" to flight exists, just a region of change in atmospheric behavior patterns.

Lift and Drag

All powered, heavier-than-air craft must be able to meet fundamentally similar aerodynamic conditions. The common parameters of all atmospheric powered flight are weight, lift, drag, and thrust. In simplified terms, the four parameters are related in the following manner: for flight that is constant in speed, altitude, and direction, *weight* must be equaled by *lift*, the upward force produced by airflow across the aircraft's lifting surfaces (principally the wings); *drag*, the resistance of air to the motion of a body through it, must be equaled by *thrust*, the propulsive output of the aircraft's engine. Aeronautical engineers distinguish between several kinds of drag, each with its own nuances. Induced drag is the result of vortices set up around a wing of finite span and is inherent in the generation of lift. For a wing of a given planform, induced drag is proportional to the ratio of wing loading to span and is inversely proportional to air density and the square of the speed. In other words, the lighter a plane's span loading and the faster it flies, the less its induced drag. Profile drag, which for laminar (smooth) flow consists of skin friction, is for a wing dependent on the wing area and is directly proportional to air density and the square of the speed. The higher the speed, the greater the profile drag. For the parts of an aircraft which perform no lifting function and are perfectly streamlined, drag consists only of skin friction and therefore behaves as profile drag on a wing. If flow around a body is not perfectly streamlined—and it never is—then turbulent flow results. The turbulent drag associated with this type of flow, sometimes called form or parasitical drag, arises from flow disruption and is proportional to the amount of deviation from streamline conditions. The quantitative difference between streamline and turbulent-flow drag is large. In addition to imperfect form, pressure alterations resulting from changing aerodynamic conditions—such as those which occur when an aerofoil (wing) stalls—also cause transition from streamline to turbulent flow. As the speed of sound is approached, drag relationships undergo radical alteration. Shock waves begin to form, since the air cannot get out of its own way. Profile and form drag increase precipitously. In general, drag coefficients (ratio of drag to lift) are fairly constant from Mach 0.1 to Mach 0.5, rise rapidly from Mach 0.5 to Mach 1.0 (Mach 1.0 = speed of sound), then decline slowly for speeds above Mach 1.0 toward a value higher than that for speeds below 0.5 Mach. Fuselage and wing shapes appropriate to subsonic and supersonic flight are entirely different. Furthermore, the range Mach 0.8 to Mach 1.2 is a transitional zone in which flow patterns are likely to become erratic.

Thrust and Propulsive Efficiency

In terms of dynamics, all aircraft propulsion systems, jet or propeller, that use atmospheric air as their operating medium are the same. Aerodynamic thrust is simply another expression for Newton's third law: that for every action there is an equal and opposite reaction. Whether an aircraft is moved by a propeller or by a turbojet is immaterial, since the propelling force is obtained from the change in momentum of a portion of the operating medium (air) acted upon by the propulsion system. The propeller or the turbojet accelerates a portion of the atmospheric air in a direction opposite (rearward) to that of the direction of the aircraft's travel. It is that "action" which engenders an equal "reaction" (thrust) that drives the aircraft forward.

One facet of thrust and of efficiency (efficiency may be defined as the ratio of output to input) is of especially critical relevance here: the way in which the thrust and efficiency of propeller and turbojet systems are related to aircraft speed and altitude. By proper design, a propeller can be produced to satisfy almost any speed and load conditions efficiently so long as the velocity of the aircraft and of the propeller blade tips do not reach the speed of sound. As aircraft speeds do approach that of sound, propeller efficiency decreases sharply and in the neighborhood of 500 mph becomes intolerably low. Similarly, the altitude at which propellers will function is limited: the higher an airplane goes, the less dense the air becomes and the less "bite" the propeller can get, so that the absolute ceiling for propeller-driven aircraft is about 55,000 feet. The turbojet demonstrates obverse characteristics. First, because of exhaust velocity of a conventional turbojet cannot be reduced much below 700 mph (1,200 mph is more common) without incurring large internal thermodynamic losses, the aircraft speeds at which a turbojet starts to be acceptably efficient begin above 400 mph. Even at 300 mph, a turbojet wastes 80 percent of its power. Second, because the less dense air at high altitudes produces less drag on the aircraft (at 40,00 feet, about one-third that at sea level), and because the much colder atmospheric temperatures at high altitude improve engine (compressor) efficiency, turbojets are increasingly efficient above 20,000 feet and do not begin to press their absolute ceiling until well over 80,000 feet. Table 6 compares two first-generation turbojets, the General Electric (Whittle) I-16 and the G.E. I-40, with two of World War II's most highly developed turbocharged radial piston engines, the Pratt and Whitney R-2800 and the Wright R-3350. The awesome advantage of the turbojets at high speed (500 mph) is obvious: for half the installed specific weight (which means less than half the induced drag) the turbojets offered equal specific fuel consumption. The same table also makes clear the turbojets' disadvantage at lower speeds (375 mph).

THE HEAT ENGINE

In all widely used powered aircraft the thrust necessary to overcome aircraft drag, as well as that required for the aircraft to take off, climb, and maneuver—all of which demand considerably more thrust than straight and level flight—has come from some form of internal combustion engine. Any internal combustion engine which uses atmospheric air as an oxidizing agent performs three distinct operations: compression, ignition, and expansion. All internal combustion engines, be they ordinary Otto cycle (automobile) engines, diesel engines, internal combustion gas turbines, or any other variation, have these three thermodynamic operations in common. In general, compression constricts air volume so that combustion will be less incomplete. Ignition, obviously, is the detonation of a combustible fuel which has been mixed with the compressed atmospheric air. Expansion is the release of the high-temperature, high-pressure gas resulting from ignition in such a way as to derive useful kinetic energy. For example, in a normal automobile engine, an air-fuel mixture at roughly atmospheric pressure is introduced from a carburetor into a cylinder. The upstroke of the piston in the cylinder compresses the air-fuel mixture, and a spark plug ignites it. The resulting hot, high-pressure gas as it expands exerts a force on the

Table 6. *Comparative Performance of Turbojet and Reciprocating Engines*

	Aircraft Gas Turbines		Reciprocating Engines	
	I-16	I-40	R-2800	R-3350
Sea-level static rating				
Military rating (bhp)	—	—	2,100	2,500
Thrust (based on 2.6. lb per bhp-lb.)	1,600	4,000	5,460	6,500
Weights (lbs.)				
Engine dry weight	815	1,850	2,315	2,769
Installed weight	1,020	2,250	4,600*	5,300*
Altitude performance				
500 mph at 35,332 ft.				
Military rating (thp)	795	2,010	1,680	2,000
Sfc	1.09	0.96	1.02	1.00
Specific installed weight†	1.32	1.12	2.74	2.64
375 mph at 35,322 ft.				
50% normal rating (thp)‡	342	815	680	840
Sfc	1.46	1.35	0.58	0.55

SOURCE: Felix Fremont, "Application and Installation Engineering of Aircraft Gas Turbines with Axial-Flow Compressors," in General Electric, *Aircraft Gas Turbine Turbine Conference, 1945* (West Lynn, Mass.: G.E., 1945), p. 121.

NOTE: bhp = brake horsepower; thp = thrust horsepower; sfc = specific fuel consumption (lbs.-fuel-per-hr.-per-thp).

*Includes turbocharger.

†Engine lb.-per-thp.

‡Fifty percent normal thrust is approximately 50 percent power.

piston, which, as it moves, converts the pressure of the expanding gas into mechanical kinetic energy, which is in turn converted into rotary motion at the crankshaft. In an internal combustion gas turbine, the three operations—compression, ignition, and expansion—occur separately rather than in one cylinder. Air is compressed by an air compressor and fed to a separate combustion chamber where fuel is added and ignited. The resulting high-temperature gas is exhausted out through a turbine, which the gas rotates, thereby directly converting temperature energy into the rotary mechanical power used to drive the compressor. In the turbojet, excess energy remains in the exhaust stream, which by its momentum (mass times its velocity) produces the thrust that drives the aircraft. The basic operations performed by a piston engine and by an internal combustion gas turbine, then, are identical, even if quite different in mechanical arrangement.

The Thermodynamic Cycle

Other things being equal, the theoretical work that can be extracted from an ideal heat engine operating on a cycle is a function only of the total difference between the maximum and minimum temperatures attained during the cycle. The maximum thermodynamic (or Carnot) efficiency for an ideal heat engine is defined to be a function of these two temperatures. Theoretically, a device cannot convert all the heat supplied at the higher temperature of the cycle into mechanical work, and unitary efficiency can therefore only be approached, not attained. If heat is rejected to ambient temperature (which for a real engine is generally much less that the minimum temperature of the cycle) further losses are incurred.[2] Since in general ambient temperature is given for propulsion systems, the only way in which theoretical work available can be raised is by raising the maximum cycle temperature. For an internal combustion engine, whether piston or turbine, maximum cycle temperature depends on three things: the temperature of the ingested working medium (ambient air at given temperature for most internal combustion engines); compression, which represents heat added in the form of mechanical work; and fuel added, which, when burned, chemically releases heat. Compression, whether done by a piston in a cylinder, by a rotary air compressor, or by some other mechanical means, requires work to be done. The larger this work for a cycle, the smaller the total net work produced by the cyclic heat engine. Such losses also reduce engine efficiency. Thus with ambient temperature fixed and with the heat of compression even ideally only just recoverable, the only way theoretically available engine work can be raised is by adding more heat in the form of fuel.

Air-Fuel Ratios and Combustion

The maximum amount of heat available from a fuel is determined by its heat of combustion, or the amount of heat chemically available by complete combustion. The highest temperature obtained from a quantity of fuel, given an initial temperature and pressure, results from burning that fuel stoichiometrically, that is, with the exact quantity of oxidant necessary to completely burn the fuel. If less oxidant than is required for a stoichiometric mixture is used, some fuel will be unburnt and its heat wasted. If more oxidant than necessary is present, the fuel will still combine stoichiometrically, but the heat released (which is fixed by the quantity of fuel) will be spread through a greater quantity of gas. For that reason, temperature rise will be less for the same total quantity of heat released. (It might be observed here that since air contains only about 20 percent oxygen, all internal combustion engines that use atmospheric air ingest a minimum of five times the gas needed for stoichiometric combustion.) Since no mechanical compression process enjoys perfect efficiency, the more air that is compressed in excess of that necessary for stoichiometric combustion, the greater the losses due to the work required for compression. Stoichiometric mixture offers the highest possible ratio between fuel input and mechanical (shaft) output. Piston engines normally employ approximate stoichiometric mixtures, about 15 parts air to 1 part fuel by weight. The actual combustion regions of turbojet engines (near the fuel injectors) also approximate a stoichiometric ratio in order to maintain combustion intensity, but the overall ratio of ingested air to fuel in turbojets is 150:1 or 200:1.[3]

Power

The theoretical power, the work per unit time, developed by a heat engine depends on the difference between the amount of heat released in the engine per unit time and the amount of heat consumed within the engine to perform required mechanical processes (chiefly compression). Thus the greater the air-flow rate through an engine, the more fuel that can be burned (stoichiometrically or otherwise), the greater the heat released, and the greater the power that can be derived. Air-flow rate is a function of three things: air density (which depends on altitude and ambient temperature), compression ratio, and engine speed (rpm). Compression ratio is the ratio of the volume of air taken in to the volume of air after compression. A compression ratio of 4:1, for example, means that the ingested air is reduced to one-fourth its atmospheric volume, with corresponding rises in temperature and pressure. For an engine—piston or turbine—turning at constant revolutions, the higher the compression ratio the higher the rate of air-mass flow through the engine, and, therefore, the greater amount of fuel that can be burned in the engine per unit time (the greater the intensity of combustion) and the higher the power developed. Obviously, for a given compression ratio, if the engine revolutions are increased, the greater the air-mass flow per unit time, the more fuel burned, and the more power produced.

The actual power obtainable from an engine is reduced by numerous other losses in addition to the work of compression. Materials limitations on permissible temperatures often compel air flow considerably in excess of stoichiometric in order to reduce gas temperatures. Such an expedient causes the work of compression to be significantly higher than that essential for complete combustion. Component efficiencies, especially for internal combustion turbines, may be low. Indeed, because of low permissible temperatures and low compressor and turbine efficiencies, most early internal combustion gas turbines were failures. Furthermore, radiation, cooling, exhaust, and mechanical (frictional) losses take a large toll of the energy developed. For example, an ordinary automobile piston engine reduces only about one-third of its fuel's heat content to mechanical power. Of the other two-thirds, one-third is lost through cylinder block radiation and the cooling system, and one-third is rejected in the exhaust. Finally, net useful output is again shrunk by the driving of accessories necessary to the engine—fuel, oil, and coolant pumps, generators, fans, and so forth. Most heat engines develop thermal efficiencies (ratio of net mechanical output to total heat, that is, fuel, input) of 25 to 35 percent.

With these fundamentals of the aeronautical environment and of the heat engine in mind, attention can be directed to specific descriptions of aero-engine types. The turbojet demands primacy because it is the subject here; the aircraft piston engine, on the other hand, was the conventional system before the turbojet, and several other types are of interest because they played important if peripheral roles in the turbojet revolution.

THE TURBOJET

If nothing else, the turbojet is the technological expression of a political ideal, for in the turbojet hot air is made to do useful work. The turbojet is a synthesis of two mechanical forms—the internal combustion gas turbine as gas generator and the reaction propulsion nozzle as thrust producer. In a turbojet, atmospheric air is ingested, compressed to several times atmospheric pressure, and introduced into a combustion chamber where fuel is added and where constant ignition is maintained. After the ingested air is heated in the combustion chamber to a temperature above 1,000° C, it is exhausted through a turbine, which provides the rotary power to drive the compressor. The air then passes out of the engine as a high-velocity gas stream, thereby producing a forward thrust. The fundamental components of the turbojet's gas generator are its compressor, its combustion system, and the turbine itself. In practice, the diffuser, which passes air from the compressor into the combustion chambers, and the turbine inlet nozzle, which directs the gas flow leaving the combustion chambers onto the turbine wheel, are equally critical.

A Technical Appendix

Compressor Types

There are two basic types of air compressors used in turbojet engines, centrifugal flow and axial flow. In centrifugal-flow compressors, air is taken in near the center of an impeller, then whirled outward, imparting to the air molecules great velocity and therefore great kinetic energy. From the rim of the impeller the air enters a diffuser where its kinetic energy is converted, by slowing the air down, into pressure or head energy. There are two subtypes of centrifugal-flow compressors: single-sided, with impeller blades on one side of the impeller disc, and double-sided, with impeller blades on both sides of the disc. The single-sided centrifugal compressor permits more efficient air flow out of the aircraft's intake ducting system; the double-sided impeller permits a smaller overall engine diameter for the same air-flow capacity. A single-stage (one impeller) centrifugal compressor circa 1940 could produce pressure ratios of from 3:1 to 4.5:1.

In an axial-flow compressor, the air flows longitudinally along the axis of the engine. An axial-flow compressor is composed of alternate sets of rotating and stationary blades ("rotor blades" and "stator vanes," respectively).[4] A set of rotor blades and a set of stator vanes together are known as a "stage." The rotor blades accelerate the ingested air both axially and radially, and, depending upon blade design, sometimes also partially compress it. The stator vanes convert a portion of the kinetic energy of the air coming off the rotor blades into a pressure head and direct the air into the next compressor stage. Since early axial-flow compressor stages (1940) could produce a maximum pressure ratio of only about 1.2:1, several stages in series were necessary to get an overall pressure rise of 3:1 or 4:1. While axial-flow compressors are more complex than centrifugal-flow units, they possess one great advantage: for equal capacity, their overall diameter is much smaller than that of a comparable centrifugal compressor, thus permitting more streamlined installation. The relative merits of centrifugal versus axial compressor turbojet engines were still very much debated until the late 1940s; but since the early 1950s, the axial-flow compressor usually has proved better able to produce the higher volumes, pressures, and efficiencies demanded by ever-more-powerful engines. Since that time most turbojet engine development has been done on axial-flow units.

Diffusers

Diffusers in centrifugal-flow compressor engines are more obvious than in the axial-flow engines, for in the centrifugal-flow unit the diffuser

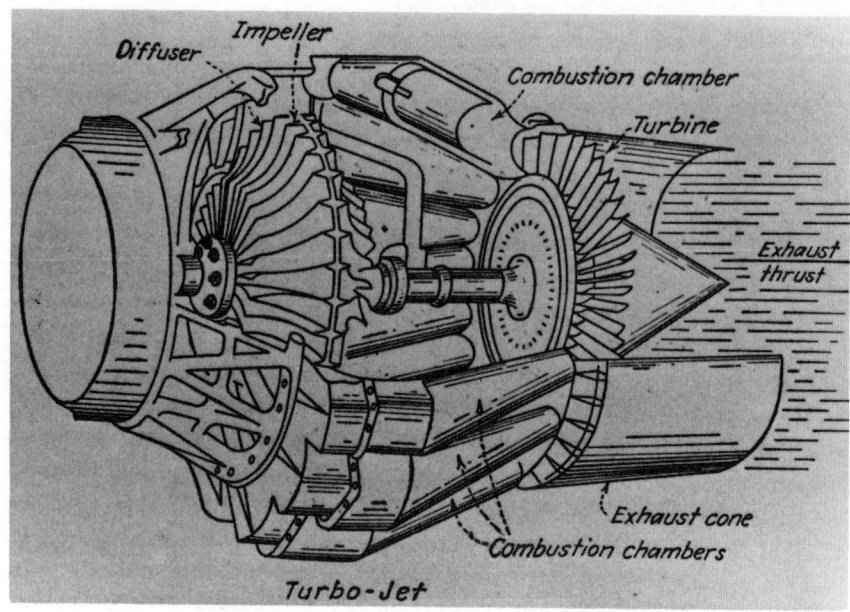

Centrifugal-flow turbojet (with double-sided impeller). (From Jack V. Casamassa, *Jet Aircraft Power Systems* [New York: McGraw-Hill, 1950].) (Wright Aeronautical Corporation.)

must not only direct the air into the combustion chambers (which in this type of engine usually involves a change in direction of the air flow of at least 90°) but also must convert the kinetic energy of the air into pressure. Thus in centrifugal-flow engines, diffuser design is of central importance. Efficiency losses due to poor diffuser configuration were among the most difficult to locate and remedy in the early turbojet engines.

In axial-flow engines, the diffuser is essentially a section of casing between the compressor and combustion chambers which incorporates stationary straightening vanes necessary for uniform air distribution into the combustion chambers. In axial engines, the function of converting kinetic into static pressure energy is performed in the compressor itself, so that the diffuser has only to insure proper air flow.

Combustion Systems

From the diffuser, ingested air is passed to a combustion chamber or chambers. Combustion chamber design is again one of the most difficult problems found in turbojet engine development because efficient jet propulsion requires a combustion intensity (in terms of BTUs released

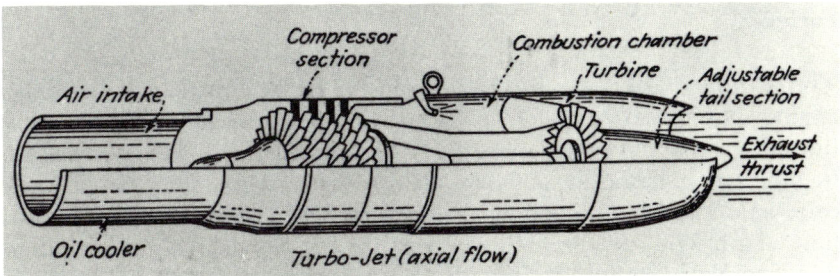

Axial-flow turbojet. (From Jack V. Casamassa, *Jet Aircraft Power Systems* [New York: McGraw-Hill, 1950].) (Wright Aeronautical Corporation.)

per unit volume of the chamber) at least 20 times greater than that of normal industrial practice and because continuous combustion must be maintained in a very-fast-moving gas stream. Furthermore, since combustion is continuous (rather than intermittent, as in a cylinder of a piston engine), cooling of the combustion chambers is especially troublesome. There are two major types of combustion chambers for turbojet engines, cylindrical "can" and annular.

A cylindrical combustion-chamber system is composed of a number of individual but interconnected can chambers. The major components of an individual cylinder are a fuel nozzle or nozzles; an inner, usually perforated, shell that serves both to contain the combustion flame and to set up combustion-supporting eddies (like a mantle in a Coleman lantern); and an outer, nonperforated shell that encloses the combustion chamber. The design of the cans usually provides that precombustion air be passed between the inner and outer shells to prevent excess heat transfer out of the combustion chamber. There are two variations of cylindrical combustion chambers: straight-through and reverse-flow. In a straight-through chamber, air enters one end and goes out the other. In a reverse-flow chamber, the air is turned 180° within the chamber and exits from the same end it entered.

An annular combustion chamber is one large axial can with an axial tube at its center through which passes the rotor shaft connecting the compressor and the turbine. The annular chamber is also composed of an inner and an outer shell, with the same functions as the similar elements in a cylindrical system. In an annular chamber, multiple fuel nozzles are used, and special inserts are often necessary to provide combustion-sustaining eddies. The annular chamber is considered more efficient than the cylindrical system, but the cans, since each represents only a fraction of the total combustion chamber volume, can be tested separately with a relatively small air flow. Because of this ease of experimentation, most early turbojets used the cylindrical system.

Turbines

The turbine diaphragm, or nozzle, and the turbine proper of a turbojet engine require performance characteristics that were thought impossible until the actual development of the turbojet. Although a centrifugal gas turbine is mechanically possible, all turbojet engines thus far put into production have employed axial-flow turbines. The function of the turbine diaphragm, or nozzle, is to direct the gas flow leaving the combustion chambers smoothly onto the buckets of the turbine wheel. The turbine wheel provides power to operate the engine compressor and also whatever accessories (fuel and oil pumps, generators, and so forth) the engine might be required to power. The turbine nozzle is the easier of the two components to design, for although it is subjected to extremely high temperatures (1,000–1,200° C) and stresses (about 300 pounds per square inch), it does remain stationary.

In terms of the state of the art before the turbojet revolution, the turbine wheel presented perhaps the most forbidding problem in the design of the gas turbine engine. The performance specifications of the turbine wheel could not initially be met by any known material. The turbine wheel is composed of two structures: the turbine disc, or hub, and the turbine buckets. The turbine disc must have sufficient rim strength to permit attachment of the buckets, yet must not become either brittle or malleable at high temperatures. The turbine buckets must withstand "heat (in the order of 900° to 1100°C.), centrifugal force (in the order of 30,000 lb. per square inch), thermostress, corrosion, vibration, and differential expansion of materials,"[5] and must meet the further requirement of reasonable ease of production. The power output of a turbojet engine turbine is phenomenal: even the early Whittle engines were designed to produce better than 3,500 h.p. out of a turbine only 19 inches in diameter. In contrast, the most powerful piston engine used during the Second World War, the 24-cylinder "H" sleeve-valve Napier-Sabre, produced only 2,400 net h.p.[6] Furthermore, turbine efficiency in a turbojet engine must approach 70 percent for the engine itself to function efficiently (that is, the turbine must not dissipate more than 30 percent of the total energy it is designed to withdraw from the gas stream passing through it).

Nozzles

Aft of the gas turbine is the exhaust nozzle, or tailpipe. It need theoretically have no moving parts (although in practice many exhaust nozzles do incorporate a moveable inner cone), and its construction would not seem to present any great problem. The shape and size of the tail cone and of the exhaust duct and nozzle, however, do have a primary effect, first, on exhaust velocity and temperature and thus on the thrust and efficiency of the whole engine, and, second, on the operating temperature of the gas turbine. For thermodynamic reasons, even nozzles with identical cross-sectional areas, but slightly different shapes, were found to produce variations in turbine-outlet temperature of several hundred degrees.[7] Later turbojet engines employed an expedient called afterburning to increase thrust briefly. Since the air through-put of a turbojet is several times that necessary to completely burn the normal fuel flow, considerable unconsumed oxygen is left in the exhaust. For afterburning, fuel is sprayed into and burned in the exhaust aft of the turbine, thus raising the exhaust temperature and hence velocity (energy) of the exhaust stream, which of course produces more thrust. Afterburning is not efficient. It is wasteful of fuel in an already fuel-hungry engine, and is used only for take-off, high-rate climb, and combat.

Finally, several other peculiarities of the turbojet engine profoundly affected its development. First, each component (compressor, combustion chambers, turbine) must be designed for some assumed set of operating conditions (aircraft speed, altitude, air density and temperature, turbine rotational speed). The whole engine operates with relative efficiency only very close to its design conditions. For instance, most turbojet engines must operate at 80 to 90 percent of their maximum designed rpm in order to be efficient. Second, in order to prevent undue internal losses, clearances in a turbojet must be very small—for example, the labyrinth lip seal around the rim of the 31-inch-diameter de Havilland Goblin II centrifugal compressor rotor had a clearance of only 0.015 inch cold and (due to differential expansion) 0.075 to 0.095 inch hot.[8] Obviously, with rotor speeds of better than 10,000 rpm, any frequent contact of parts eventually would be fatal to the engine. Manufacturing tolerances therefore must be extremely precise. Third, turbojet engines, for their size and weight, must handle very large quantities of air—air flow through the early Whittle engines was on the order of 20 pounds per second, or 1,200 pounds per minute, which is equal to approximately 16,000 cubic feet of air per minute.[9] A turbojet engine consumes about "the volume of one standard railroad tank car of air per second."[10] It takes a given air molecule about 0.02 second to pass completely through a turbojet engine.[11]

While the turbojet engine was and is mechanically simple, the phenomena which constitute its operation, both in terms of physical and chemical complexity and in terms of rapidity of occurrence, are difficult to manage properly. To sustain and to integrate the processes of turbojet operation is an engineering accomplishment of the first magnitude.

Essentials of the Aeronautic Environment

PISTON AERO-ENGINES

A reciprocating piston engine driving some sort of propeller was the conventional aircraft propulsion system from the beginning of manned, powered flight until the turbojet revolution. In a reciprocating engine, the rotary drive shaft motion which the propeller converts into aerodynamic thrust is produced by the action on a crankshaft of pistons operating within cylinders. Virtually all automobile gasoline engines are reciprocating piston engines, and the type is therefore well known. Aircraft piston engines may be in-line, that is, with cylinders longitudinally behind one another ("V" engines, with two "banks" or lines of cylinders, are in-line engines), or radial, that is, with cylinders arranged radially around a single crankshaft (each concentric cluster of cylinders is a row; radial engines may have as many as four rows).[12] Aircraft engines may be liquid-cooled (like most large in-line engines) or air-cooled (like most radial engines); they may be spark-ignition Otto cycle (4-cycle, found in almost all gasoline engines) or compression-ignition Diesel (2-or-4 cycle) in which ignition is caused by the heat generated in compression of the air within the cylinder, rather than by a spark plug. All piston engines used extensively in aircraft have employed a propeller.

Propellers

Not unlike the components of turbojet engines, a propeller must be designed for specific operating conditions—rpm, aircraft speed, altitude, and so forth. In order to overcome limitations on engine speed imposed by physical limitations on propeller-tip speed, most high-performance engines by the late 1930s drove their propellers through reduction gearing. In addition, the development of variable-pitch (as opposed to fixed-pitch) propellers allowed the propeller to meet a more diverse range of operating requirements. Contra-rotating propellers, or two propellers rotating in opposite directions on concentric shafts, often driven by the same engine, permitted the propeller to absorb more power without a larger overall diameter. Nevertheless, as indicated, there are definite physical constraints to the use of propellers beyond an aircraft speed of 500 mph.

Supercharging

Among other factors, the shaft power produced by a piston engine is directly dependent upon the amount of air that can be introduced into the engine cylinders. Since air density declines rapidly as altitude in-

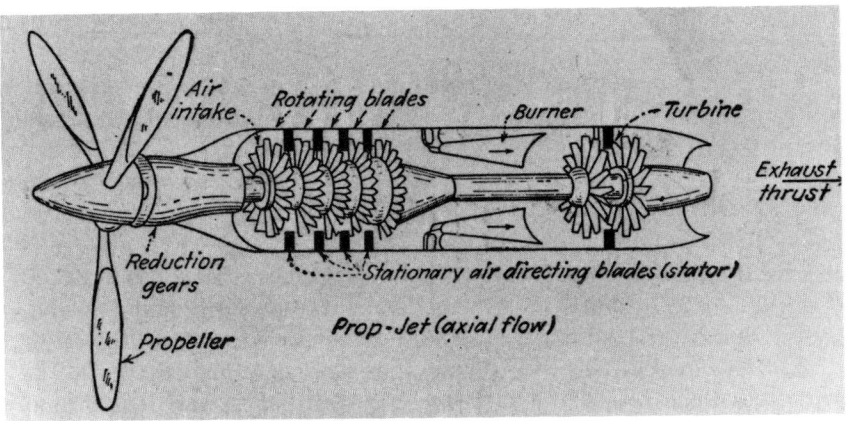

Turbo-prop. (From Jack V. Casamassa, *Jet Aircraft Power Systems* [New York: McGraw-Hill, 1950].) (Wright Aeronautical Corporation.)

creases, normally aspirated piston engine power also falls off rapidly with altitude. Engine designers solved this problem by compressing the air before it entered the cylinders, usually in some form of centrifugal air compressor. When the air compressor used for this purpose is geared directly to the engine drive-shaft, it is called a supercharger. When the air compressor is driven by a turbine which is in turn driven by the engine exhaust gases (which contain otherwise lost energy), it is called a turbocharger or a turbosupercharger. In some turbocharged piston engines, the exhaust-driven turbine produces more power than is needed to operate the air compressor; this excess power is sometimes fed to the propeller by gearing the turbocharger turbine to the engine drive-shaft. Such an engine is called a turbocompound, or simply a compound, engine.

While the fully developed reciprocating piston engine did achieve high efficiency, a low power-to-weight ratio, and excellent fuel economy, it could compete with the turbojet engine in none of these respects at high speeds (over 500 mph) and altitudes (over 30,000 feet). To obtain high power from piston engines requires complex mechanisms that are not only difficult and expensive to construct but costly to maintain. Although the piston engine is still used for low-speed, low-altitude applications (usually on small planes requiring low-output, low-cost power systems), the gas turbine engine is virtually predominant in all other fields of aircraft propulsion.

OTHER ENGINE TYPES

There are several other types of engines which are sometimes referred to as providing jet propulsion and which to a greater or lesser extent did complete with the turbojet. The more important systems are discussed here.

A *turboprop* is essentially a turbojet engine in which the turbine, in addition to supplying power for the compressor, also provides rotary power to turn a propeller. Often, separate turbines (but operating in the same gas stream) are used to power the compressor and the propeller. The turboprop engine actually was seriously discussed before the development of the turbojet. At medium speeds (300–400 mph) and altitudes the turboprop is generally more efficient than a pure turbojet. The turboprop's main disadvantage is that it carries with it all the weight and complexity of mechanism's characteristic of any propeller system.

A *turbofan* (also called a ducted fan, fan-jet, or bypass engine) is a variant of the turbojet which combines qualities of the pure turbojet and the turboprop. In a turbofan engine, a very large multistage axial-air compressor is employed. Only a portion of the compressed air is passed through the combustion system and the turbine which powers the compressor; the rest of the compressed air is "bypassed" around the combustion chambers and turbine to rejoin the exhaust stream behind the turbine. In the ducted fan engine, part of the compressed air is blown directly into the atmosphere in a manner similar to the thrust stream generated by a propeller. By either method, a turbofan increases mass flow and reduces exhaust velocity, thereby raising propulsive efficiency at moderate speeds.

A *pulse jet* is a form of compressorless jet engine. Air is taken into a combustion chamber through a one-way shutter valve, which closes after each "gulp" of air. Fuel is added and the mixture ignited. The hot, expanding gas produced exits through a rearward-facing nozzle, thus creating thrust. The rapid evacuation of the combustion chamber produces a partial vacuum, which causes the forward shutter valve to open to admit another charge of air. Fuel is sprayed in constantly, and the heat of the chamber is sufficient to sustain ignition once the engine is started. Combustion is intermittent, producing the characteristic "buzz" that identified the most famous of the vehicles using the pulse-jet engine, the German V.1 flying bomb. Although simple, cheap, and lightweight, the pulse jet is inefficient and therefore has very limited application.

A *ram-jet* is a jet engine without a mechanical compressor. The air is compressed by ram-effect. At high speeds a carefully shaped convergent-divergent duct converts part of the velocity of oncoming air (relative to the aircraft) into pressure within the duct by retarding the velocity of the stream passing through it. Fuel is added in a constant combustion process. The resulting hot gas expands out the rear of the engine, thereby producing thrust. The ram-jet is the simplest aircraft jet engine (with

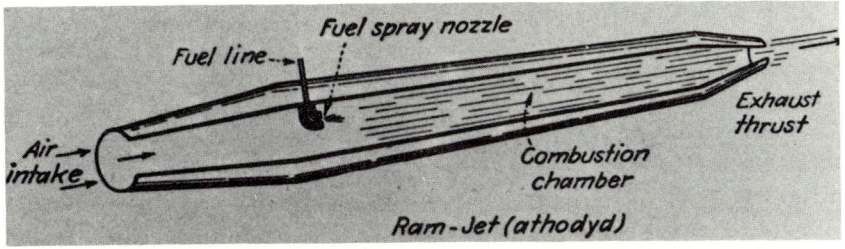

Ram-jet. (From Jack V. Casamassa, *Jet Aircraft Power Systems* [New York: McGraw-Hill, 1950].) (Wright Aeronautical Corporation.)

no moving parts), but is also, at velocities lower than several thousand miles per hour, the least efficient.

A *rocket* is a jet engine only in the sense that it produces thrust by the expansion and rearward ejection of a gas (or liquid). Rockets carry both their fuel and any necessary oxidizing agent, and thus require nothing from their operating medium. Rockets may have liquid or solid propellants. Intercontinental ballistic missiles, baking-soda-and-water toy submarines, and balloons inflated and released are all forms of rockets.

The turbojet engine is distinguished from other forms of aircraft propulsion in that the turbojet (1) uses atmospheric air both as its operating medium and as the oxidizing agent for its fuel; (2) uses a gas turbine to power an air compressor; and (3) produces a propulsive thrust solely by the ejection of a high-velocity gas stream through a nozzle. These three characteristics define a turbojet engine.

EFFICIENCY

Efficiency is an expression that recurs time and again in any discussion of the turbojet revolution. In simplest terms, efficiency is the ratio of energy output to energy input, usually expressed as a percentage. All stated efficiencies result from some measurements by some measuring apparatus according to some more or less arbitrary definition. They are thus abstractions rather than real quantities. Overall engine efficiency is the product of individual component efficiencies: if a turbojet compressor has an efficiency of 80 percent and the turbine 70 percent (and only these two components are considered), the engine will have an efficiency of $.80 \times .70$, or 56 percent. Small errors in component-efficiency estimates thus have a large effect on overall performance expectations.

In order to enhance their power to analyze the complex phenomena that make up the heat engine, engineers employ a spectrum of efficiency

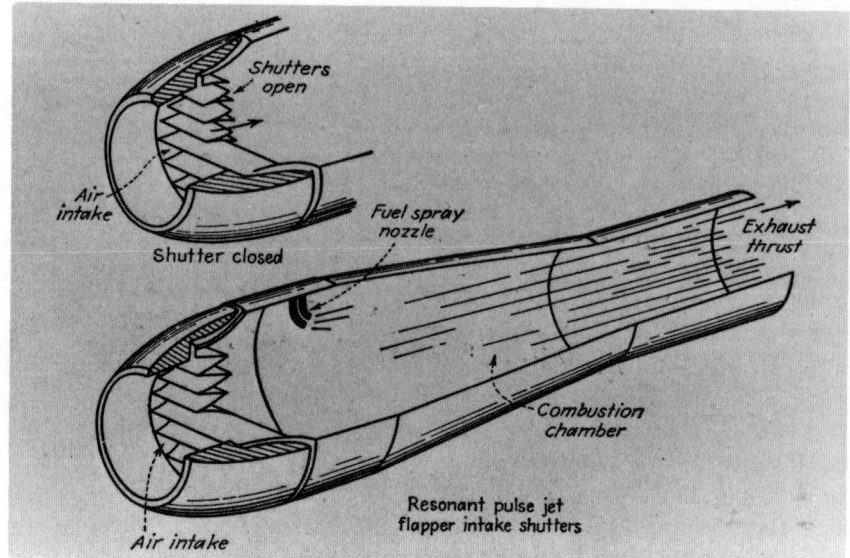

Pulse-jet. (From Jack V. Casamassa, *Jet Aircraft Power Systems* [New York: McGraw-Hill, 1950].) (Wright Aeronautical Corporation.)

definitions. All are basically input-output ratios. The broadest and most common definition of efficiency is thermal efficiency, the ratio of net useful output to total energy (heat content of fuel) input. Thermal efficiencies of 25 to 40 percent are normal for heat engines. Combustion efficiency compares total heat chemically available in fuel with heat actually released in combustion in an engine. Mechanical efficiency compares thrust actually generated with theoretically available power. There are many other specialized efficiency definitions. For each major component of a turbojet engine there may be several different ways of indicating its efficiency. And even a seemingly passive object like an air duct may have a crucial efficiency—pressure or velocity losses through it may adversely affect the whole engine.

There is one special definition of an efficiency that is perhaps more basic to the history of the turbojet than any other: Froude, or propulsive, efficiency. Propulsive efficiency is an expression applicable to any vehicle operating in a viscous medium (gas or liquid) and compares actual work done by the vehicle's propulsion system with the work theoretically required if the vehicle were towed at the same velocity by a rope. The mathematical statement for Froude efficiency is

$$E_F = \frac{2}{V_2/V_1 + 1}$$

A Technical Appendix

where V_1 is the velocity of the vehicle and V_2 is the final velocity of that portion of the surrounding medium whose momentum is changed to create a propulsive reaction. Clearly, propulsive efficiency is at a maximum (that is, is unitary) when $V_1 = V_2$, which is why the turbojet had little hope of competing with the piston engine at aircraft speeds below 500 mph. As much as the word *efficiency* is encountered in this study, only rarely, for obvious reasons, is it explicitly defined; for the nontechnician, statements of efficiency are best just accepted.

Here indeed, much is accepted without the rigorous definition common in engineering. Efficiency and the turbojet itself are described qualitatively rather than quantitatively. Yet the turbojet is and was for its creators a composite of mathematically expressed concepts. There are now hundreds of complicated mathematical models for various phases of turbojet design. Even during the gestation of the turbojet, design was quite naturally expressed mathematically. Yet, for all its dependence upon mathematical tools, if turbojet engine development were nothing more than algebraic puzzle solving, it would not be the extraordinary technological revolution it is. To understand multitudinous equations is necessary to understanding construction of a turbojet engine; but to perceive men in their confrontation with radical technology is essential to comprehending the turbojet revolution. It is ultimately a human revolution, a revolution in human thought, in human conception, and in human commitment; it is only expressed in machines.

A Note on Bibliography

Because a formal bibliography would merely replicate the information already presented in the footnotes to this study, none has been included. Since the study is rigorously organized topically, and, within each topic, chronologically, sources on specific innovations, persons, or periods should be relatively easy to locate. Again because of the structure of the study, virtually all methodological sources are discussed and cited in chapter 1.

Because this inquiry focuses on middle-level technological change, almost all sources are in the public domain and most are published. Most important have been original technical publications, although retrospective papers by original authors and contemporary textbook accounts also have been valuable. A few secondary or historical accounts have been essential, notably Robert Schlaifer's and Miller's and Sawer's books. A wide range of methodological sources have been utilized. Of central importance are the works of Thomas Kuhn, Donald T. Campbell, and Herbert A. Simon.

Sir Frank Whittle, Dr. Hans von Ohain, Professor Herbert Wagner, and Mr. Clarence L. Johnson each very generously and graciously answered written inquiries from the author. Dates of their responses were: Whittle, 20 May 1970; von Ohain, 31 July 1970; Wagner, resumé, 18 September 1970, and letter, 26 October 1970; and Johnson, letters, 25 September and 24 October 1978. In addition, Dr. John D. Stanitz of Cleveland Ohio, very generously provided the author with a typescript of his 1964 Society for the History of Technology convention paper, "Aerodynamic Considerations in Early Turbojet Engine Development."

List of Abbreviations

B.I.O.S.	British Intelligence Objectives Sub-Committee (Great Britain)
C.I.O.S.	Combined Intelligence Objective Sub-Committee (Supreme Headquarters Allied Expeditionary Forces)
Jl. RAeS	*Journal of the Royal Aeronautical Society* (Great Britain)
NACA TM	National Advisory Committee for Aeronautics, Technical Memorandum (U.S.; translations of foreign technical publications)
NACA TN	National Advisory Committee for Aeronautics, Technical Note (U.S.; research notes)
***Proc. Inst. M.E.* (London)**	*Proceedings of the Institution of Mechanical Engineers* (London)
Trans. ASME	*Transactions of the American Society of Mechanical Engineers*

Notes

Chapter 1

1. H. Roxbee Cox, British Air Ministry, in an address to the Institute of Aeronautical Sciences, "British Aircraft Gas Turbines," *Journal of the Aeronautical Sciences* 13 (1946): 53-83.

2. Probably the most competent account of the invention of the turbojet engine is contained in Robert Schlaifer's *Development of Aircraft Engines* (Cambridge, Mass.: Harvard University Press, 1950). Schlaifer's study is a Harvard Business School-backed *ex post* inquiry into the relative efficacy of various governmental development funding approaches. While Schlaifer presents a serviceable narrative of the creation of the first turbojets (some of which, indeed, will be drawn upon here), he does not confront several basic historical issues. Those lacunae are perhaps understandable given his interest. Nevertheless, it is disappointing to find an author perfectly aware that the turbojet was founded upon radical assumptions about aircraft speed and yet not questioning the origin of those assumptions. Schlaifer pulls back from investigating science-technology interaction. As a result, his explanation of the origin of the turbojet is catastrophically inadequate: "The relatively late date at which work began in the United States is simply the result of historical accident: Whittle, von Ohain, and Wagner were not Americans" (p. 489).

3. A. L. Kroeber, "The Superorganic," *American Anthropologist*, n.s. 19 (1917): 163-214; W. F. Ogburn and D. Thomas, "Are Inventions Inevitable?" *Political Science Quarterly* 37 (1922): 83-98; W. F. Ogburn, *Social Change*, (New York: B. W. Huebsch, 1922); S. C. Gilfillan, *The Sociology of Invention* (Chicago: Follett, 1935), p. 6. See also S. C. Gilfillan, *Supplement to the Sociology of Invention* (San Francisco: San Francisco Press, 1971); and Edward W. Constant II, "On the Diversity and Co-Evolution of Technological Multiples: Steam Turbines and Pelton Water Wheels," *Social Studies of Science* 8 (1978): 183-210.

4. *Project Hindsight* (1969), Office of the Director of Defense Research and Engineering; *TRACES* (1968), IIT Research Institute for NSF. For a highly critical analysis of *Hindsight*, see Karl Kreilkamp "Hindsight and the Real World of Science Policy," *Science Studies* 1 (1971): 43-66.

5. See Jacob Schmookler, *Invention and Economic Growth* (Cambridge, Mass: Harvard University Press, 1966). For a devastating analysis of Schmookler's book, see Barkev S. Sanders, "Commentary on *Invention and Economic Growth*, *Idea* 10 (1966): 587-608.

6. John Jewkes, David Sawers, and Richard Stillerman, *The Sources of Invention* (London: Norton, 1958).

7. D. Hamberg, "Invention and the Industrial Laboratory," *Journal of Political Economy* 71 (1963): 95-115.

8. National Bureau of Economic Research, *The Rate and Direction of Inventive Activity: Economic and Social Factors* (Princeton: Princeton University Press, 1962).

9. Stillman Drake, in his *Galileo Studies: Personality, Tradition, and Revolution* (Ann Arbor: University of Michigan Press, 1970), argues for the primacy of biographical reconstruction in the history of science. He attributes the eminence of even so well established a tradition as the history of scientific ideas to "a gracious custom of polite inattention to everything it really involves" (p. 12).

10. Donald T. Campbell, "Evolutionary Epistemology," in P. A. Schlipp, ed., *The Philosophy of Karl Popper,* vol. 14-I of *The Library of Living Philosophers* (LaSalle, Ill.: Open Court, 1974), pp. 413-63.

11. E. H. Carr, *What Is History?* (New York: Knopf, 1963).

12. Max Weber, "Objectivity in Social Science," in *The Methodology of the Social Sciences* (1904; reprint ed., New York. Free Press, 1964); William Todd, *History as Applied Science: A Philosophical Study* (Detroit: Wayne State University Press, 1972), esp. pp. 162-81.

The distinction for Weber is apparently between ideal-typical models, which are acknowledgedly artificial constructs and are justified solely because of their efficacy in analyzing social reality, and ideal-deterministic models of the Hegelian or Marxian sort, which claim, in the Platonic formal sense, to capture the ultimate reality underlying social appearances. Ideal-typical models might also be distinguished from hypothetico-deductive models in the physical sciences, which at least some positivists in Weber's time thought to portray an objective, external, real world. Some circumspect contemporary philosophers of science, however, might well describe hypothetico-deductive mathematical models of physical phenomena in terms reminiscent of Weber's ideal-typical constructs.

13. Weber (1904), "Objectivity," p. 71.

14. The following summary is based upon my reading of Campbell's development of his ideas. See especially Donald T. Campbell, "A Phenomenology of the Other One: Corrigible, Hypothetical, and Critical," in Theodore Mischel, ed., *Human Action: Conceptual and Empirical Issues* (New York: Academic Press, 1969), pp. 41-69; "Objectivity and the Social Locus of Scientific Knowledge," Presidential Address to the Division of Social and Personality Psychology of the American Psychological Association, 1969; "Variation and Selective Retention in Socio-Cultural Evolution," in Herbert R. Barringer et al., eds., *Social Change in Developing Areas* (Cambridge, Mass.: Schenkman, 1966); "Blind Variation and Selective Retention in Creative Thought as in Other Knowledge Processes," *Psychological Review* 67 (1960): 380-400; "Methodological Suggestions for a Comparative Psychology of Knowledge Processes," *Inquiry* 2 (1959): 152-84.

15. Campbell, "Evolutionary Epistemology," p. 421.

16. For introspectionists' interpretations see Paul Souriau, "Theory of Invention," trans. E. L. Clark (n.p.) from *Theorie de l'invention,* (Paris, Hachette: 1881); Ernst Mach, "On the Part Played by Accident in Invention and Discovery," *The Monist* 6 (1896): 161-75; Henri Poincaré, "Mathematical Creation", in *The Foundations of Science* (New York: Science Press, 1921), pp. 383-94; Jacques Hadamard, *An Essay on the Psychology of Invention in the Mathematical Field* (Princeton; Princeton University Press, 1945); see also Campbell, "Blind Variation," for a review of other relevant literature. For the alternative view, see Herbert A. Simon, *The Sciences of the Artificial* (Cambridge, Mass.: MIT Press, 1969). The historical evidence used in this study is ordinarily not of a character to permit assessment of arguments concerning internal or, especially, subconscious processes, and these issues, together with those raised by the much larger literature on creativity in general, will be neglected.

17. The relationship between scientific conjecture and experimental technology, which does affect the environment, is not now clear. New findings may vitiate this interpretation. See Derek Price, *Science since Babylon* (New Haven: Yale University Press, 1961).

18. The technological testing of scientific theory may provide secular protection against communal solipsism.

19. Fritz Zwicky, in *Morphological Astronomy* (Berlin: Springer, 1957), and in "Tasks We Face," *Journal of the American Rocket Society* 84 (1951): 3-20, presents an interesting program for self-conscious "blind-variation." Supposedly, his "morophological approach," using matrices, will generate all possible system variants. Unfortunately, the appropriate matrix entries are obvious only *ex post*.

20. Layton sees basic differences in purpose and approach between science and technology: "We might restate the matter by noting that the laws of science refer to nature and the rules of technology refer to human artifice. The function of technological rules is to provide a rational basis for design, not to enable man to understand the universe. The difference is not just one of ideas but of values. 'knowing' and 'doing' reflect the fundamentally different goals of the communities of science and technology" (Edwin T. Layton, "Technology as Knowledge," *Technology and Culture* 15 [1974]: 40). Layton also notes essential differences in rigor between the pure and the "engineering" sciences and the existence of hierarchies of technical design methods:

> The expanded range of application of the engineering sciences was accompanied by a tendency away from analytic solutions, a reliance on approximations, and, to some extent, a lessening of mathematical rigor. A given problem in the strength of materials might be solved rigorously by the theory of elasticity, or it might be treated by less rigorous graphical methods.
>
> The development of hierarchies of methods of variable rigor, along with the importance of economic factors in determining their use, served to distinguish the engineering sciences from physics, where only the most rigorous methods were normally admitted.

(Edwin Layton, "Mirror-Image Twins: The Communities of Science and Technology in Nineteenth-Century America," in George Daniels, ed., *Nineteenth-Century American Science: A Reappraisal* [Evanston. Northwestern University Press, 1972], pp. 224-25.)

21. Some of the best work on the history and sociology of technology is ostensibly concerned with science: Merton on science in seventeenth-century England or Robert Schofield on the Lunar Society. Some sociologists have tried to identify an engineering "community"; see, for example, Robert Perrucci and Joel E. Gerstl, *Profession without Community: Engineers in American Society* (New York: Random House, 1969). Unfortunately, Perrucci and Gerstl's determination to impose an *a priori* conception of profession, combined with a like determination not to sully their discipline with gritty details of what engineers actually do, render them oblivious to the community structure of technological practice. For an ideal analysis of communities of practitioners, in science, see Diana Crane, *Invisible Colleges* (Chicago: University of Chicago Press, 1972).

22. Part of this ambiguity in defining communities of practitioners would seem to derive from the fact that an individual's community membership is usually mediated by formal organizational membership. For example, a piston-engine designer ordinarily would be a practitioner (individual community membership) who works for a firm or corporation (organizational membership) which is also an element in a community of practitioners at a differently defined hierarchical level. If John Kenneth Galbraith is correct in locating the essential determinants of practice or decision-making in complex corporations in their technostructures, then it is appropriate to speak of communities of practitioners as composed interchangeably of individuals or firms. Notably, within the broad span of all of aeronautics, the community membership of the men who created the

turbojet is defined only as not piston engine practitioners and not members of firms manufacturing aircraft piston engines. Beyond that, the turbojet protagonists' institutional affiliations are remarkably diverse.

23. "Traditions of practice" is used here to denote what communities of technological practitioners hold in common, rather than Thomas Kuhn's "paradigm," for two reasons. First, "paradigm" is controversial and not well defined. More importantly, it is not clear that a technological tradition of practice comprises a set of specific exemplars in the same sense that a scientific paradigm, in its most narrow and precise usage, does. An aspiring early-twentieth-century piston-engine designer probably would have learned about piston engines by way of the Carnot cycle and a single-cylinder laboratory engine, plus some exposure to contemporary practice. The relationship of those experiences to his subsequent design efforts, however, would not seem quite to correspond to the role envisioned for scientific paradigms-as-exemplars in the work of a scientist. Thus "tradition of practice" and "conventional system" are thought to capture more accurately the community focus of technological practice.

24. Layton sees science and technology as separate but similar and interacting social systems—"mirror-image twins." He suggests a sophisticated model for linkage: "The passage of information from one community to the other often involves extensive reformulation and acts of creative insight. This requires men who are in some sense members of both communities. These intermediaries might be called 'engineer-scientists' or 'scientists-engineers,' depending on whether their primary identification is with engineering or with science. Such men play a very important role as channels of communication between the communities of science and technology" (Layton, "Mirror-Image Twins," p. 227).

25. Hugh G. J. Aitken, "Science, Technology, and Economics: The Invention of Radio as a Case Study," in Wolfgang Krohn, Edwin T. Layton, Jr., and Peter Weingart, eds., *Sociology of the Sciences Yearbook, 1978* (Boston: Reidel, 1978), pp. 89–111.

26. See Herbert A. Simon, *The Sciences of the Artificial* (Cambridge: MIT Press, 1969), esp. the last chapter, "The Architecture of Complexity."

27. Quoted in J. R. T. Hughes, *The Vital Few* (New York: Houghton Mifflin, 1966), p. 327.

28. Thomas S. Kuhn, *The Structure of Scientific Revolutions* (Chicago: University of Chicago Press, 1962), p. 23.

29. Nathan Rosenberg, "Technological Change in the Machine Tool Industry, 1840–1910," *Journal of Economic History* 23 (1963): 414–46.

30. The critical notion here is random search at a specific level within a complex hierarchical system. See Thomas P. Hughes, "The Electrification of America: The System Builders," *Technology and Culture* 29 (1979): 124–61.

Another source of invention, not really relevant to the complex systems of interest here, may lie in the autonomous intuition, or sense of play, of individuals. Someone may simply have a bright idea not directly related to any particular problem in any existent technology. Novel inventions such as the yo-yo or the hula-hoop may fall into this category. Often, perhaps, these inventions result from the application of new materials or production processes to ancient contrivances: the kinship between barrel hoops and hula-hoops or pie-pans and Frisbees is obvious.

There likewise seems to be considerable inventive activity inspired by the desire to find new applications for existent proprietary products, materials, or processes. Hyping return on investment in R&D and in embodied capital is a major motivation in these instances. Such activities, however, would appear to be largely subsumed under the articulation processes discussed in the text.

31. Erich R. Pianka, *Evolutionary Ecology* (New York: Harper & Row, 1973), p. 175. For a charming and thorough account of some of the empirical phylogenetic evidence on

coevolution, see Verne Grant and K. A. Grant, *Pollination in the "Phlox" Family* (New York: Columbia University Press, 1965); for an analysis of some of the theoretical issues involved, see Andrew J. Beattie, "Floral Evolution in Viola," *Annals of the Missouri Botanical Garden* 61 (1974): 781-93.

32. Thomas P. Hughes, "The Science-Technology Interaction: The Case of High-Voltage Power Transmission Systems," *Technology and Culture* 17 (1976): 646-62; and idem, *Elmer Sperry: Inventor and Engineer* (Baltimore: Johns Hopkins Press, 1971).

33. The individual does not design a new "tradition"; he designs a new device. He may be fully aware that his device is different from the conventional system, but his goal is a thing, not a tradition of practice. The individual's subjective belief that a new system is a real possibility is inferred from his objective behavior in designing such a device, which inference is also made for objective anomaly-provoked change.

34. See R. M. MacIver, *Social Causation* (New York: Harper & Row, 1942), esp. pp. 239-65; and Thomas S. Kuhn, "Energy Conservation as Simultaneous Discovery," in Marshall Clagett, ed., *Critical Problems in the History of Science* (Madison: University of Wisconsin Press, 1959), esp. p. 323 and p. 345, n. 9, on the nature of central causes in complex social systems.

35. Ballisticians had been aware of the fundamental differences between subsonic (hydrodynamic) flow and high supersonic flow since the work of Ernst Mach in the last half of the nineteenth century. It was not until the mid-twenties, however, that aerodynamicists were able to establish that virtually none of their conjectures about subsonic flow held for transsonic speeds and that the ballisticians' generalizations about high supersonic speeds did not apply either.

36. This characteristic of presumptive anomaly was pointed out to me by Professor Donald Campbell.

Moreover, aerodynamics itself might be defined as an applied science or, perhaps, as an engineering science. While the knowledge sought by such science is generalized, it may be constrained by a set of relevance selectors or problem and solution definitions different from natural science. See Gernot Böhme, Wolfgang Van Den Daele, and Wolfgang Krohn, "The Scientification of Technology," and Peter Weingart, "The Relation Between Science and Technology—A Sociological Explanation," both in Krohn, Layton, and Weingart, eds., *Sociology of the Sciences Yearbook, 1978;* see also Walter G. Vincetti, "Control-Volume Analysis: A Difference in Thinking Between Engineering and Physics," *Technology and Culture* (forthcoming).

While such discussions do raise critical issues concerning the possibly contrasting natures and values of science, engineering science, and technological practice, they do not appear to alter the relationship between theory and technological presumptive anomaly protrayed here.

37. For a fuller treatment of the factors governing the diffusion of innovations, see Gerald Zaltman and Nan Lin, "On the Nature of Innovations," *American Behavioral Scientist* 14 (1971): 651-73.

38. Simon, *Sciences of the Artificial*, p. 97.

39. A new system, however perfect its scientific justification, however excellent its execution, can still fail because of technological obsolescence or irrelevance. For example, after laminar flow became relatively well understood during the early 1920s, a new system for sailing-ship design was proposed and tried. A German engineer's proposal promised both better efficiency and greater ease of handling. The new sails were simply large, hollow, rigid, aluminum-skinned cylinders which were rotated around the vertical "masts" by small electric motors. The sails were rotated at a slightly greater speed than the velocity of the wind. The rotation caused each cylinder to carry on its surface a portion of the laminar layer, thus forming a high-pressure region on one side of the sail and a low-pressure area

on the other. The new system was tried on several ships and found quite successful. But by 1930, for economic purposes, advances in sail technology as a prime-mover system were irrelevant. With the development of low-speed diesels and geared steam turbines, sails were obsolete, and the new proposal, whatever its intrinsic merits, useless. Since it was obsolete, the new sail attracted no coterie of adherents and therefore, by our definition, did not constitute a technological revolution. The case of the rotorships is a good example of conventional systems' (diesels and steam turbines) developing rapidly enough to nullify a potential rival. See *Scientific American,* January 1925, p. 12, and March 1927, p. 199.

40. David Bloor, "Essay Review: Two Paradigms for Scientific Knowledge?" review of I. Lakatos and A. Musgrave, eds., *Criticism and the Growth of Knowledge,* in *Science Studies* 1 (1971): 101–15; G. R. Simonson, "The Demand for Aircraft and the Aircraft Industry, 1907–1958," *Journal of Economic History* 20 (1960): 361–82. Bloor notes the possible "constant evolution and change" of the social basis of a scientific specialty, and Simonson describes the emergence of electronic firms, rather than aircraft firms, as prime contractors for guided missile development, thereby defining a new community of aerospace practitioners.

41. This ideal first emerged from a talk by Professor Carl W. Condit and later discussion with him. The way I have construed the evidence is, however, my own doing.

42. See Arthur P. Modella and Nathan Reingold, "Theorists and Ingenious Mechanics: Joseph Henry Defines Science," *Science Studies* 3 (1973): 323–51. See also the seminal articles by Charles C. Gillispie, "The Natural History of Industry," *Isis* 48 (1957): 398–407, and Robert E. Schofield, "The Industrial Orientation of Science in the Lunar Society of Birmingham," ibid., pp. 408–15. As Kuhn points out, each reaches opposite conclusions concerning the relationship between science and technology, because for Gillispie science is theory and for Schofield science is scientists and method (Thomas S. Kuhn, "The Relations between History and the History of Science," in *The Essential Tension* [Chicago: University of Chicago Press, 1977], p. 144, n. 10, reprinted from *Daedalus* 100 [1971]: 271–304). Robert K. Merton, *Science, Technology, and Society in Seventeenth-Century England* (Chicago: University of Chicago Press, 1938), is also relevant, as is Edwin T. Layton, "Millwrights and Engineers: Science, Social Roles, and the Evolution of the Turbine in America," in Krohn, Layton, and Weingart, eds., *Sociology of the Sciences Yearbook, 1978,* pp. 61–87.

43. George H. Daniels, *American Science in the Age of Jackson* (New York: Columbia University Press, 1968), chap. 3.

44. Thomas P. Hughes, "Electrification."

45. See, for example, the index volumes to the *Transactions of the American Society of Civil Engineers,* or the *Journal of the Franklin Institute.*

46. The evolution of those traditions of technological testability may be analogous in some sense to Nathan Rosenberg's concept of technological convergence. Technological convergence is said to occur when technologies developed in one sector or industry prove adaptable to a wide variety of other industries or processes: for example, precision metalworking machinery developed to manufacture sewing machines and bicycles was directly applied to the manufacture of automobiles. Similarly, as noted, the Prony brake, or friction dynamometer, was applied to all prime movers, not just to water turbines or wheels. More significantly, and going somewhat beyond Rosenberg's concept, the ideological or normative component of this tradition of technological testability went on to infuse virtually all modern technological practice. See Nathan Rosenberg, "Technological Change in the Machine Tool Industry, 1840–1910," *Journal of Economic History* 23 (1963): 414–46. According to C. W. Thompson, of the Department of Industrial Engineering, Northwestern University, currently about half of the Institution of Electrical and Electronic Engineers annual list of the two hundred top innovations is devoted to testing equipment.

47. Ian I. Mitroff, "Norms and Counter-Norms in a Select Group of the Apollo Moon Scientists: A Case Study of the Ambivalence of Scientists," *American Sociological Review* 39 (1974): 579-95.

48. Ibid., p. 592.

49. Donald T. Campbell, "Objectivity and the Social Locus of Scientific Knowledge." On the efficacy of experimental tests as compared to natural selection, Campbell notes: "Experimentation, even if nature answers in the subjective language of the experimenter rather than in the pure tones of the *Ding an Sich,* has a selective efficiency far above natural selection. We can learn the value of rauwolfia for high blood pressure through three thousand years of Dravidian custom selection, but we get it with a lot of irrelevant and ineffective pharmacological lore that it would take another thousand years to weed out. Experimentation can achieve a winnowing of equivalent purity in one decade or less." See also Donald T. Campbell, "A Tribal Model of the Social System Vehicle Carrying Scientific Knowledge," *Knowledge* 1 (1979): 181-201.

50. The term is Kenyon B. DeGreene's, *Sociotechnical Systems* (Englewood Cliffs, N.J.: Prentice-Hall, 1973).

The concept of technological testability discussed here would seem to encompass Walter G. Vincetti's more limited but more precise notion of "parameter variation" as an especially central and powerful methodology in the generation of technological knowledge. See his "The Air-Propeller Tests of W. F. Durand and E. P. Lesley: A Case Study in Technological Methodology," *Technology and Culture* 20 (1979): 712-51.

51. Simon, *Sciences of the Artificial,* p. 74.

52. D. L. Marples, "The Decisions of Engineering Design," *IRE Transactions on Engineering Management,* EM-8 (1961): 55-71.

53. P. J. Booker, "Principles and Precedents in Engineering Design," *The Engineering Designer,* September 1962, p. 30. See also Thomas J. Allen, "Studies of the Problem Solving Process in Engineering Design," *IEEE Transactions on Engineering Management,* EM 13 (1966): 72-83.

54. Ibid., p. 24.

55. Ibid., p. 30.

56. This hierarchical structure may account for the perception by some that innovation involves continuous movement back and forth along a spectrum of activities from research through invention to marketing. See the papers collected in Nathan Reingold and Arthur Mollela, eds., "The Interaction of Science and Technology in the Industrial Age," *Technology and Culture* 17 (1976): 620-742. In fact, as noted below, movement is often "vertical" through the system's hierarchy as well as "horizontal" through developmental phases.

57. Burton Klein and William Meckling, "Application of Operations Research to Development Decisions," *Operations Research,* May-June 1958, pp. 352-63, corroborate the views expressed by Marples and Booker and argue that even sophisticated operations research techniques cannot be expected to yield reliable cost or performance estimates for systems undergoing major developmental change.

58. For a superb critical review of simplistic "market-pull" analyses of technological change, see David Mowery and Nathan Reingold, "The Influence of Market Demand upon Innovation: A Critical Review of Some Recent Empirical Studies," *Research Policy* 8 (1979): 102-53. For the most radical innovations of concern here, Mowery and Rosenberg conclude: "Also of importance is the fact that all of the studies which attempt to rank innovations or research events by the importance of these occurrences found that the most radical or fundamental ones were those least responsive to 'needs.' Even within the flawed conceptual and methodological framework of these empirical studies, then, the 'demand-pull' case is admittedly weakest for the most significant innovations. Does an explanatory

schema retain much usefulness when it is contradicted by the most important occurrences in the set of events which it purports to illuminate?" (p. 140).

59. Each author has his own vocabulary for describing these functional stages; most differences seem trivial.

60. See "Pure Technology," in Albert Teich, ed., *Technology and Man's Future* (New York: St. Martin's Press, 1977), pp. 30–41; and perhaps more convincingly, Kurt Vonnegut, *Player Piano* (New York: Dell, 1952).

61. Thomas P. Hughes, "Electrification" and Reese V. Jenkins, *Images and Enterprise* (Baltimore: Johns Hopkins University Press, 1975).

62. Thomas P. Hughes, *Elmer Sperry*.

63. Thomas P. Hughes, "Electrification."

64. See also Robert L. Perry, "Innovation and Military Requirements: A Comparative Study," Rand Corporation RM-5182-PR (1967). It should be noted that while war conditions usually lead to substantial expansion of funds available for military research, development, and procurement, the same economic consideration—anticipated cost effectiveness—governs allocation of even lavish resources. Given the intrinsic military premium on even small margins of comparative performance superiority, however, resources expanded beyond subsistence often are partially used to support high-risk projects (such as the turbojet) which can promise potentially high-performance returns.

65. In addition to other cost factors, the relative cost of a particular device is also a function of its cost of production. Usually the adoption of a new system implies the introduction of new or substantially modified production techniques.

66. The cost overruns that plague development of almost all new systems may be partially attributable to the inherent difficulties of prerevolutionary evaluation of new systems. The more radical the technology, the less likely realistic cost prediction. In this context, it may well be that transparadigm systems analysis is dysfunctional.

67. See W.E.G. Salter, *Productivity and Technical Change* (London: Cambridge University Press, 1960).

68. See Robert W. Fogel, "Railroads as an Analogy to the Space Effort: Some Economic Aspects," in Bruce Mazlich, ed., *The Railroad and the Space Program* (Cambridge: MIT Press, 1965).

69. Thomas P. Hughes, "Technological Momentum in History: Hydrogenation in Germany, 1898-1933," *Past and Present* 44 (1969): 106-32.

70. John Kenneth Galbraith, *The New Industrial State* (New York: Houghton Mifflin, 1967).

71. See Schlaifer, *Development of Aircraft Engines*.

72. Frank Whittle, *Jet: The Story of a Pioneer* (London: Frederick Muller, 1953).

73. Merton, *Sociology of Science;* Mitroff, "Norms and Counter-Norms in a Select Group of Apollo Moon Scientists" pp. 579-95; Joseph Ben-David, "Role and Innovation in Medicine," *American Journal of Sociology* 65 (1960): 557-68; Diana Crane, "Social Structure in a Group of Scientists: A Test of the 'Invisible College' Hypothesis," *American Sociological Review* 34 (1969): 335-52; S. Cole and J. R. Cole, "Scientific Output and Recognition: A Study of the Reward System in Science," *American Sociological Review* 32 (1967): 377-90.

74. Reinhold Niebuhr, *Moral Man and Immoral Society* (1932; reprint ed., New York: Scribners, 1960), notes in a slightly different context: "The inertia of society is so stubborn that no one will move against it, if he cannot believe that it can be more easily overcome than is actually the case. And no one will suffer the perils and pains involved in the process of radical social change, if he cannot believe in the possibility of a purer and fairer society than will ever be established. These illusions are dangerous because they justify fanaticism; but their abandonment is perilous because it inclines to inertia" (p. 221).

Chapter 2

1. Norman A. F. Smith, "The Origins of the Water Turbine and the Invention of Its Name," in Rupert Hall and Norman Smith, eds., *History of Technology* (London: Mansell, 1977), pp. 215-59.
2. Terminology and concepts in nineteenth-century turbine practice were extraordinarily confused and confusing. See Edwin T. Layton, "Millwrights and Engineers: Science, Social Roles, and the Evolution of the Turbine in America," in Wolfgang Krohn, Edwin T. Layton, Jr., and Peter Weingart, eds., *Sociology of the Sciences Yearbook, 1978* (Boston: Reidel, 1978), pp. 67-87; and Arthur T. Stafford and Edward P. Hamilton, "The American Mixed-Flow Turbine and Its Setting," *Transactions of the American Society of Civil Engineers* 85 (1922): 1235-1356. For the modern concepts of impulse and reaction as used in steam turbine practice, see Walter Hossli, "Steam Turbines," in *Scientific Technology and Social Change: Readings from "Scientific American"* (San Francisco: W. H. Freeman, 1969), pp. 158-68.
3. Abbot Payson Usher, *A History of Mechanical Invention* (Cambridge: Harvard University Press, 1929), pp. 382-85; John Muendel, "The Horizontal Mills of Medieval Pistora," *Technology and Culture* 15 (1974): 194-225; and Louis C. Hunter, *Industrial Power in America* (Charlottesville: University of Virginia Press, 1979). Dr. Hunter very generously made available to me proof copies of his manuscript.
4. J. F. d'Aubuisson de Voisin, *A Treatise on Hydraulics for the Use of Engineers*, trans. Joseph Bennett (Boston: Little, Brown, 1852), pp. 407-16.
5. Smith, "Origins of the Water Turbine"; Hunter, *Industrial Power*; Layton, "Millwrights and Engineers"; and Edwin T. Layton, "Scientific Technology, 1845-1900: The Hydraulic Turbine and the Origins of American Industrial Research," *Technology and Culture* 20 (1979): 64-89. Professor Layton generously provided me with an earlier version of his paper.
6. Thomas S. Kuhn, "Energy Conservation as an Example of Simultaneous Discovery," in Marshall Clagett, ed., *Critical Problems in the History of Science* (Madison: University of Wisconsin Press, 1959), pp. 321-56; see esp. pp. 348-51, pp. 45-53, and Erwin Hiebert's critique, pp. 394-95. As Kuhn notes, an adequate, comprehensive historical account of these eighteenth- and early nineteenth-century conceptual transformations, which indeed were to prove essential to the evolution of the energy conservation laws, has yet to be written.
7. Hunter Rouse and Simon Ince, *History of Hydraulics* (New York: Dover, 1963), p. 62.
8. Ibid., p. 62.
9. I.H.B. Spiers and A.G.H. Spiers, *The Physical Treatises of Pascal* (New York: Columbia University Press, 1937).
10. D.S.L. Cardwell, *From Watt to Clausius: The Rise of Thermodynamics in the Early Industrial Age* (Ithaca: Cornell University Press, 1971), pp. 68-69; John Smeaton, *Experimental Enquiry Concerning the Natural Powers of Wind and Water* (London: The Royal Society, 1759, 1794), pp. 74-75; and Smith, "Origins of the Water Turbine," which contains an excellent summary of seventeenth- and eighteenth-century experiments.
11. Smeaton, *Experimental Enquiry*, p. 2; see also Terry S. Reynolds, "Scientific Influences on Technology: The Case of the Overshot Waterwheel, 1752-1754," *Technology and Culture* 20 (1979): 270-95.
12. Smeaton, *Experimental Enquiry*, p. 1.
13. Ibid., pp. 6-7.
14. Ibid., pp. 25-26.
15. See Cardwell, *From Watt to Clausius*, pp. 68-70.

16. Like most Englishmen at the time, Smeaton was a confirmed Newtonian and therefore would have little use for the continental concept of kinetic energy.

17. Clifford Ambrose Truesdell, "Rational Fluid Mechanics, 1687-1765," editor's introduction to *Leonardi Euleri Opera Omnia*, 2nd ser. 12 (1954): xlvi-xlvii.

18. Rouse and Ince, *History of Hydraulics*, p. 107.

19. Truesdell, "Rational Fluid Mechanics," p. xlvvi; Jakob Ackeret, preface to *Leonhardi Euleri Opera Omnia*, 2nd ser. 15 (1957): xlvii-xlix.

20. Rouse and Ince, *History of Hydraulics*, p. 125.

21. Cardwell, *From Watt to Clausius*, notes: "When the potential or latent *vis viva* becomes mgh, the *vis viva* must be $\frac{1}{2}$ mv^2 and not mv^2 as before. Thus the relationship between the two forms of *energy* was determined by the progressive recognition that mg^h was a basic and universally applicable measure in technology, mechanics and physics" (p. 167). Kuhn points out, however, that it was G. Coriolis (1829) who first insisted that *vis viva* be $\frac{1}{2}$ mv^2 because it should be equal to the work it could produce (Kuhn, "Energy Conservation").

22. See Ellwood Morris, "Remarks on Reaction Water Wheels Used in the United States; and on the Turbines of M. Fourneyron, an Hydraulic Motor, recently used with great success on the continent of Europe," *Journal of the Franklin Institute* 34 (1842): 224, on attempts to compare turbines in terms of grain ground.

23. See, for early wheels, Marcel Crozet-Fourneyron, *Invention de la turbine* (Paris and Liege: Library Polytechnique, 1924); Abbott Payson Usher, *A History of Mechanical Inventions* (Cambridge: Harvard University Press, 1954); Irving P. Church, *Hydraulic Motors* (New York: John Wiley, 1911); and d'Aubuisson, *Treatise on Hydraulics*. Also Calvin Wing, "Specification of a patent for the making of Reaction Water Wheels...," *Journal of the Franklin Institute* 11 (1831): 85-91; and Robert Eastman, "Specification of a patent for an improvement in the Reacting Water Wheel," *Journal of the Franklin Institute* 15 (1833): 320-23.

24. D'Aubuisson, *Treatise on Hydraulics*, p. 365.

25. For this reason, Smith, "Origins of the Water Turbine," considers Poncelet's wheel a form of "pressureless" turbine.

26. Usher, *Mechanical Inventions*, p. 388.

27. D'Aubuisson, *Treatise on Hydraulics*, p. 444.

28. Crozet-Fourneyron, *Invention de la turbine*, p. 25. Burdin is a puzzling figure, having proposed, early and unsuccessfully, water turbines, gas turbines, and submarines. See *Dictionaire de Biographie Française* (1956), s.v. "Burdin, Claude."

29. Frederick W. Keator, "Benoit Fourneyron (1802-1867)," *Mechanical Engineering* 61 (1939): 295-301.

30. Benoit Fourneyron, "Memoire on the application of the Prony brake to the measurement of power of large motors," *Bulletin de la société Industrielle de Mulhausen*, 1829, no. 6.

31. Arthur Morin, "Experiments on water wheels having a vertical axis, called turbines," trans. Ellwood Morris, *Journal of the Franklin Institute* 36 (1843): 234-46+.

32. M. de Prony, "Note sur un Moyen de mesurer l'effect dynamique des machines de rotation," *Annales de Chimie et de Physique* 19 (1822): 165-73. On Prony, see Rouse and Ince, *History of Hydraulics*, p. 141.

33. M. Prony, "Rapport: Machines A Vapeur du Gros-Cailou: Deuxieme Note: Sur un moyen de mesurer l'effect dynamique des machines de rotation," *Annales des Mines*, 1st ser. 12 (1826): 91-99.

34. Ellwood Morris, "On the Frictional Dynamometer, or Brake, of M. de Prony, a cheap, simple, ard effective instrument, for measuring the actual power developed by Machines," *Journal of the Franklin Institute* 35 (1843): 229.

35. Fourneyron, "Application of the Prony brake." Garland Allen, in his recent "In-

troduction of *Drosophila* into the Study of Heredity and Evolution: 1900-1910," *Isis* 66 (1975): 322-33, voices what seems to be an increasingly common view among historians of science, that "advances in theory often depend upon the introduction of new methods, techniques, and 'favorable' material." For turbine technology, the testability techniques centered on the Prony brake would seem to have had an analogous role in advancing (or creating) a progressive tradition of practice. In his case study, Allen examines the social origins of *Drosophila* experimentation and stresses the importance of "a particular community of workers" in the development of a new theory (p. 333). In contrast, the device celebrated eponymously as the Prony brake was apparently the singular suggestion of Riche de Prony, although a meticulous search of late-eighteenth-century and early-nineteenth-century sources might alter that belief. Nevertheless, what is crucial about the Prony brake parallels what Allen finds crucial about *Drosophila* experimentation: rigorous, detailed development and application by an emergent community of practitioners.

36. Fourneyron, "Application of the Prony brake," and "Memoire in four parts...," Bulletin, *Société d'Encouragement pour l'Industrie Nationale*, January-March 1834.

37. No information is readily available on industrial-machinery market structure in early-nineteenth-century France. Membership lists published in the *Bulletin de la Société Industrielle de Mulhausen* (1829), in which Fourneyron first described use of the Prony brake, do not give professional or occupational data. The administrative council lists for the Société d'Encouragement pour l'Industrie Nationale, formed in 1802 under the auspices of the minister of commerce and public works, are more informative. In 1834, the year in which Fourneyron published a complete description of his turbine in the society's bulletin, the president, Baron Thenard, was also a member of the Royal Academy of Sciences; and four of five other major officers were government officials. Of seventeen members, associate members, and honorary members on the Comité des Arts Mechaniques, two were members of the Academy of Sciences, two held academic appointments, and eight were government officials, three in the Corps des Ponts et Chaussées, two in the Corps des Mines (*Bulletin*, 1834, pp. 292-93). One of the honorary members was Baron de Prony, member of the Academy of Sciences and inspector general of Ponts et Chaussées. In 1839, of sixteen members of the same committee, three listed academic appointments, one was a member of the Academy, and seven held governmental positions, three relating to mining, two to bridges and roads (*Bulletin*, 1839, pp. 281-82). Other committees seem to have been equally dominated by academics and government officials. In contrast, of seventy-one new members admitted to the society in 1835 (*Bulletin*, 1834, pp. 481-82), only two held government jobs and only two others listed themselves as Ecole alumni. The other sixty-seven listed various private occupations, mostly related to manufacturing. Although the Société d'Encouragement pour l'Industrie Nationale thus does not seem to have limited its membership to those with formal scientific training, its administrative structure and its critical prize committees clearly seem to have been dominated by a scientific and bureaucratic elite. Statistical analysis of complete data, however, might revise or considerably alter this interpretation of the society's structure.

Ironically, for all Fourneyron's accomplishments in advancing scientific technology, and for all Arthur Morin's praise of him, Morin was selected over Fourneyron for membership in the Academy of Sciences in 1843—apparently because Fourneyron "was not a polytechnician" (Keator, "Benoit Fourneyron," p. 300). On the pecking order of nineteenth-century French schools, see Frederick B. Artz, *The Development of Technical Education in France, 1500-1850* (Cambridge: MIT Press, 1966).

38. Thomas S. Kuhn, "Mathematical versus Experimental Traditions in the Development of Physical Science," in *The Essential Tension* (Chicago: University of Chicago Press, 1977), pp. 31-65, esp. pp. 52-58.

39. Cardwell, *From Watt to Clausius*, pp. 156-57.

40. Ibid., p. 129.
41. Morin, "Experiments on water wheels."
42. On linkage roles in technology, see Ronald G. Havelock et al., *Planning for Innovation through Dissemination and Utilization of Knowledge* (Ann Arbor: University of Michigan Press, 1971).
43. Elwood Morris, "Experiments on the useful effect of Turbines in the United States," *Journal of the Franklin Institute* 36 (1843): 378.
44. Irving P. Church, *Hydraulic Motors* (New York: John Wiley, 1911), pp. 122-24.
45. Ibid., pp. 120-21.
46. See Norman Smith, "Origins of the Water Turbine," and *Man and Water: A History of Hydro-Technology* (London: Scribners, 1975), and esp. Hunter, *Industrial Power*, and Layton, "Millwrights and Engineers," for details on the richness of the nineteenth-century water turbine tradition.
47. American Society of Civil Engineers, *Biographical Dictionary of American Civil Engineers* (1972), s.v. "James B. Francis."
48. Hunter, *Industrial Power*, pp. 292-93; Layton, "Millwrights and Engineers," pp. 75-76.
49. James B. Francis, *Lowell Hydraulic Experiments* (Boston: Little Brown, 1855), pp. 1-7.
50. R. H. Thurston, "The Systematic Testing of Turbine Waterwheels in the United States," *Trans. ASME* 8 (1887): 359-420.
51. Francis, *Lowell Hydraulic Experiments*, pp. 14-19. Cardwell, *From Watt to Clausius*, presents John Smeaton's intensive but empirical development of the Newcomen steam engine in the third quarter of the eighteenth century as the first instance of "systematic evolutionary improvement" (p. 33) through rigorous *ceteris paribus* testing of successive small design variations.
52. Francis, *Lowell Hydraulic Experiments*, pp. 41-42.
53. Rouse and Ince, *History of Hydraulics*, p. 166.
54. The virtue of Howd's design remains a point of dispute. Layton, "Millwrights and Engineers," and Robert E. Horton (*Trans. ASCE* 85 [1922]), in discussion of Stafford and Hamilton, "American Mixed-Flow Turbine," rate Howd relatively highly, while Smith, *Man and Water*, who is followed here, does not. The Proprietors did pay Howd $1,200 for his patent (Hunter, *Industrial Power*, p. 319), although it is not clear whether that tidy sum was recompense for the value of his invention or tribute to his capacity for making trouble.
55. Layton, "Millwrights and Engineers"; idem, "Scientific Technology, 1845-1900."
56. Thurston, "Systematic Testing of Turbine Waterwheels," pp. 363-68.
57. Samuel Webber, "Water Power: Its Generation and Transmission," *Trans. ASME* 17 (1896): 41-57.
58. Arnold Pfau, "Wheels of the Pressure Type," *Transactions of the International Engineering Congress, 1915*, vol. 2, *Electrical Engineering and Hydroelectric Power Development*, paper no. 148, pp. 443-502; W. A. Doble, "The Tangential Water-Wheel," *Transactions of the American Institute of Mining Engineers* 29 (1899): 852-94; and idem, "Water Wheels of the Impulse Type," *Transactions of the International Engineering Congress, 1915*, 2, paper no. 149: 503-59. See also Hunter, *Industrial Power*.
59. Webber, "Water Power," p. 48.
60. Hunter, *Industrial Power*, pp. 350-52.
61. Pfau, "Wheels of the Pressure Type."
62. Chester W. Larner, "Characteristics of Modern Hydraulic Turbines," *Proceedings of the American Society of Civil Engineers* 35 (1909): 723-63.
63. Lewis F. Moody, "The Present Trend of Turbine Development," *Mechanical Engineering* 43 (1921): 235-44.

64. William M. White, "American Hydraulic Turbines," *Mechanical Engineering* 52 (1930): 393.

65. B. E. Smith, "The Kaplan Adjustable-Blade Turbine," *Trans. ASME* 52.1 (HYD 52-8) (1930): 137.

66. R. V. Terry, "Development of the Automatic Adjustable-Blade-Type Propeller Turbine," *Trans. ASME* 63 (1941): 395-409. Francis-type runners, however, are still used, in single-unit sizes as large as 200,000 h.p.; see *Encyclopedia Britannica* (1970), s.v. "Turbine, Water."

67. See Larner, "Modern Hydraulic Turbines," especially the discussion by John C. Parker (*Proc. ASCE* 35 [1909]), pp. 1287-89.

68. See below, Chapter 4.

69. Layton, "Scientific Technology, 1845-1900," p. 87.

70. Smith, *Man and Water.*

71. Larner, "Modern Hydraulic Turbines."

72. See ibid., pp. 759-63.

73. William F. Durand, "The Pelton Water Wheel," *Mechanical Engineering* 61 (1939): 447-54.

74. S. N. Knight, quoted in ibid., p. 449.

75. Durand, "The Pelton Water Wheel," p. 453.

76. F. G. Hesse, quoted in ibid., p. 453. Both Hesse and Durand apparently confused the Jonval wheel (France, 1841) with the Girard wheel (France, 1850). Hesse has his theory straight but his names wrong. The Jonval was an axial-flow, pure reaction wheel. The Girard was an impulse wheel manufactured in both axial- and radial-flow configurations, and is the appropriate antecedent to the Pelton design. Jonval, however, also invented the draft tube and introduced the practice of running two turbines in a common casing to obviate end thrust. Hesse seems merely to have conflated the contributions of Jonval and Girard. See Irving P. Church, *Hydraulic Motors* (New York: Wiley, 1911), pp. 73, 121-24, on Girard and Jonval; and F. C. Lea, *Hydraulics for Engineers and Engineering Students* (London: Arnold, 1916), pp. 368-69, on Girard.

77. Durand, "The Pelton Water Wheel," pp. 453-54.

78. Lester G. Pelton, quoted in ibid., p. 454.

79. Hamilton Smith, Jr., "Water Power with High Pressure and Wrought-Iron Water Pipe," *Transactions of the American Society of Civil Engineers* 13 (1884): 15-40. I am deeply indebted to an anonymous referee for *Social Studies of Science* for pointing out this source to me.

80. Robert K. Merton, "On Multiple Discoveries as a Strategic Research Site," in *The Sociology of Science* (Chicago: University of Chicago Press, 1973), pp. 371-83.

81. For a short discussion of the equifinality of open systems, see Kenyon B. DeGreene, *Sociotechnical Systems: Factors in Analysis, Design, and Management* (Englewood Cliffs, N.J.: Prentice-Hall, 1973), pp. 37, 43-45; and Merton, "On Multiple Discoveries as a Strategic Research Site," pp. 371-83.

82. Doble, "The Tangential Water-Wheel."

83. Doble, "Water Wheels of the Impulse Type."

84. Robert K. Merton, "The Normative Structure of Science," in *The Sociology of Science,* pp. 267-80.

85. Cardwell, *From Watt to Clausius,* pp. 79-84.

86. Arthur M. Green, *Pumping Machinery* (New York: Wiley, 1913), pp. 13-14.

87. Ibid., pp. 45-50.

88. L. E. Harris, "Some Factors in the Early Development of the Centrifugal Pump, 1689 to 1851," *Transactions of the Newcomen Society* 28 (1953): 187-202.

89. Edward Hopkinson and Alan E. L. Chorlton, "The Evolution and Present Development of the Turbine Pump," *Proc. Inst. M.E.* (London) 67 (1912): 59.
90. Harris, "Development of the Centrifugal Pump"; Greene, *Pumping Machinery*, p. 43; Hopkinson and Chorlton, "Development of the Turbine Pump," p. 8.
91. Harris, "Development of the Centrifugal Pump."
92. R. L. Daugherty, *Centrifugal Pumps* (New York: McGraw-Hill, 1915).
93. Harris, "Development of the Centrifugal Pump," pp. 193-95.
94. Ibid., pp. 198-99.
95. Hopkinson and Chorlton, "Development of the Turbine Pump," p. 42.
96. William O. Webber, "Some Types of Centrifugal Pumps," *Trans. ASME* 26 (1905): 764.
97. Daugherty, *Centrifugal Pumps*, p. 11.
98. Hopkinson and Chorlton, "Development of the Turbine Pump," p. 59.
99. A. H. Gibson, *Osborne Reynolds and His Work in Hydraulics and Hydrodynamics* (London: The British Council, 1948), pp. 1-8.
100. Ibid., p. 20.
101. See Church, *Hydraulic Motors*, pp. 119-21.
102. The science Reynolds used may not have been all that old or that unoriginal. Bernoulli's theorem as stated here and as portrayed in modern mathematical terms is not quite the way Bernoulli stated it. Giovanni Venturi (1797) made notable contributions concerning the flow of fluids in tubes of changing diameter (Rouse and Ince, *History of Hydraulics*, pp. 136-38). Rouse and Ince credit Julius Weisbach with "popularizing the application of the Bernoulli equation in its presently accepted form" (p. 164). Weisbach's work, although published in German in 1850, was not translated into English until 1877, two years after Reynolds' turbine pump patent. Thus, Reynolds may have done more purely theoretical development than at first seems apparent.
103. Gibson, *Osborne Reynolds*, pp. 20-21.
104. Ibid.
105. J. F. Schubeler, in Hopkinson and Chorlton, "Development of the Turbine Pump," p. 44.
106. Hopkinson and Chorlton, "Development of the Turbine Pump."
107. Ibid., pp. 59-60.

Chapter 3

1. Abbott Payson Usher, *A History of Mechanical Invention* (Cambridge: Harvard University Press, 1929), p. 392.
2. J. G. Keenan, *Elementary Theory of Gas Turbines and Jet Propulsion* (London: Geoffrey Cumberlege, 1946), p. 4; Jack V. Casamassa, *Jet Aircraft Power Systems* (New York: McGraw-Hill, 1950), p. 2.
3. James Watt, quoted in Charles A. Parsons, "Presidential Address to the British Association" (1919), in *The Scientific Papers and Addresses of the Hon. Charles A. Parsons*, ed. G. L. Parsons (Cambridge: Cambridge University Press, 1934), p. 133.
4. *Encyclopaedia Brittanica* (1964), s.v. "Turbine, Steam."
5. Charles A. Parsons, "The Steam Turbine on Land and Sea" (1906), in *Papers*, p. 65, Rollo Appleyard, *Charles Parsons* (London: Constable, 1933), pp. 73-81.
6. Parsons, "The Steam Turbine on Land and Sea."
7. Lester G. French, *Steam Turbines* (New York: McGraw-Hill, 1907), p. 63.

8. This account of Tournaire's work and its significance is taken from Henry Harrison Suplee, *The Gas Turbine* (Philadelphia: Lippincott, 1910). Suplee reproduces (in English translation) the complete text of Tournaire's remarkable memoir; see pp. 19-20.

9. See William Todd, *History as Applied Science* (Detroit: Wayne State University Press, 1972), on the use of counterfactual assumptions in assessing historical reconstruction.

10. Joseph H. Keenan and A. H. Shapiro, "History and Exposition of the Laws of Thermodynamics," *Mechanical Engineering* 69 (1947): 915-21.

11. Ibid.; see also D.S.L. Cardwell, *From Watt to Clausius: The Rise of Thermodynamics in the Early Industrial Age* (Ithaca: Cornell University Press, 1971).

12. *Encyclopedia Britannica* (1954), s.v. "Turbine, Steam."

13. George H. Gibson, "A Pioneer in High Pressure Steam: Carl Gustav Patrick De Laval," *Power* 68 (1928): 762-64.

14. W. Bergh, "Centrifugal Separator for Liquids," *Proc. Inst. M.E.* (London) 37 (1882): 519-27.

15. Gibson, "Pioneer in High Pressure Steam: De Laval," p. 762.

16. A similar Hero reaction turbine and centrifugal separator, for sugar syrup, had been exhibited by Henry Bessemer at the Crystal Palace Exposition in 1851 (L. E. Harris, "Some Factors in the Early Development of the Centrifugal Pump, 1689 to 1851," *Transactions of the Newcomen Society*, 28 [1953]: 187-202).

17. Parsons, "The Rise of Motive Power and the Work of Joule" (1922), in *Papers*, p. 156, and idem, "The Steam Turbine on Land and Sea," p. 63.

18. Parsons, "Motive Power and High-Speed Navigation: Steam Turbines" (1900), in *Papers*, p. 28; idem, "The Rise of Motive Power," p. 156. For a lucid description of modern turbine types, see Walter Hossli, "Steam Turbines," in *Scientific Technology and Social Change: Readings from "Scientific American"* (San Francisco: W. H. Freeman, 1969), pp. 158-68.

19. Parsons, "Motive Power and High-Speed Navigation," p. 28.

20. Parsons, "The Steam Turbine on Land and Sea," p. 64.

21. Parsons, "The Expansive Working of Steam in Steam Turbines" (1909), in *Papers*, p. 76.

22. Gibson, "Pioneer in High Pressure Steam: De Laval," p. 762.

23. Lord Rayleigh, "Some Personal Reminiscences of Sir Charles Parsons" (1934), in Parsons, *Papers;* Appleyard, *Charles Parsons*, pp. 18-26.

24. Appleyard, pp. 18-23.

25. Ibid., 23-26.

26. Rayleigh, "Reminiscences of Parsons," pp. xx-xxi.

27. Appleyard, *Charles Parsons*, pp. 27-30.

28. Parsons, "The Steam Turbine" (1911), in *Papers*, p. 108.

29. Parsons, "The Compound Steam Turbine and Its Theory, As Applied to the Working of Dynamo-electric Machines" (1887), in *Papers*, p. 3.

30. Robert D. Napier, *On the Velocity of Steam and Other Gases and the True Principles of the Discharge of Fluids* (London: E & F. N. Spon, 1866). Although Napier's experimental results, within their limits, were accurate, his theoretical explanation, which involved an attempt to apply purely Newtonian notions to compressible fluids, was erroneous.

On the state of theoretical knowledge when Parsons began his turbine work, Alexander Richardson (*The Evolution of the Parsons Steam Turbine* [London: Offices of "Engineering," 1911]) notes:

> When the subject of the steam turbine was taken in hand, in 1884, very little was known as to the properties of steam in rapid motion beyond the fact that the resistance to the flow of steam in jets bore an approximate relationship to that of water at the same velocity for

small amounts of expansion, and little *as to the amount of the conversion of potential into kinetic energy when steam expanded through divergent nozzles.* (p. 18).

See also Sanford A. Moss, *Superchargers for Aviation* (New York: National Aeronautics Council, 1942), p. 9.

31. Parsons, "Motive Power and High Speed Navigation," p. 29.
32. Appleyard, *Charles Parsons*, pp. 69-89.
33. R. H. Parsons, *The Steam Turbine and Other Inventions of Sir Charles Parsons* (London: The British Council, 1944), p. 7.
34. Appleyard, *Charles Parsons*, p. 26. Richardson, writing in 1911 (*The Evolution of the Parsons Steam Turbine*), portrays Parson's radial-flow turbines as abject failures; other sources mark them as qualified successes. The parallel-flow type was clearly superior, but it appears that Richardson was somewhat overenthusiastically towing the then current company line in his damnation of the radial-flow type.
35. R. H. Parsons, *The Steam Turbine*, pp. 7-8.
36. Such large turbines are complex: for example, each might require over a million individual turbine blades in its construction. Present critical-heat regenerative-cycle steam turbine generator units produce up to 600,000 kW. at efficiencies approaching 40 percent.
37. Charles A. Parsons, "Motive Power and High-Speed Navigation," pp. 32-34.
38. Idem, "The Steam Turbine on Land and Sea," and Richardson, *The Evolution of the Parsons Steam Turbine,* pp. 70-72.
39. *New York Times,* 18 July 1897.
40. Charles A. Parsons, "The Steam Turbine on Land and Sea," and Appleyard, *Charles Parsons.*
41. R. H. Parsons, *The Steam Turbine,* p. 12.
42. Thomas G. Marx, "Technological Change and the Theory of the Firm: The American Locomotive Industry, 1920-1955," *Business History Review* 50 (1976): 1-24.
43. Richard Hough, *Dreadnought: A History of the Modern Battleship* (New York: Macmillan, 1964), pp. 20-21.
44. Henry Harrison Suplee, "Auguste C. E. Rateau, 1863-1930," obituary notice, *Mechanical Engineering* 52 (1930): 571; *Grand Larousse Encyclopedeque* (1964), s.v. "Rateau, Auguste." On the order and quality of French technical schools, see Frederick B. Artz, *The Development of Technical Education in France, 1500-1850,* (Cambridge, Mass.: M.I.T. University Press, 1966), esp. chap. 4.
45. A. Rateau, "Different Applications of Steam-Turbines," *Proc. Inst. M.E.* (London) 67 (1904): 744-45.
46. See Thomas Parke Hughes, *Elmer Sperry: Inventor and Engineer* (Baltimore: Johns Hopkins Press, 1971), for the concept of "inventor-entrepreneur," esp. pp. 66-71, 283-84.
47. *National Cyclopedia of American Biography* (1958), s.v. "Curtis, Charles Gordon."
48. W.L.R. Emmet, "The Curtis Steam Turbine," *Proc. Inst. M.E.* (London) 67 (1904): 716-17.
49. Kendall Birr, *Pioneering in Industrial Research* (Washington, D.C.: Public Affairs Press, 1957), esp. pp. 28-33.
50. Sanford A. Moss, "Gas Turbines and Turbosuperchargers," *Trans. ASME* 66 (1944): 351-71.
51. William J. Kearton, *Steam Turbine Theory and Practice* (New York: Pitman, 1931), p. 357.
52. Aurel Stodola, *Steam and Gas Turbines* (New York: McGraw-Hill, 1927), 1: 608. A turbine similar to the Reidler-Stumpf was later used to power the fuel pumps of German rocket engines. The British unsuccessfully tried a similar design for the same purpose; see A. D. Baxter, "Aircraft Rocket Motors," *Aircraft Engineering* 19 (1947): 249-56.
53. Theodore Baumeister, *Fans* (New York: McGraw-Hill, 1935), p. 16.

54. "Roots Low-Pressure Blower: History," The P. H. & F. M. Roots Co. Bulletin no. 1011 (1925), p. 2.
55. A. Rateau, "High-Speed Centrifugal Pumps and Fans," *The Engineer* 93 (1902): 240.
56. Ibid.
57. A. Rateau, "High-Pressure Centrifugal Fans," *Engineering* 84 (1907): 248–51+.
58. Rateau, "High-Speed Fans," p. 240; and "Action of Centrifugal Pumps and Fans," *Engineering* 84 (1907): 690.
59. Rateau, "High-Speed Fans," p. 230.
60. Ibid.
61. Ibid., anonymous commentator, p. 240.
62. Rateau, "High-Pressure Fans," p. 249.
63. Ibid., p. 288. W. Reavell, "Reciprocating Air Compressors and Turbo Compressors for High Pressures," *Engineering* 84 (1907): 3.
64. Rateau, "High-Pressure Fans," p. 288.
65. Ibid., p. 287.
66. Ibid., p. 250.
67. *Engineering* 84 (1907): 3–5; Reavell, "Air-Compressors," ibid., pp. 29–30.
68. Rateau, "High-Pressure Fans," pp. 250–51.
69. Ibid.
70. Reavell, "Air-Compressors."
71. A. R. Howell, "Fluid Dynamics of Axial Compressors," *Proc. Inst. M.E.* (London) 153, War Emergency Issue no. 12 (1945): 441–52; A. Carnegie, discussion of "Reciprocating Air Compressors and Turbo Compressors for High Pressures," *Engineering* 84 (1907): 2.
72. Carnegie, p. 2.
73. A. I. Ponomareff, "Principles of the Axial-Flow Compressor," *Westinghouse Engineer* 7 (1947): 40–46. Ponomareff was manager of the Pump and Blower Engineering Section, Westinghouse Electric.
74. Ibid. pp. 40–41.
75. "Experimental Gas Turbine Plant," *The Engineer* 182 (1946): 51–52.
76. See Stodola, *Steam and Gas Turbines*, 2: 1217–35, for some of the many proposed systems.
77. Ibid., p. 1237. See E. D. Meyer, "The Combustion Gas Turbine," *Proc. Inst. M.E.* (London) 141 (1939): 200.
78. Meyer, "The Combustion Gas Turbine," p. 200.
79. Ibid.
80. W. J. Stern, "The Internal Combustion Turbine," Great Britain, Aeronautical Research Council (ARC) Engine Sub-Committee Reports, no. 54 (1920), p. 64.
81. Dugald Clerk, "Notes on the Gas-Turbine," *Engineering* 94 (1912): 367–68.
82. Stodola, *Steam and Gas Turbines*, 2: 1235.
83. Moss, "Gas Turbines and Turbosuperchargers," p. 355.
84. Stodola, *Steam and Gas Turbines*, 2: 1235.
85. Rene Armengaud, "The Gas Turbine: Practical Results with Actual Operative Machines in France," *Cassiers' Magazine* 31 (1907): 196.
86. Rateau, "High-Pressure Fans," p. 250.
87. P. R. Sidler, comment on Moss, "Gas Turbines and Turbosuperchargers," *Trans ASME* 66 (1944): 355.
88. Armengaud, "The Gas Turbine."
89. Clerk, "Notes," p. 367.
90. Armengaud, "The Gas Turbine."
91. Stern, "The Internal Combustion Turbine," pp. 58–59.

92. This account of the Moss–General Electric projects is based upon Moss, "Gas Turbines and Turbosuperchargers."
93. Interview with Glenn B. Warren by Ilan Kusiatin, Harvard Business School, in Schenectady, N.Y., 18 September 1975. I am indebted to Dr. George Wise of the General Electric Corporate Research and Development Laboratories, Schenectady, for providing me with a transcript of this interview.
94. Clerk, "Notes on the Gas-Turbine," p. 367.
95. See Chapter 5 below. Notably, Charles Parsons too had entertained the notion of a gas turbine, and, as Richardson (*The Evolution of the Parsons Steam Turbine*) reported in 1911:

> The success of the steam turbine has encouraged the engineering profession to investigate the problems of the internal combustion turbine which would dispense with boilers, and many auxiliary engines. One of the seemingly insurmountable difficulties is that the high temperatures of the gas would burn the turbine blades. It was suggested that this temperature might be reduced by the expansion of the flame in a divergent nozzle. Mr. Parsons and Mr. Stoney made some experiments with a flame flowing under pressure through such a divergent nozzle. A platinum wire was placed at each end, and the temperature effect on it observed through glass windows. It was found that the two wires attained almost identical temperatures, both showing a bright red heat. The gases had cooled on expansion, but the heat was restored by the impact of the gas against the wire. The same result would occur with blades, and thus there seems an insurmountable obstacle to the practical success of an internal combustion turbine.

96. "The Efficiency of the Gas Turbine," *Engineering* 92 (1911): 706.
97. Moss, "Gas Turbines and Turbosuperchargers," pp. 356, 371.

Chapter 4

1. R. Giacomelli and E. Pistolesi, "Historical Sketch," in W. F. Durand, ed., *Aerodynamic Theory* (Berlin: Springer, 1934), pp. 310-12.
2. W. R. Durand, "Historical Sketch of the Development of Aerodynamic Theory," *Trans. ASME* 51, Aer-51-3 (1929): 13-19.
3. Ibid.
4. Ibid.
5. The term "Reynolds' Number" was coined by Arnold Summerfield in 1908.
6. J. L. Pritchard, "The Dawn of Aerodynamics," *Jl. RAeS* 61 (1957): 152-56.
7. Theodore von Karman, *Aerodynamics* (Ithaca: Cornell University Press, 1954). Most of this treatment of aerodynamics, except where otherwise noted, is based on von Karman's history.
8. "Mach Number" (to indicate velocity relative to the local speed of sound), "Mach Cone," and "Mach Angle" were not used until after 1935, when Jacob Ackeret introduced the term "Mach Number" at the Volta High Speed Conference.
9. See Karl Popper, *Conjectures and Refutations* (New York: Harper & Row, 1962).
10. "Prandtl," *The Guggenheim Medalists* (New York: Daniel Guggenheim Fund, 1964).
11. Von Karman, *Aerodynamics*, p. 88.
12. Gernot Böhme, Wolfgang van den Daele, and Wolfgang Krohn, "The 'Scientification' of Technology," *Sociology of the Sciences Yearbook, 1978* (Boston: Reidel, 1978), p. 242.
13. Von Karman, *Aerodynamics*, p. 88.

14. Again, see Popper, *Conjectures and Refutations,* for demarcation between science and nonscience.

15. Von Karman, *Aerodynamics,* pp. 166-73.

16. R. C. Pankhurst, "Aerodynamics at NPL, 1917-1970," *Nature* 238 (1972): 375-80. Actual choice of a propeller for a given application still required recourse to empirically derived data. See Walter G. Vincetti, "The Air-Propeller Tests of W. F. Durand and E. P. Lesley: A Study in Technological Methodology," *Technology and Culture* 20 (1979): 712-51.

17. J. L. Nayler, "Aeronautical Research at the National Physical Laboratory," *Jl. RAeS,* Centenary Issue, 70 (1966): 83.

18. B. M. Jones, "The Importance of 'Streamlining' in Relation to Performance," Great Britain, Aeronautical Research Committee Reports and Memorandum (R&M), no. 1115 (1927).

19. B. M. Jones, "The Streamline Aeroplane," *Jl RAeS* 33 (1929): 357-85.

20. Von Karman, *Aerodynamics,* p. 69.

21. To be discussed in Chapter 6.

22. Theodore von Karman. "The Problem of Resistance in Compressible Fluids," in *Volta High Speed Conference* (1935), p. 256 (hereafter cited as *VHSC*).

23. G. P. Douglas, "Research on Model Airscrews at High Speed," in *VHSC,* pp. 446-72; Hugh L. Dryden, "Supersonic Travel within the Last Two Hundred Years," *The Scientific Monthly* 78 (1954): 289-95.

24. G. I. Taylor, "Well-Established Problems in High-Speed Flow," in *VHSC,* pp. 186-203.

25. E. N. Jacobs, "Methods Employed in America for the Experimental Investigation of Aerodynamic Phenomena at High Speeds," in *VHSC,* pp. 356-87.

26. Von Karman, "The Problem of Resistance," pp. 241, 243.

27. Taylor, "Problems in High Speed Flow," pp. 186-203.

28. W. F. Hilton, *High-Speed Aerodynamics* (New York: Longmans, Green, 1951).

29. See von Karman, "The Problem of Resistance," p. 236, 241; Taylor, "Problems in High-Speed Flow," p. 201.

30. A. A. Rubbra, "Alan Arnold Griffith," *Biographical Memoirs of the Fellows of the Royal Society of London, 1964* (1964), pp. 117-36.

31. Frank Nixon, "Aircraft Engine Developments during the Past Half Century," *Jl. RAeS* 70, Centenary Issue (1966): 165. The De Havilland Comet, Lockheed Electra, and Douglas DC-10 notwithstanding.

32. See A. R. Howell, "Griffith's Early Ideas on Turbomachinery Aerodynamics," *Aeronautical Journal,* December 1976, pp. 521-29, for a full discussion of Griffith's original paper; also F. W. Armstrong, "The Aero Engine and Its Progress: Fifty Years after Griffith," *Aeronautical Journal,* December 1976, pp. 499-520. I am indebted to Mr. P. R. Mann of the National Aeronautical Collection, Science Museum, South Kensington, for calling these sources to my attention.

33. R. G. Harris and R. A. Fairthorne, "Wind Tunnel Experiments with Infinite Cascades of Aerofoils," Great Britain, Aeronautical Research Committee Reports and Memorandum (R&M), no. 1206 (1928).

34. A. A. Griffith "The present position of the internal combustion turbine as a powerplant for aircraft," (n.p., 1929), quoted in Rubbra, "Alan Arnold Griffith," p. 122.

35. Robert Schlaifer, *The Development of Aircraft Engines* (Cambridge: Harvard University Press, 1950), pp. 332-33.

36. Rubbra, "Alan Arnold Griffith."

37. Dryden, "Supersonic Travel," pp. 289-95; Jacob Ackeret, "Aeronautical Education and Research at the Swiss Institute of Technology in Zurich," NACA TM no. 616 (1930);

Jacob Ackeret, "High-Speed Wind Tunnels," NACA TM no. 808 (1935) (translation of Ackeret's Volta High Speed Conference paper).

38. C. Seippel, "The Development of the Brown Boveri Axial Compressor," *Brown Boveri Review* 27 (1940): 108-13.

39. Ibid.

40. M. M. Postan, D. Hay, and J. D. Scott, *The Design and Development of Weapons* (London: H.M. Stationery Office, 1964), p. 187.

41. J. W. Adderly, "German Gas Turbine Developments during the Period 1939-1945," B.I.O.S. Final Report, no. 12, (1949), pp. 32-33; P. R. Price, "Gas Turbine Development by B.M.W.," C.I.O.S., item no. 5, file no. XXVI-30, pp. 26-28B.

42. Albert Betz, "Diagrams for Calculations of Aerofoil Lattices," NACA TM no. 1022 (July 1942), originally published in *Ingenieur-Archiv* (1931); Albert Betz, "Axial Superchargers," NACA TM no. 1073 (August 1944), originally published in *Jahrbuch 1938 der Deutschen Luft Fahrforschung;* W. Encke, "Investigations on Experimental Impellers for Axial Blowers," NACA TM no. 1123 (April 1947); Postan, *Design and Development of Weapons.*

Chapter 5

1. A. H. Roy Fedden, "Aircraft Power Plant: Past and Future," *Jl. RAeS* 48)1944): 335-88.

2. Ibid.

3. Ibid.

4. Ibid.

5. F. M. Owner, "Random Reminiscences in Centenary Year," *Jl. RAeS* 70, Centenary Issue (1966): 181.

6. Sterry B. Freeman, "Marine Oil Engines," *Proc. Inst. M.E.* (London) 129 (1935): 203-9. "Scavenging" refers to the complete extraction of the burned mixture from a cylinder after the power stroke, a necessary condition for efficient operation.

7. E. T. Vincent, *Supercharging the Internal Combustion Engine* (New York: McGraw-Hill, 1948), pp. 1-2.

8. James D. Roots, comment on A. Rateau, "The Use of the Turbo-compressor for Attaining the Greatest Speeds in Aviation," *Proc. Inst. M.E.* (London) 116 (1922): 1038-39.

9. Alfred J. Buchi, "Personal Reflections on Forty Years of Diesel Engine Development," *Mechanical Engineering* 61 (1939): 213-16.

10. Alfred J. Buchi, "Turbo-charging and Gas Turbines," *Journal of the American Society of Naval Engineers* 60 (1948): 261-91.

11. A. Rateau, "The Use of the Turbo-compressor forAttaining the Greatest Speeds in Aviation," *Proc. Inst. M.E.* (London) 116 (1922): 809-10.

12. Vincent, *Supercharging,* pp. 1-2.

13. Sanford A. Moss, "Gas Turbines and Turbosuperchargers," *Trans. ASME* 66 (1944): 349-50.

14. A. L. Berger and Opie Chenoweth, "The Turbosuperchargers," *Society of Automotive Engineers Journal* 29 (1931): 281.

15. Interview with Glenn B. Warren by Ilan Kusiatin, Harvard Business School, in Schenectady, N.Y., 18 September 1975.

16. A.H.R. Fedden, "Aircraft Engines," *Proc. Inst. M.E.* (London) 129 (1935): 197-203.

17. Robert Schlaifer, *The Development of Aircraft Engines* (Cambridge: Harvard University Press, 1950), p. 196. This account of Heron's development of the sodium-cooled valve is taken from Schlaifer's chap. 7, especially the Appendix, pp. 196-98.

18. Ibid., p. 197.

19. This account of the development of the Bristol sleeve-valve engine is based upon Roy Fedden, "The First 25 Years of the Bristol Engine Department," *Jl. RAeS* 65 (1960): 332-52.
20. Ibid., p. 346.
21. See Chapter 6 for an account of the Schneider Trophy races.
22. This account of the development of the Rolls Royce R engine is taken from H. E. Wimperis, "The British Technical Preparation for the Schneider Trophy Contest, 1931," *Volta High Speed Conference* (1935), pp. 23-29.
23. F. R. Banks, "Ethyl: Some Information on the Use and Advantages Gained by the Employment of Tetraethyl in Fuels for Aviation Engines," *Jl. RAeS* 38 (1934): 309-72.
24. See Ronald Miller and David Sawers, *The Technical Development of Modern Aviation* (London: Routledge & Kegan Paul, 1968), chap. 3: "Creating the Economic Airplane," pp. 47-97, esp. pp. 53-71.
25. J. C. Hunsaker, "Research in Aeronautics," in U.S. Presidential, National Resources Planning Board, *Research: A National Resource*, 2 (Washington, D.C.: Government Printing Office, 1940): 129-43.
26. Miller and Sawers, *Technical Development*, pp. 64-66.
27. H. B. Howard, "Aircraft Structures," *Jl. RAeS* 70; Centenary Issue (1966): 54; B. S. Shenstone, "Hindsight Is Always 100 Per Cent," ibid., p. 131; A. J. Murphy, "Materials in Aircraft Structures," ibid., pp. 114-20; I. J. Gerald, "Monocoque Construction," *Jl. RAeS* 41 (1937): 467-92.
28. Hunsaker, "Research in Aeronautics," p. 138.
29. Paul H. Wilkinson, *Aircraft Diesels* (New York: Pitman, 1940), p. 17.
30. This account of aircraft diesels is largely based upon Wilkinson's book.
31. William Green, *War Planes of the Second World War,* vol. 10, *Bombers* (New York: Doubleday, 1968), pp. 45-51.
32. Wilkinson, *Aircraft Diesels*, pp. 149-51.
33. These are dry weights exclusive of cooling systems, oil coolers, and so on. Comparative weights are tricky, since so much depends on installation. See Schlaifer, *Development of Aircraft Engines*, pp. 676-78. Nevertheless, for consistent weight definitions, diesels do suffer by approximately the ratios indicated.
34. Fedden, "Aircraft Power Plant," p. 362.
35. Harry R. Ricardo, "Piston Aero-Engines," *Proc. Inst. M.E.* (London) 157, War Emergency Issue no. 29 (1947): 194-96.
36. *Encyclopaedia Britannica* (1954), s.v. "Rockets."
37. G. Geoffrey Smith, *Gas Turbines and Jet Propulsion for Aircraft* (New York: Philosophical Library, 1944), p. 6.
38. This account of early rocketry activity is based upon Ley (New York: Viking, 1968) and Dornberger (New York: Viking, 1954) unless otherwise noted.
39. See photographs in Ley, *Rockets, Missiles, and Men in Space*, plates 2-4.
40. Smith, *Gas Turbines and Jet Propulsion*, p. 22.
41. Ibid., pp. 21-25.
42. Edgar Buckingham, "Jet Propulsion for Airplanes," *NACA Ninth Annual Report*, Technical Report no. 159 (1923).
43. Eastman N. Jacobs and James M. Shoemaker, "Tests on Thrust Augmentors for Jet Propulsion," NACA Technical Report no. 431 (1928).
44. G. B. Schubauer, "Jet Propulsion with Special Reference to Thrust Augmentors for Jet Propulsion," NACA TN no. 442 (1934).
45. R. H. Goddard, "A New Turbine Rocket Plane for the Upper Atmosphere," *Scientific American* 146 (1932): 148-49.
46. W. J. Stern, "The Internal Combustion Turbine," Great Britain, Aeronautical Research Committee, Engine Sub-Committee Reports, no. 54 (1920).

47. Claude Seippel, "Gas Turbines in Our Century," *Trans. ASME* 75 (1953): 122.
48. Ibid.
49. C. Seippel, "The Development of the Brown Boveri Axial Compressor," *Brown Boveri Review* 27 (1940): 108–13; idem, "Henry Thomas and the Birth of the Gas Turbine," *Brown Boveri Review* 54 (1967): 132; E. D. Meyer, "The Combustion Gas Turbine: Its History, Development, and Prospects," *Proc. Inst. M.E.* (London) 150 (1939): 1–10. (E. D. Meyer is apparently a misidentification of Adolf Meyer.)
50. Curt Keller, "The Escher Wyss–AK Closed-Cycle Turbine: Its Actual Development and Future Prospects," *Trans. ASME* 68 (1946): 791–822; W. M. Crim, J. R. Hoffman, and G. B. Manning, "The Compact AK Process Nuclear System," *Trans. ASME* 88 (1966): 127–38.
51. James Calderwood, "Research on Internal Combustion Prime Movers," *The Engineer* 181 (1946): 428–29+.
52. Lysholm attempted to use a centrifugal compressor with no less than eleven stages. His design was thus closer to that of Armengaud and Lemale than to later turbojet practice and could not have offered the great power for extremely low weight that would be the hallmark of the turbojet.
53. A.J.R. Lysholm, "A New Rotary Compressor," *Proc. Inst. M.E.* (London) 150 (1943): 11–16; idem, "A Contribution to the Solution of the Gas Turbine Problem," *Proc. Inst. M.E.* 157, War Emergency Issue no. 36 (1945): 498–523.
54. Adolf Meyer, "Recent Developments in Gas Turbines," *Mechanical Engineering* 69 (1947): 273–77.
55. Christian Lorenzen, "The Lorenzen Gas Turbine and Supercharger for Gasoline and Diesel Engines," *Mechanical Engineering* 52 (1930): 665–72; R. Tom Sawyer, *The Modern Gas Turbine* (New York: Prentice-Hall, 1947), pp. 39–40.
56. "The Jendrassik Combustion Turbine," *Engineering*, 1939, pp. 186–88.
57. H. R. Ricardo, "High-Altitude Engines: Thermodynamics and Carburetion," *VHSC* (1935), pp. 557–69.
58. Ibid.

Chapter 6

1. R. Smelt, "A Critical Review of German Research on High-Speed Airflow," *Jl. RAeS* 50 (1946): 900.
2. E. N. Jacobs, "Methods Employed in America for the Experimental Investigation of Aerodynamic Phenomena at High Speeds," *Volta High Speed Conference* (1935), pp. 357–58.
3. Ibid., pp. 360–61.
4. Ibid., p. 376.
5. Ibid., p. 380.
6. John T. Sinette, Jr., Oscar W. Schey, and J. Austin King, "Performance of NACA Eight-Stage Axial-Flow Compressor Designed on the Basis of Airfoil Theory," NACA Report no. 758 (1943), pp. 81–82.
7. See note number 42, Chapter 4.
8. Reale Accademia D'Italia, Fondazione Allesandro Volta, *Convegno di Scienze Fisiche, Matematiche e Naturali*, Theme: "High Speeds in Aviation," 2nd ed. (Rome, 1940) (original edition 1935).
9. Missing is Herman Glauert, killed in 1934 in a freak accident. While watching workmen dynamite a stump, he was struck by a flying splinter.
10. This statistical analysis includes a chi-square (χ^2) analysis of the distribution of the

90 citations and construction of 99 percent confidence intervals for the German and American figures. The 90 citations contain multiple citations to some papers. Multiple citation is ordinarily and, in this context, legitimately taken as an indicator of the relative quality of different papers. In this sense, the 90 citations are a mixed indicator of quality and quantity. The citations' validity as a gross indicator of theoretical aerodynamic output, quantitative and qualitative, in respective countries is not thereby impugned.

The chi-square analysis depends upon specification of a null hypothesis involving the following assumptions: all countries in the universe (5) are in fact equal in qualitative and quantitative aerodynamic output; the 90 citations represent a true random selection from among that equally distributed population; therefore, the expected value of the citations to each country's work is the mean value, $\bar{x} = \frac{90}{5} = 18$ citations. Assuming the 90 citations are a representative random sample, as external qualitative evidence suggests, chi-square analysis asks what the chance of the observed distribution appearing is if the null hypothesis above is in fact true. In this case, the chance is less than 5 in 1,000, $\chi^2 = 17.1$, for a probable value of less than .005.

The preeminence of non-U.S. sources does not result from authors' citing only works of their own nationality: excluding the Douglass paper on airscrews (Great Britain), which cites only three British theoretical sources, the five remaining papers which have footnotes (two German [counting von Karman as a German], one American, and two Italian) contain a total of 90 citations. Construction of a 3 x 2 contingency table (German, U.S., and Italian papers, by own versus other country citations) and application of a test for dependence yields a $\chi^2 = 1.81$ with 2 degrees of freedom, which provides no support for rejecting the hypothesis that there is no relation between nationality of paper and nationality of citations.

Finally, a difference of true means test applied to the 26 German and 12 American citations found in the six theoretical papers supports the conclusion that Germany was producing on the order of twice as many contributions to theoretical high-speed aerodynamics as the United States. From observed means of $\bar{x}_{GER} = 4.3$ and $\bar{x}_{US} = 2.0$ for the six papers, true population means are indicated to differ $\mu_{GER} - \mu_{US} = 2.3 \pm 1.2$ at a 95 percent confidence level (Student's t).

11. G. V. Lachmann, "Aerodynamic and Structural Features of Tapered Wings," *Jl. RAeS* 41 (1937): 162-237.

In his remarkably perceptive examination of W. F. Durand's and E. P. Lesley's propeller experiments, Walter G. Vincetti notes:

From the mid-1920s, work in *NACA* laboratories was noted for the comprehensive empirical study of aerodynamic problems by the method of parameter variation. Perhaps the most important (but far from the only) example of this was the study of the performance of airfoils as it depends on the parameters that define the airfoil's shape, a study that continues in the laboratories of NACA's successor, NASA. The work of Durand and Lesley was the first aerodynamic study under NACA auspices to employ this method. Because of the work's comprehensiveness and quality, and because of Durand's prestige and membership on the committee, we may suppose that the Durand-Lesley effort was an important influence in establishing the empirical pattern of NACA research. To the extent that this is true, the methodological effects of the Stanford propeller work went far beyond propeller technology. (Walter G. Vincetti, "The Air-Propeller Tests of W. F. Durand and E. P. Lesley: A Case Study in Technological Methodology" *Technology and Culture* 20 (1979): 740).

As Vincetti also observes, the purpose of NACA as defined in the organic act was "to supervise and direct the scientific study of the problems of flight, with a view to their practical solution.... (p. 721)."

Note to Page 156

12. This section on wind tunnels is based on information drawn from Jacob Ackeret, "Aerodynamical Education and Research at the Swiss Institute of Technology in Zurich," NACA TM no. 616 (1930); idem, "High-Speed Wind Tunnels," NACA TM no. 808 (1935; translation of Ackeret's Volta Conference paper); George W. Gray, *Frontiers of Flight: The Story of NACA Research* (New York: Knopf, 1948); W. F. Hilton, "British Aeronautical Research Facilities," *Jl. RAeS* 70, Centenary Issue (1966): 103-7; R. C. Pankhurst and D. W. Holder, *Wind-Tunnel Technique* (London: Pitman, 1952); Ludwig Prandtl and O. G. Tietjens, *Applied Hydro- and Aeromechanics* (New York: Dover, 1934).

13. See Smelt, "Critical Review," for a relatively complete survey of German high-speed research during the war.

14. Hilton, "British Aeronautical Research Facilities," p. 103.

15. Ibid., p. 104.

16. Robert R. Gilruth, "From Wallops Island to Project Mercury, 1945-1958," in *Essays on the History of Rocketry and Astronautics: Proceedings of the Third through Sixth History Symposia of the International Academy of Astronautics*, 2 (Washington, D.C.: NASA, 1977): 445-74.

17. Gray, *Frontiers of Flight*, p. 332.

18. J. C. Hunsaker, "Research in Aeronautics," in U.S. Presidential, National Resources Planning Board, *Research: A National Resource* (Washington, D.C.: Government Printing Office, 1940), p. 138.

19. Theodore von Karman, *The Wind and Beyond* (New York: Little, Brown, 1967), p. 224.

20. Charles Mathews, quoted in Ellis Rubinstein, "Dollars vs. Satellites," *IEEE Spectrum* 13, no. 10 (1976): 74-80.

21. Figures are from R.E.G. Davies, *A History of the World's Airlines* (London: Oxford University Press, 1964), table 52.

22. Hunsaker, "Research in Aeronautics," p. 132, fig. 25.

23. H. S. Hele-Shaw, and T. E. Beacham, "The Variable Pitch Airscrew," *Jl. RAeS* 32 (1928): 525-54.

24. Lloyd Morris and Kendall Smith, *Ceiling Unlimited* (New York: MacMillan, 1953).

25. Robin Higham, *Britain's Imperial Air Routes, 1918 to 1939* (London: Foulis, 1960).

26. Morris and Smith, *Ceiling Unlimited*, p. 292.

27. Ibid.; and Frederick C. Thayer, Jr., *Air Transport and National Security* (Chapel Hill: University of North Carolina Press, 1965).

28. American figures are from Hunsaker, "Research in Aeronautics." British figures are from Higham *Imperial Routes*, p. 346. All figures have been rounded to the nearest $1,000. Conversion value of the pound is arbitrarily chosen to be $4.80. German figures have not been sought because German currency restrictions and physical priority allocations after 1933 vitiate any dollar-value comparison. See Arthur Schweitzer, *Big Business in the Third Reich* (Bloomington: Indiana University Press, 1964).

29. See Nathan Rosenberg, *Technology and American Economic Growth* (New York: Harper, 1972), for a general statement of the role played by the economic environment in selecting technology for adoption, adaptation, or development.

30. *New York Times*, 14 November 1926.

31. J. L. Nayler, "The Aeronautical Research Council," *Jl. RAeS* 70, Centenary Issue (1966): 79-81; J. S. Buchanan, "The Schneider Cup Race, 1925," *Jl. RAeS* 30 (1926): 434-52; F. R. Banks, "Five Decades of the Aero Engine," *Jl. RAeS* 66 (1962): 672-82; idem, "Fifty Years of Engineering Learning," *Jl. RAeS* 72 (1968): 209-28.

32. P. A. Ralli, "Design of Airscrews for Schneider Trophy Race, 1927," *Jl. RAeS* 32 (1928): 767.

33. H. R. Ricardo, "The Development and Progress of the Aero Engine," *Jl. RAeS* 34 (1930): 1001-2.

34. "Crazy Boy," *Newsweek* 3 November 1934, pp. 25-26.
35. See John P. Heinmuller, *Man's Fight to Fly* (New York: Oxford University Press, 1944), for a series of sketches of spectacular aeronautical feats.
36. K. E. Bailes, "Technology and Legitimacy: Soviet Aviation and Stalinism in the 1930's," *Technology and Culture* 17 (1976): 55-81.
37. Howard Hughes, quoted in Heinmuller, *Man's Fight to Fly*, p. 221.
38. George H. Daniels, *Science in American Society* (New York: 1971).
39. See Thomas Kuhn, "Mathematical versus Experimental Traditions in the Development of Physical Science," and, especially, "Objectivity, Value Judgment, and Theory Choice," particularly p. 333, n. 8, both in *The Essential Tension* (Chicago: University of Chicago Press, 1977).
40. See the somewhat controversial argument advanced by Paul Foreman, "Weimar Culture, Causality, and Quantum Theory, 1918-1927: Adaptation by German Physicists and Mathematicians to a Hostile Intellectual Environment," *Historical Studies in the Physical Sciences* 3 (1971): 1-115.

Chapter 7

1. As mentioned in Chapter 4 Isaac Newton designed a steam carriage with a boiler feeding directly to a nozzle.
2. Gohlke, "Thermal-Air Jet-Propulsion," *Aircraft Engineering* (London) 14 (1942): 32-39 (translated from *Flugsport*, vol. 31 [January-February 1939]).
3. Now Air Commodore Sir Frank Whittle. Names and titles will appear as they were at the time of the events described.
4. This summary of Whittle's early life, and much else in this section on his work, is drawn from his book, *Jet: The Story of a Pioneer* (London: Frederick Muller 1953). See also his "The Birth of the Jet Engine in Britain," in *The Jet Age: Forty Years of Aviation*, ed. Walter J. Boyne and Donald S. Lopez (Washington, D.C.: National Air and Space Museum, Smithsonian Institution, 1979), pp. 3-24.
5. F. W. [Frank Whittle], "Speculation," *RAF Cadet College Magazine*, Fall 1928, pp. 106-10. Subsequent references are cited by page number in the text.
6. Whittle, *Jet*, pp. 20-21.
7. See also Whittle "Evolution of the Internal Combustion Turbine Engine," *Gas Turbine International*, March-April 1976, pp. 16-19.
8. Frank Whittle, letter to the author, 20 May 1970.
9. Whittle, *Jet*, pp. 24-25.
10. Frank Whittle, "The Early History of the Whittle Jet-Propulsion Gas Turbine," *The Aeroplane* (London), October-November 1945 (the first James Clayton Lecture to the Institution of Mechanical Engineers).
11. Whittle, *Jet*, pp. 24-27.
12. This issue was raised in John D. Stanitz's paper presented at the 1964 convention of the Society for the History of Technology, entitled "Aerodynamic Considerations in Early Turbojet Engine Development." Dr. Stanitz has very generously provided me with a draft of his paper.
13. Roy Fedden, "The First 25 Years of the Bristol Engine Department," *Jl. RAeS* 65 (1960): 348.
14. Robert Schlaifer, *The Development of Aircraft Engines* (Cambridge: Harvard University Press, 1950), p. 335.
15. Whittle, letter.

16. Ibid.

17. Schlaifer, *The Development of Aircraft Engines;* M. M. Postan, D. Hay, and J. D. Scott, *Design and Development of Weapons* (London: H. M. Stationery Office, 1964); Robert L. Perry, "Innovation and Military Requirements," Rand Corporation Memorandum RM-5182-PR (1967).

18. Whittle, *Jet,* p. 39; idem, "The Turbo-Compressor and the Supercharging of Aero Engines," *Jl. RAeS* 35 (1931): 1047–74.

19. Stanitz, "Early Turbojet Development."

20. Whittle, *Jet,* p. 53.

21. Whittle, "Early History," quoted in Stanitz, "Early Turbojet Development."

22. F. W. Armstrong, "The aero-engine and its progress—fifty years after Griffith," *Aeronautical Journal* (December 1976): 499–520.

23. A. A. Griffith, "Report on the Whittle jet propulsion system," Report No. E. 3545 (ARC 2897; A. M. Ref.: 572886136, R.A.E. Ref. El2025) (Farnborough: Royal Aircraft Establishment, February 1937, n.p.). I am indebted to the librarian, Aeronautical Research Council, National Physical Laboratory, Teddington. Middlesex, for providing me with a copy of Griffith's report.

24. Ronald W. Clark, *Tizard* (Cambridge, Mass.: M.I.T. University Press, 1965), p. 105; see also Derek Wood, *The Narrow Margin* (New York: Paperback Library, 1969).

25. Clark, *Tizard,* pp. 104–63.

26. Schlaifer, *The Development of Aircraft Engines,* p. 433.

27. Whittle, "Early History."

28. Whittle, *Jet,* p. 88–89.

29. Ibid., p. 96.

30. C.N.H. Lock, "Problems of High-Speed Flight as Affected by Compressibility," *Jl. RAeS* 42 (1938): 192–228.

31. Hans von Ohain, letter to the author, 31 July 1970; Stanitz, "Early Turbojet Development."

32. Ernst Caspari, "The Lectures of Professor Robert Pohl in Göttingen," *American Journal of Physics* 19 (1951): 61–63.

33. Von Ohain, letter.

34. Ibid. Roger Allan Williams, "The Development of Luftwaffe Aircraft in the Nazi Era" (Ph.D. diss., University of Minnesota, 1971), offers a different interpretation of von Ohain's conversion to the turbojet: "When von Ohain joined the Heinkel Company in 1936, he thought of his invention only as an auxiliary engine to assist the take-off of an aircraft. As a trained physicist, he knew that a jet unit would be a suitable primary engine only for an aircraft which would fly much faster and higher than any at that time. Siegfried Guenter, Heinkel's brilliant young designer, acquainted him with the latest estimates of the high speed and altitude capabilities of future aircraft, given a power plant superior to the conventional piston engine and propeller."

35. Hans von Ohain, "The Evolution and Future of Aeropropulsion Systems," in *Jet Age,* p. 29.

36. Ibid.

37. Ibid.

38. Von Ohain, letter.

39. This paragraph, and the two that follow, are based on von Ohain, "Evolution of Aeropropulsion Systems," pp. 29–32.

40. Ernst Heinkel, *Stormy Life* (New York: Dutton, 1956), p. 111. This book may or may not be a dependable source, since it was written with the cooperation of a journalist. The selections used here seem valid in light of other evidence and events recorded elsewhere. See a review of the book under its English title, *He 1000,* in *Jl. RAeS* 58 (1954): 580.

41. Heinkel, *Stormy Life*, p. 111.
42. Von Ohain, "Evolution of Aeropropulsion Systems," p. 32.
43. Stanitz, "Early Turbojet Development."
44. Schlaifer, *Development of Aircraft Engines*, pp. 377-78.
45. This section is based largely on a resumé Dr. Wagner supplied the author, dated 18 September 1970, and a letter to the author, dated 26 October 1970.
46. Donald Miller and David Sawers, *The Technical Development of Modern Aviation* (London: Routledge & Kegan Paul, 1968), pp. 64-67, which is based, so a note indicates, on "information supplied by Professor H. Wagner."
47. Wagner, resumé and letter; *Who's Who in Germany* (1964), p. 1811; *Wer ist Wo? Luftfahrtwissenschaft und -technik*, p. 115; Adolf Rohrbach, "Large All-Metal Seaplanes," *Jl. RAeS* 28 (1924): 655-75; Herbert Wagner, "Flat-Sheet Metal Girders with Very Thin Metal Web," NACA TM no. 604 (1931); idem, "Torsion and Buckling of Open Sections," NACA TM no. 807 (1936).
48. Wilhelm Hoff, "Research Work of the 'DVL,'" *Jl. RAeS* 35 (1931): 771-816.
49. Wagner letter.
50. Ibid.
51. Schlaifer, *Development of Aircraft Engines*, pp. 379-81.
52. Wagner, letter.
53. Perry, "Innovation and Military Requirements," p. 23.
54. Half of the compression was accomplished by energy transfer in the rotor blades themselves.
55. Schlaifer, *Development of Aircraft Engines*, p. 379.
56. Ibid., p. 383.
57. S. T. Robinson, "Interrogation of Dipl. Ing. Helmut Schelp" (1945), C.I.O.S., items no. 5 and 26, file no. XXXII-46, Library of Congress, PB 6678 (1945). The current Federation International Aeronautique (F.I.A.) world speed record for a piston-engined, propeller-driven aircraft is 483 mph, established in 1969.
58. Schlaifer, *Development of Aircraft Engines*, p. 384. Most current subsonic turbojet aircraft have a maximum speed between 500 and 600 mph, so Schelp's estimate was rather good.
59. Ibid., p. 385; and Clifford F. Maier, "A History of the Development of the World's First Operational Turbo-jet Airplane, the Me-262, to June, 1943" (Ph.D. diss., University of Washington, 1971). Maier's account closely follows Schelp's own "Survey of Special Engine Development in Germany," written 25 May 1945 (Maier, p. 43).
60. Schlaifer, *Development of Aircraft Engines*, pp. 391-92.

Chapter 8

1. Ernst Heinkel, *Stormy Life* (New York: Dutton, 1956), pp. 228-230.
2. Robert Schlaifer, *The Development of Aircraft Engines* (Cambridge: Harvard University Press, 1950), p. 407.
3. J. W. Adderley, "German Gas Turbine Developments during the Period 1939-1945," B.I.O.S. Report no. 12.
4. The He 011 decision points up an interesting question with regard to the whole German war production effort. Much criticism has been made of German diversion of resources and talent from immediate projects to more advanced ones, such as the He 011, with the implication that they should have concentrated exclusively on immediate needs. If, however, it is assumed that the Germans expected to continue the war indefinitely

(which at least dominant personalities in the High Command did), it would have been irrational to divert all resources to immediate problems, only to lag catastrophically behind if present difficulties were in fact overcome. The Germans had a genuine Hobson's choice.

5. William Green, *War Planes of the Second World War,* vol. 1, *Fighters* (New York: Doubleday, 1960).

6. Schlaifer, *Development of Aircraft Engines,* p. 419; see also Anselm Franz, "The Development of the 'Jumo 004' Turbojet Engine," in *The Jet Age: Forty Years of Jet Aviation,* ed. Walter J. Boyne and Donald S. Lopez (Washington, D.C.: National Air and Space Museum, Smithsonian Institution, 1979), pp. 69–74.

7. D. R. Maguire, "Enemy Jet History," *Jl. RAeS* 52 (1948): 75–84.

8. Schlaifer, *Development of Aircraft Engines,* p. 426.

9. H. Roxbee Cox, "British Aircraft Gas Turbines," *Journal of the Aeronautical Sciences,* 13 (1946): 53–83.

10. Hayne Constant, quoted in A. A. Rubbra, "Alan Arnold Griffith," *Biographical Memoirs of the Fellows of the Royal Society of London* (1964), p. 124. A photograph of the original Griffith contrarotating test compressor is in F. W. Armstrong, "The Aero Engine and Its Progress: Fifty Years after Griffith," *Aeronautical Journal,* December, 1976.

11. Rubbra, "Alan Arnold Griffith," pp. 125–28.

12. Ironically, Griffith's materials research now offers hope for a new revolution in ceramic turbine materials. See Tom Alexander, "Hot Prospects for the New Ceramics," *Fortune,* April 1976, pp. 153–58+.

13. D. M. Smith, "The Development of an Axial-Flow Gas Turbine for Jet Propulsion," *Proc. Inst. M.E.* (London) 157, War Emergency Issue no. 36 (1945): 471–82.

14. Cox, "British Aircraft Gas Turbines"; E. S. Moult, "The Development of the Goblin Engine," *Jl. RAeS* 51 (1947): 655–85.

15. Interview with Glenn B. Warren by Ilan Kusiatin, Harvard Business School, in Schenectady, N.Y., 18 September 1975; Glenn B. Warren "Jet-Propelled Airplanes," *General Electric Review* 55 (1952): 34–36.

16. Warren, "Jet-Propelled Airplanes," p. 36.

17. Ibid.

18. Schlaifer, *Development of Aircraft Engines,* p. 484.

19. Sanford A. Moss, letter to Dale Streid, 20 September 1939, quoted in draft manuscript by W. R. Travers, General Electric Aircraft Engine Group. I am indebted to Mr. Travers for providing me with pages from his forthcoming history of General Electric aircraft engines.

20. W. R. Travers, letter to the author, 17 August 1978.

21. Warren, interview.

22. Sanford A. Moss. "Airplane Gas Turbines with Propeller or Jet Propulsion," *G.E. Bulletin* No. 816, File 1-E, November 1940 (Revised edition of Original Bulletin dated January 1940), unpublished.

23. Schlaifer, *Development of Aircraft Engines,* p. 466.

24. Warren, interview.

25. Alan Howard and C. J. Walker, "An Aircraft Gas Turbine for Propeller Drive," *Mechanical Engineering* 69 (1947): 827–35.

26. Letters to the author from Clarence L. Johnson, senior advisor to Lockheed Aircraft Corporation, 25 September and 24 October 1978. Mr. Johnson very generously provided most of the material upon which this account of Lockheed's L-133/L-1000 proposal is based.

27. Lockheed Aircraft Corporation, Inter-Departmental Communication, N. C. Price to C. L. Johnson, 22 September 1943.

28. Lockheed L-1000 Project Jet Propulsion Engine Technical Data, submitted to AAF Material Command, 31 August 1943.

29. Robert Gross, quoted in *Of Men and Stars: A History of Lockheed Aircraft Corporation* (1957), chap. 8, p. 6. I am indebted to Wayne Pryor, Lockheed Corporate Communications, for providing me with this source.

30. Schlaifer, *Development of Aircraft Engines*. Much of this material on the several unsuccessful American projects is drawn from the Schlaifer book.

31. Macon C. Ellise and Clinton E. Brown, "NACA Investigation of a Jet-Propulsion System Applicable to Flight," NACA Report no. 802 (1949; originally written in 1944).

32. Fred T. Jane, *Jane's All the World's Aircraft* (New York: MacMillan, 1948), p. 46d.

33. Joseph Szydlowski, "Design and Development of Small Aircraft Gas Turbines in France," *Jl. RAeS* 69 (1965): 459–66; Jane, *All the World's Aircraft* (1948), p. 47d; "The Szydlowski-Planiol Centrifugal Supercharger," *Jl. RAeS* 43 (1939): 114.

34. P. Destival, "French Turbo-Propeller and Turbo-Reaction Engines," *Jl. RAeS* 53 (1949): 111–39.

35. See papers in the NASA-published *Essays on the History of Rocketry and Astronautics: Proceedings of the Third through Sixth History Symposia of the International Academy of Astronautics*, esp. Vergeny S. Shchetinkov, "Main Lines of Scientific and Technical Research at the Jet Propulsion Research Institute (RNII), 1933–42," 2 (1977): 43–58.

36. Richard E. Stockwell, *The Soviet Air and Rocket Forces* (New York: Praeger, 1959), pp. 229–40; Robert A Kilmarx, *A History of Soviet Air Power* (New York: Praeger, 1962); Asher Lee, *The Soviet Air Force* (New York: Harper, 1950).

37. Neville Duke, *Sound Barrier* (New York: Philosophical Library, 1955), p. 29.

38. F. E. Pickles, "Caproni-Campini Aircraft and Allied Developments in Italy" (1944), C.I.O.S., item no. 5, file no. XII-24, Library of Congress.

39. Gohlke, "Thermal-Air Jet-Propulsion," *Aircraft Engineering* (London) 14 (1942): 32–39; Noel Daum "The Griffon Aircraft and the Future of the Turbo-Ram-Jet Combination in the Propulsion of Supersonic Aeroplanes," *Jl. RAeS* 63 (1959): 327–40.

40. Willy Ley, *Rockets, Missiles, and Men in Space* (New York: Viking, 1968), pp. 425–27; E. Sänger and I. Bredt, "A Ram-Jet for Fighters," NACA TM no. 1106; Irene Sänger-Bredt, "The Silver Bird Story: A Memoir," in *Essays on the History of Rocketry and Astronautics*, 1 (1977): 195–228.

41. Mano Ziegler, *Rocket Fighter* (New York: Doubleday, 1963).

42. See D. N. Walker, "Techniques of Testing Gas Turbine Engines," *Proc. Inst. M.E.* (London), War Emergency Issue no. 12 (1945): 472–83; and E. E. Stoeckly, "Testing of Gas-Turbine Jet-Propulsion Engines," in General Electric, *Aircraft Gas Turbine Engineering Conference, 1945* (West Lynn, Mass., General Electric, 1945), pp. 227–33, on the development of testing techniques for turbojets.

43. Frank Whittle, *Jet* (London: Frederick Muller, 1953), p. 224.

44. Robert L. Perry, "Innovation and Military Requirements," Rand Corporation Memorandum RM-5182-PR (1967); Schlaifer, *Development of Aircraft Engines*; and Pierre Closterman, *The Big Show* (New York: Ballantine, 1951).

45. Data on virtually all aircraft mentioned in this section are drawn from Fred Jane's *Jane's All the World's Aircraft* (New York: MacMillan, 1945, 1947, 1948), and from William Green's series, *War Planes of the Second World War* (1960-). Additional data on the Me 262 comes from Closterman, *The Big Show*, and from R. S. Hirsch and Uwe Feist, *Messerschmitt 262* (Fallbrook, Calif.: Aero, 1967).

46. Eugene M. Emme, *Hitler's Blitzbomber* (Maxwell Air Force Base: Air University, 1951).

47. See Clifford F. Maier, "A History of the Development of the World's First Oper-

ational Turbo-Jet Airplane, the Me-262, to June, 1943" (Ph.D. diss., University of Washington, 1971), and Roger Allan Williams, "The Development of Luftwaffe Aircraft in the Nazi Era" (Ph.D. diss., University of Minnesota, 1971), for greater detail on the development of German jet aircraft.

48. The Germans also worked on two other nonturbojet reaction propulsion projects. One, the Me 328, was of wooden construction and was powered by two Argus pulse-jets similar to those used on the V-1. The other, the Bachem Natter, was a wooden interceptor powered by a Walther rocket motor. Neither saw service. The two represented desperation designs in the closing months of the war.

49. R. Smelt, "A Critical Review of German Research on High-Speed Airflow," *Jl. RAeS* 50 (1946): 917–19.

50. See Perry, "Innovation and Military Requirements," for a study of the influence of the different operational requirements of the various countries.

51. Schlaifer, *Development of Aircraft Engines*, p. 400.

52. See Richard P. Hallion, *Supersonic Flight: The Story of the Bell X-1 and Douglas D-558* (New York: MacMillan, 1972), for an account of the American transonic aircraft program after the war.

53. See Donald Miller and David Sawers, *The Technical Development of Modern Aviation* (London: Routledge & Kegan Paul, 1968), on the great cost effectiveness of the turbojet transports.

54. Theodore von Karman, *Aerodynamics* (Ithaca: Cornell University Press, 1954).

55. See the revealing comments of W. G. Lundquist, representing the Wright Aeronautical Division of Curtiss-Wright, in General Electric, *Aircraft Gas Turbine Engineering Conference*, p. 214.

Chapter 9

1. Such an in-depth search process, using prespecified criteria, is one variety of heuristic employed in artificial intelligence programs.

2. President's Opening Remarks, *Proc. Inst. M.E.* (London) 102, pt. 2 (1922): 700.

3. A. Rateau, "The Use of the Turbo-Compressor for Attaining the Greatest Speeds in Aviation: Specific Pressures and Weights of Air under Normal Atmospheric Conditions," *Proc. Inst. M.E.* (London) 102, pt. 2 (1922): 795–831.

4. William Reavell, discussion of Rateau's paper, *Proc. Inst. M.E.* (London) 102, pt. 2 (1922): 1030.

5. *Mechanical Engineering* 52 (1930): 259.

Appendix

1. Neville Duke and Edward Lanchbery, *Sound Barrier* (New York: Philosophical Library, 1955), p. 85.

2. A steam engine (piston or turbine) running with a vacuum condenser may reject heat at a temperature equivalent less than ambient, but the conditions inside the condenser are the product of negative work. Overall efficiency for the particular engine may be improved, but the engine is still subject to the same thermodynamic laws as any other engine.

I am indebted to Mr. Ronald Lasser of Carnegie-Mellon University for helping write this section.

3. F. R. Banks, "Five Decades of the Aero Engine," *Jl. RAeS* 66 (1962): 672-82.

4. The Society of Automotive Engineers terminology for turbojet engine components is as follows:

bucket—a part rotated by hot gas, such as a turbine bucket.
blade—a rotating member used to compress gas; such as a rotor blade.
vane—a stationary member used to change gas direction, such as a stator vane in a compressor or nozzle diaphragm vane.

(Jack V. Casamassa, *Jet Aircraft Power Systems* [New York: McGraw-Hill, 1950], p. 41).

5. Ibid., p. 41.

6. These figures can be misleading. The 3,500 h.p. out of the Whittle turbine is used for the work of driving the compressor. The 2,400 h.p. for the Napier-Sabre is net work. Still, that an experimental design should generate such power as the Whittle turbine is extraordinary.

7. Casamassa, *Power Systems*, p. 129.

8. Leslie E. Neville and Nathaniel F. Silsbee, *Jet Propulsion Progress* (New York: McGraw-Hill, 1948), p. 86.

9. Frank Whittle, "The Early History of the Whittle Jet-Propulsion Gas Turbine," in *The Aeroplane* (1945), p. 544.

10. Casamassa, *Power Systems*, p. 109.

11. Neville and Silsbee, *Jet Propulsion*, p. 60.

12. The First World War also saw the extensive use of another type of piston engine, the "rotary." In this type, the cylinders were arranged radially around the drive shaft, but the drive shaft was attached to the aircraft firewall and the propeller attached to the engine block. Thus the "engine" rotated and the "drive shaft" remained stationary. The Le Rhone rotary was one of the most famous of this type and was used in the French Nieuport and the English Sopwith Camel.

During the Second World War, other variations in cylinder arrangement—"flat," "H," "X," and so forth—were tried anew. Other basic engine types, such as tandem and opposed piston, were also tried.

Index

Ackert, Jakob, 38, 106, 109-10, 114-16
Aerodynamics, 99-116
    and development of high speed, 175-77
    in England during World War II, 208-13
    in Germany during World War II, 213-18
    historical development of, 100-116
    impact of, 116
    from 1920-1940, 152-60
    at supersonic speeds, 219, 240
Aero-engine manufacturers, 9, 31, 206
Aerofoil lattices, 115
Aerofoil theory of axial compressors, 111
Aeronautical Research Council (ARC), 107, 111, 143, 190, 231-41
Aeronautics, 247-50
    Great Britian's Advisory Committee for, 102
Aerostructures, revolution in, 129-33
Aerothermodynamics, 240
Aircraft performance, 161-66
Aircraft piston engines, 118-29
Aircraft structures, 124-33
Airframe design philosophy, 239
Air-fuel ratios and combustion, 252-53

Air mail, 168-69
Air Ministry, 190-94, 198
Aitken, Hugh G., 2, 11, 30
AK turbine, 146-47
Allis-Chalmers, 61, 146, 222-23, 225
Allison Division of General Motors, 9
American aviators, 173-75
American Rocket Society, 140
American Society of Civil Engineers, 53
American Society of Mechanical Engineers, 246
Antz, Hans, M., 206-7, 211
Appold, John, 58
Archimedean screw, 57
Armengaud, René, 90-93, 97
Armstrong-Siddley, 185, 214, 217
    Jaguar engine of, 124
Asiatic Petroleum Company (Shell), 193
Avery, William, 64
Avon and Conway (Rolls Royce Turbojets), 214
Axial flow test compressor, 214-15
Axial flow turbojet, 179, 214

Baconian
    methodology, 21, 36
    sciences, 43
Barber, Elinor, 23

Barber, John, 89
Barker's mill reaction wheels, 34
Bavarian Motor Works (B.M.W.), 9, 115, 206
Beacham, T. E., 121
Berlin Academy of Sciences, 37
Bernoulli, Daniel
    study of hydrodynamics by, 101-2
    theorem of, 60, 101-2
Bernoulli family, 36, 101
Bethune compressor, 85
Betz, Albert, 106, 110, 115-16, 205
Bi-fuel arrangement, 149
Biggin Hill experiments, 191
Birmann, Rudolph, 225
"Blitzbombers," 231
Boeing, 163-64
Booker, P. J., 25-26, 30
Borda, Charles, 38-39
Boyd, T. A., 120
Boyden, Uriah, 45-47
Bramo, 206
Branca, Giovanni, 64
Brazil Straker, 126-27
Breguet airline distance formulation, 181
Briggs, L. J., 109
Bristol Aeroplane Company, 218
Bristol Aircraft, 125, 185
British Interplanetary Society, 140
British Thomson-Houston Company (B.T.H.), 185, 188
Brown Boveri, 93, 114-15, 124, 145-46, 226, 243
Buchi, Alfred, 122
Buchi-Duplex turbocharging system, 122
Burdin, Claude, 34, 39, 64
Busemann, Adolf, 109, 156
Byron Jackson Company, 61

California Institute of Technology, 106, 108, 223
Camerer, R., 51
Campbell, Donald, T., 6-7, 14, 20-21
Campini, Secondo, 182-83
    ducted fan design by, 182, 225-27

"Can combustors," 217
C. A. Parsons and Company, 72, 214
Cardwell, Donald, 44
Carnot, Lazare, 38
Carnot, Sadi, 67
Carr, E. H., 4
Cayley, Sir George, 103
Centaurus (Bristol sleeve-valve engine), 127
Centrifugal blower, 58
Clerk, Dugald, 122
Clothier, W. C., 111
Co-evolution
    biological, 14
    technological, 13-15, 50-51, 54, 56, 83-84, 117, 120, 129, 240, 243-44
Coleman, Nicholas J., 52
Combustion systems, 256-57
Commercial aviation, 160-69, 176-77
Community of practitioners, 8-10, 12, 16, 20, 23, 32, 41, 43, 48, 57, 240
Compressor types, 95-96, 255
Condit, Carl, 2
Constant, Hayne, 192, 214-16
Convergent-divergent nozzle, 61, 68-69, 71, 92, 109, 139, 141, 219
Counterfactual argument on role of science, 66-67
Cracker catalytic turbocompressors, 146
Craft technologies, 23, 58
Craft traditions, 23, 34, 47
Crocco, G. A., 103
Crystal Palace Exhibition of 1851, 58
Cukor, Coloman, 148
Curtis, Charles G., 68, 79, 81
Curtiss-Wright Company, 124, 240

d'Alembert, Jean, 37, 101-2
Darrieus, M., 114
Daugherty, R. L., 59
Dayton Electric Company (DELCO), 126
de Belidor, Bernard Forest, 35

Index

De Laval, Carl Gustav Patrick, 61, 68–71, 77, 81
De Laval Company, 124, 139, 225
Deparcieux, C. A., 35
de Prony, Baron Riche, 41
Desaguliers, J. T., 35
Destival, P., 226
Deutsche Versuchsanstalt für Luftfortforschung (DVL), 202
Diamler-Benz, 9, 206
Diesels, 134–38
Diffusers, 255–56
Dimensional analysis, 50
Doble, William A., 54
Dornberger, General Walter, 140
Douglas, Donald, 164
Douglas, G. P., 109
Dryden, Hugh L., 109
Ducted-fan system, 225
Duralumin ("dural"), 132
Durand, W. R., 123, 222
Dynamometer, 22, 34, 41, 57
    hydraulic, 50, 129
    Prony, 44, 46
Dynamos, 63, 73
    development of, 14, 50

Earhart, Amelia, 174
Economic factors, 27–30
Edgar, Graham, 120, 126
Edison, Thomas, 21, 27–29
Efficiency, 263–65
Egyptian tympanum, 57
Eiffel, A. G., 103, 156
Electric Manufacturing Company, 79
Electric strain gauge, 121
Electro-Mecanique Cie., 114, 226
Elliot Corporation, 147
Ellor, James E., 124
Emerson, James, 48
Encke, W., 106, 110, 115–16, 205
Engineer, 84
Engineering design, 24–27
Engine types. See *specific types*
    miscellaneous, 261–63
Escher-Wyss, 146–47

Euler, Johann Albrecht (son), 35
Euler, Leonhard, 37–39, 47, 50, 58, 101–2
Evolutionary epistemology, 6–8, 12

Fedden, H. Roy, 126–27, 135
Ferguson, Eugene, 2
"Firebolts," 138–39
Firth, Vickers, 127
Flow theory, 108
Flume, testing, 48, 49
"Flying Fortress," 164
Fokker, 132
Forced invention, 26
Ford, Henry, 12, 131
Ford Tri-motor, 161
Fourneyron, Benoit, 34, 38–39, 41, 46, 48, 51, 57, 59, 66
France, 226
Francis, James B., 34, 45–48, 51, 57, 71
Franz, Anselm, 210
"free-vortex" blading, 111, 114, 191
Freidrich, Rudolph, 204
French, Lestor, 64
French Academy of Sciences, 43–44
Froude, R. E., 102, 106
Fuel chemistry, 120
Functional failure, 12–13, 17

Galbraith, John Kenneth, 30
Galileo, 35, 43, 57, 101
Gas turbine revolution, 97–98
Gay-Lussac gas law, 44
General Electric Company, 79, 95–96, 124, 158, 218–23, 226, 240
German Aerodynamic Institute, 202
German aero-engine manufacturers, 206
German Rocket Society, 140
German V-2 ballistic missile, 140
Gilfillan, S. C., 2
Glauert, Hermann, 106, 109
Gloster, Aircraft, 192–93
Goddard, Robert, 140, 143
Goldstein, Sydney, 107
Griffith, A. A., 110–16, 184–85, 190–91, 213–15, 218

Guenther, Siegfried, 198
Guenther, Walter, 198
Guillaume, Charles, 179, 186–244
Gundermann, Wilhelm, 199
Gwynne, John, 58

Hachette, J. H. P., 38
Hahn, Max, 198–99
Halford, Frank B., 216–17
Hamburg, David, 2
Hamilton-Standard, 121
Harding, Warren, 168
Harris, H. S., 141, 182
Heat engine, 250–54
Heinkel, Ernst, 197–200, 206–7
Heinkel aircraft, 140, 164, 201, 209–10
Hele-Shaw, H. S., 121
Heron, Sidney D., 125–27
Hero of Alexandria's steam-powered aeolipile, 37–38, 63
Herschel, Clemens, 48
Hesse, F. G., 52–55, 94
Hierarchical design and decomposition, 24–27
Hierarchical structure of selection and retention, 29–31
Hindsight, 2
Holyoke Water and Power Company, 48, 50
Holzworth, Hans, 90–91, 97
Hook gauge, 46
Hoover, H., 168
Houndry catalytic cracking process, 145–46
Howd, Samuel B., 47
Hughes, Howard, 175
Hughes, Thomas P., 2, 14, 29
Hull, C. F., 109
Hydraulic machinery, 38–39
Hydraulic power transmission, 90
Hydraulic regulator, 46
Hydrodynamic (aerodynamic) flow similarity
  law of, 59
  theory of, 105

Imperial Airways, 165
Internal combustion gas turbine revolution, 144–48, 219
Internal water-cooling system, 92
International Academy of Astronautics, 139
International competition, 169–77

Jacobs, Eastman N., 142, 152–53, 225–26
Japan, 227
Jendrassik, George, 148
Jenkins, Reese, 2
Jet engine design, 141
Jet propulsion, 182, 219
Jewkes, John, 2
Johnson, Clarence, 223
Johnston, W. H., 58
Jones, B. M., 107–8, 129, 157, 183, 187
Jonval, 45, 52
Joukowski, Nikolai E., 103–4
Journal of the Franklin Institute, 41, 44–45
Journal of the Royal Aeronautical Society, 186
Junkers, and Rohrbach, 131
Junkers, Hugo, 96–97, 115, 134–35
Junkers Aircraft Company, 201, 203
Junkers Engine Company, 134, 203, 206
Junkers turbojet, 204

Kalitinsky, Andrew, 225
Kaplan, Victor, 49–50
Karavodine system, 96
Keller, Curt, 146
Kestrel V5 engine, 197
Knight, S. N., 52
Koppenberg, Heinrich, 203
Kroeber, A. L., 2
Kuhn, Thomas S., 6, 10, 20, 43
Kutta, M. Wilhelm, 104
Kutta-Joukowski theorem, 104–5, 109

Lagrange, Joseph Louis, 101–2
Lakatos, Imre, 20

Lanchester, Frederick, 106
Lasley, R. E., 225
Layton, Edwin T., 2, 8, 11, 47
Leduc, René, 227
Lemale, Charles, 90–93, 97
Lewis, George W., 153–58
Lift and drag, 248–49
Lilienthal, Otto, 104
Lindbergh, Charles, 174
Lippisch, Alexander, 227
Ljungstrom, Berger, 59, 80
Ljungstrom Steam Turbine Company, 147
Lock, C. N. H., 194
  use of propellor theory by, 107
Lockheed, 174, 223–24
Lorenzen, Christian, 147–48
Lorin scheme, 141
Lowell hydraulic experiments, 45, 47, 53, 67, 71
Lubbock, I., 193
Lufthansa, Deutsche, 167, 197
Lusser, Robert, 200, 206
Lysholm, A. J. R., 147

"Mach angle" and velocity study, 104
Mach, Ernst, 108
  use of Schlieren technique by, 104
MacLaurin, Colin, 35
Mader, Otto, 203, 206, 209–10
Manchester University, 59
Mariotte, Edme, 35
Marples, David L., 25–26, 30
Massachusetts Institute of Technology, 225
Mather and Platt, 61
Mauch, Hans A., 205–6
Maxim, Hiram, 103
Melot "thrust augmentor," 141
Merlin (Rolls Royce), 197
Merton, Robert, 23, 43, 57
Messerschmitt, Willy, 198, 207
Metropolitan-Vickers Electrical Company, 215–17
Meyer, Adolf, 146
Midgeley, Thomas, 120
Millikan, Clark, 223

Ministry of Air Production (Great Britain), 193
Mitchell, Billy, 168
Mitroff, Ian, 23
Model
  ideal-typical, 5–6, 24
  information-transformation, 11, 30
  use of, 4–5
Monocoque, 131–33
Moore, Joseph, 52
Morin, Arthur, 40, 44
Morize, O., 141
Morize ejector, 141
Moss, Sanford, 80, 90, 94–95, 97–98
  invention of gas turbine by, 94, 123–24, 218, 220–21
Motion, fluid, 101
Müller, Max Adolf, 203–4, 209, 211
Munk, Max, 106

Nagler, Forrest, 49
Napier, Robert D., 71
National Advisory Committee for Aeronautics (NACA), 123–24, 153, 157–58, 215–16, 222–25
National Air Service, 168
National Bureau of Economic Research, 2
National patterns of aeronautic development, 151–77, 230–37, 237–39
  cultural influences on, 177, 244
National Physical Laboratory (Great Britain), 194
Navier, C. L. M. H., 102
Newcastle and Direct Electric Lighting Company, 76
New Deal, 168
Newton, Sir Isaac, 101, 139
New York Times, 194
Nimonic, 80, 193
Normal technology, 10–11, 16, 19–20, 22, 32, 73, 77, 80, 87, 117–18, 148–50, 176, 208–40, 244–45

Northrop, John K., 133, 225
Northrop aviation, 225
Nozzles, 258–59

Oberth, Hermann, 140
Ogburn, William F., 2
Owner, Frank M., 121, 127, 185

Papin, Denis, 58
"Paraffin turbine" naval torpedo, 93
Parent, Antoine, 35
Paris Air Show, 227
Parker Brothers, 47
Parsons, Charles, 59, 64, 68–77, 81, 87
Parsons Marine Turbine Company, 76
Pascal, Blaise, 35
Patents
  role of, defined, 31
  systems of, 243–44
Pavlecka, Vladimir H., 225
Pelton, Lester G., 53, 57
Pelton water wheel, 15, 51–54, 56
"Phantom" (McDonnell FD-1), 237
Phillips, Horatio, 103
Piston aero-engine manufacturers, 9, 11
Piston engines, 118–29, 260–61
  internal combustion, 142
  superchargers, 33
  two-stroke, 122
Piston pump, 57
Piston steam engine, 38
Pitot traverse method, 108
Planiol, M., 226
Pohl, Robert W., 195, 197
Poisson, Denis Simeon, 102
Poncelet, J. V., 39, 47
Popper, Karl, 6, 20
Post, Wiley, 175
Power, 253–54
Power Jets, Ltd., 187–89, 191–93, 217
Prandtl, Ludwig, 103–5, 109–10
Prandtl subsonic theory, 105–6
Pratt and Whitney, 124, 225, 240

Presumptive anomaly, 15, 16–18, 20, 27, 32, 99, 116, 176–77, 196, 204, 207, 242–44
Price, Derek, 2
Price, Nathan, 224
Prony brake, 22, 34, 40–43, 46–49, 51, 53, 56–57, 94
Propeller cavitation, 74, 103
Propeller-piston engine, 1
Propeller runner, 49
Propellers, 138, 260
Propeller-tip compressibility "burble," 109–10, 138, 158
Pump
  centrifugal, 58–61
  four-stage turbine, 61
  multistage, 58–61
  rotary, 57–58
  three-stage, 61
  turbine, 59–62
Pye, David R., 192

Radial-conical diffuser, 41
Ralli, P. A., 171
Ramelli, 58
Rankine, W. J. M., 67, 102, 106
Rateau, Auguste, 22, 61, 68, 78–79, 81, 84–87, 92–93, 103, 123–24, 245
Reaction propulsion, 138–43
Reaction propulsion nozzle, 139
Reavell, William, 245
Research
  classical, 26
  corporate, 30–31
  mission-oriented, 26–27
Reynolds, Osbourne, 50, 59–61, 102–3, 139
Reynolds' Number, 59, 102–3, 105, 158
Ricardo, H. R., 120, 135, 149, 171–72
Risdon Iron Works, 52–53
Robins, Benjamin, 103
Rocket propulsion, 182
Rockets, 139–40, 219, 226, 227
Rogers, Will, 174

Rohrbach Metal Aeroplane Company, 201-2
Rolls Royce, 128-29, 197, 214, 240
    Welland I, 193-94
Roots Brothers, 58, 83
    blower of, 58, 124
    compressors of, 147
Rosenberg, Nathan, 2
Rotary air compressors, 33, 58
Rover Company, 193
Royal Aeronautical Society, 103, 194
Royal Aircraft Establishment (RAE), 124, 152, 157, 192
    turboprop project of, 213, 217
Royal Air Force (RAF), 180
Royal Dutch Shell Oil Company, 120
Rynin, Nikolai, 140

Sänger, Eugene, 227
Sawers, David, 2
Schelp, Helmut, 179, 204-7, 209, 211, 213, 242
Schlichting, Herman, 109
Schmidt pulse-jet, 205
Schmookler, Jacob, 2
Schneider Trophy contests, 128-29, 151, 169-73
Schubauer, G. B., 142
Science, classical, 43
Science and technology, 20-21
Selective retention function, 7-8, 15, 21
Sherbondy, E. H., 124
Sherman, C. F., 109
Shoemaker, James M., 142
Simon, Herbert, 12, 25, 30
Smeaton, John, 35-36, 39, 103
Smelt, Ronald, 152
Smith, Hamilton, 53
"Smokejack," 63, 89
Société des Turbomoteurs, 92-93
Société Rateau, 226
Société Turbomeca, 226
Sociocultural factors, 31-32
Southwell, R. V., 107
Soviet aircraft, 174

Soviet Society for "reaction motion," 140
Soviet Union, 226
Stack, John, 109, 158
Stanton, T. E., 103, 109, 156
"Starting jets," 219
Steam engine, 41, 59-60
    Cornish, 44
    Newcomen, 35
Stern, W. J., 143-44
Stillerman, Richard, 2
Stodla, Aurel, 109, 145
Stokes, George G., 102
Stolze, F., 89
"Stratified supercharging" system, 149-50
Stream tube, 38, 46
Storer, Norman, 23
Stressed-skin construction, 131, 164, 202, 229
Subsonic aerodynamics, 105-7
Sulzer Brothers, 61
Sun Oil Company, 146
Superchargers, 122-25
Supercharging, 260-61
    definition of, 122
    stratified, 149-50
Supersonic aerodynamics, 104-15, 240
Supersonic research, 152-54
Supplee, Henry Harrison, 66
Swain, A. M., 47
Swiss Federal Institute of Technology, 114
Swiss National Exhibition, 146
Szydlowski, Joseph, 226

Taylor, G. I., 109-10
Technological
    change, 2-3, 24, 27
    competition, 16-20
    ecology, 54
    revolution, 16-20, 22, 32
        effects of patents on, 31
    testability and testing, 20-23, 24, 32, 47, 49, 51, 54, 56, 62, 83, 228-29, 244
    traditions, 244

Technology, 6–8. *See* Normal technology
Terry, R. V., 50
Tetraethyllead, 120
Theory of Gravitation, 101
"Theory of Rupture," 110
Thermodynamic
  cycle, 252
  theory, 67
Thomas, Dorothy, 2
Thomas, Henry, 146
Thomson-Houston (G.E. affiliate), 124
Thomson, James (brother), 45
Thomson, William, 59
Thomson Trophy races, 173
Thrust and propulsive efficiency, 249–50
Thrust augmentation, 141–42
Tinling, J. C. B., 187
Tizard, Henry, 190–92, 213, 216–17
Tournaire, 64, 66–67
TRACES, 2
Traditions of practice, 8, 10, 19
Transcontinental and Western (TWA), 165
Transport aircraft performance, 161–66
Tufting, 108
Turbinia, 74–76
Turbine, 258
  AK, 146–47
  axial-flow partial-admission impulse, 45
  axial-flow reaction, 45
  constant-pressure gas, 90
  constant-volume gas, 90
  definition of, 34
  double-rotation radial out-flow reaction, 80
  "explosion," 96
  gas, 16, 18, 28, 33, 62, 94, 115, 122, 216, 225
  hot air, 89
  hydraulic (Nagler), 49–50
  internal combustion gas, 63–98, 144–48
  inward-flow radial, 45–46
  Mannheimer, 91
  multistage reaction, 70, 73
  propeller, 49
  Reidler-Stumpf, 80
  steam, 14–15, 18, 28, 33, 56, 59–61, 63–98, 226
  "Tremont," 71
  velocity-staged, 79
  water, 14, 22, 28, 33–34, 37–39, 43, 45, 47, 49–51, 53, 56, 59, 62, 66, 83, 89
Turbine manufacturers, 81–82
Turbine-pump theory, 86
Turbo-air compressors, 83–89
  cracker catalytic, 146
  manufacturers of, 85
  parameters of, 87
Turbo-Engineering Corporation, 225
Turbojet, 89, 99, 129, 218–26, 254–59
  air compressors of, 61, 63
  airframes of, 229–37
  axial flow of, 214
  contraflow ducted-fan of, 214
  definition of, 63
  development of, 178–207
  development of, since World War II, 239–40
  practitioners, 204
  projects involving, 226
  propulsion of, 184
  use of, 1
Turbojet aircraft, 230–37
Turbojet revolution, 3, 15, 18–19, 59, 82, 98–100, 109–10, 116–17, 125, 129, 139, 148, 175, 177, 178–207, 213, 237, 241–46
Turboprop engine, 111–16, 143–44, 182
Turbosuperchargers, 33, 78, 96, 98, 120, 122–25, 226

United States Air Corps, 124
United States turbojets, 218–26
University of California, Berkeley, 52

Values, 31-32
Valve designs, 125-27
Variable-pitch propellors, 120-21, 161
Variation-retention function, 7-8, 21
"Velox" boiler, 145
Versailles Treaty, 198
Victoria's Diamond Jubilee, 75
Vogt, Adolf, 96
Voigt, Woldemar, 207
Voigt gas turbines, 122
Volta High Speed Conference, 143, 153-54, 225
von Braun, Wernher, 140
von Helmholtz, Hermann, 102
von Karman, Theodore, 105, 107-9, 129, 152, 159, 223
    flow theory of, 108
    supersonic theory of, 109
von Ohain, Hans, 179, 194-99, 201-2, 204, 208-10, 213, 242
von Opel, Fritz, 140
von Segner, Johann Andreas, 37
Vortex motion, 102
Vortex street, 108

Wagner, Herbert, 131, 179, 201, 203-4, 208-9, 211, 213, 242
Wagner's beam theory, 202
Walther, Helmut, 227
Warren, Glenn B., 96, 218-19, 227
Wasielewski, Eugene W., 153
Water wheel, 36, 38-40
    "hurdy-gurdy," 52-53
    Jonval, 52
    Pelton, 15, 51-53

Watt, James, 35, 38, 57-58, 64
Weber, Max, 5
Weir, 48, 56
Wenham, Herbert, 103
Westinghouse Electric Corporation, 88, 222-23
"Wetting," 126
"Whirling arm" experiments, 103
Whittle, Frank, 27, 31, 127, 179-94, 201-2, 204, 207, 213, 216, 218, 226, 242, 244-46
Wilkinson, Paul H., 134
Williams, R. Dudley, 187
Wilm, Alfred, 132
Wilson, Robert, 64
Wind tunnel
    cascade experiments, 111
    closed-channel propeller, 157
    closed circuit, 156-57
    compressed air, 157
    development of, 156-59, 175
    low-speed, 156
    open-return, 157
    open-throat, 156
    slotted-throat, 158
    supersonic air, 156
    variable density, 157
Wood, R. McKinnon, 106
World's Columbian Exposition, 94
Worthington, 61
Wright Brothers, 103
Wright Engine Company, 126

Ziolkovsky, Konstantin, 140
Zurich Technical University, 38